STATISTICS FOR PHYSICAL SCIENCES

AN INTRODUCTION

STATISTICS FOR PHYSICAL SCIENCES

AN INTRODUCTION

B. R. MARTIN
Department of Physics and Astronomy
University College London

AMSTERDAM • BOSTON • HEIDELBERG • LONDON
NEW YORK • OXFORD • PARIS • SAN DIEGO
SAN FRANCISCO • SINGAPORE • SYDNEY • TOKYO
Academic Press is an imprint of Elsevier

Academic Press is an imprint of Elsevier
225 Wyman Street, Waltham, MA 02451, USA
525 B Street, Suite 1900, San Diego, CA 92101-4495, USA
The Boulevard, Langford Lane, Kidlington, Oxford OX5 1GB, UK
Radarweg 29, PO Box 211, 1000 AE Amsterdam, The Netherlands

First edition 2012

Copyright © 2012 Elsevier Inc. All rights reserved.

No part of this publication may be reproduced, stored in a retrieval system or transmitted in any form or by any means electronic, mechanical, photocopying, recording or otherwise without the prior written permission of the publisher Permissions may be sought directly from Elsevier's Science & Technology Rights Department in Oxford, UK: phone (+44) (0) 1865 843830; fax (+44) (0) 1865 853333; email: permissions@elsevier.com. Alternatively you can submit your request online by visiting the Elsevier web site at http://elsevier.com/locate/permissions, and selecting Obtaining permission to use Elsevier material

Notice
No responsibility is assumed by the publisher for any injury and/or damage to persons or property as a matter of products liability, negligence or otherwise, or from any use or operation of any methods, products, instructions or ideas contained in the material herein. Because of rapid advances in the medical sciences, in particular, independent verification of diagnoses and drug dosages should be made

Library of Congress Cataloging-in-Publication Data
Application submitted

British Library Cataloguing in Publication Data
A catalogue record for this book is available from the British Library

For information on all **Academic Press** publications
visit our web site at elsevierdirect.com

Printed and bound in USA
12 13 14 15 10 9 8 7 6 5 4 3 2 1

ISBN: 978-0-12-387760-4

Working together to grow
libraries in developing countries

www.elsevier.com | www.bookaid.org | www.sabre.org

ELSEVIER   BOOK AID International   Sabre Foundation

Contents

Preface   ix

1. Statistics, Experiments, and Data   1

1.1. Experiments and Observations   2
1.2. Displaying Data   4
1.3. Summarizing Data Numerically   7
    1.3.1. Measures of Location   8
    1.3.2. Measures of Spread   9
    1.3.3. More than One Variable   12
1.4. Large Samples   15
1.5. Experimental Errors   17
    Problems 1   19

2. Probability   21

2.1. Axioms of Probability   21
2.2. Calculus of Probabilities   23
2.3. The Meaning of Probability   27
    2.3.1. Frequency Interpretation   27
    2.3.2. Subjective Interpretation   29
    Problems 2   32

3. Probability Distributions I: Basic Concepts   35

3.1. Random Variables   35
3.2. Single Variates   36
    3.2.1. Probability Distributions   36
    3.2.2. Expectation Values   40
    3.2.3. Moment Generating, and Characteristic Functions   42
3.3. Several Variates   45
    3.3.1. Joint Probability Distributions   45
    3.3.2. Marginal and Conditional Distributions   45
    3.3.3. Moments and Expectation Values   49

3.4. Functions of a Random Variable   51
    Problems 3   55

4. Probability Distributions II: Examples   57

4.1. Uniform   57
4.2. Univariate Normal (Gaussian)   59
4.3. Multivariate Normal   63
    4.3.1. Bivariate Normal   65
4.4. Exponential   66
4.5. Cauchy   68
4.6. Binomial   69
4.7. Multinomial   74
4.8. Poisson   75
    Problems 4   80

5. Sampling and Estimation   83

5.1. Random Samples and Estimators   83
    5.1.1. Sampling Distributions   84
    5.1.2. Properties of Point Estimators   86
5.2. Estimators for the Mean, Variance, and Covariance   90
5.3. Laws of Large Numbers and the Central Limit Theorem   93
5.4. Experimental Errors   97
    5.4.1. Propagation of Errors   99
    Problems 5   103

6. Sampling Distributions Associated with the Normal Distribution   105

6.1. Chi-Squared Distribution   105
6.2. Student's t Distribution   111
6.3. F Distribution   116
6.4. Relations Between χ^2, t, and F Distributions   119
    Problems 6   121

7. Parameter Estimation I: Maximum Likelihood and Minimum Variance   123

7.1. Estimation of a Single Parameter   123
7.2. Variance of an Estimator   128
    7.2.1. Approximate methods   130
7.3. Simultaneous Estimation of Several Parameters   133
7.4. Minimum Variance   136
    7.4.1. Parameter Estimation   136
    7.4.2. Minimum Variance Bound   137
    Problems 7   140

8. Parameter Estimation II: Least-Squares and Other Methods   143

8.1. Unconstrained Linear Least Squares   143
    8.1.1. General Solution for the Parameters   145
    8.1.2. Errors on the Parameter Estimates   149
    8.1.3. Quality of the Fit   151
    8.1.4. Orthogonal Polynomials   152
    8.1.5. Fitting a Straight Line   154
    8.1.6. Combining Experiments   158
8.2. Linear Least Squares with Constraints   159
8.3. Nonlinear Least Squares   162
8.4. Other Methods   163
    8.4.1. Minimum Chi-Square   163
    8.4.2. Method of Moments   165
    8.4.3. Bayes' Estimators   167
    Problems 8   171

9. Interval Estimation   173

9.1. Confidence Intervals: Basic Ideas   174
9.2. Confidence Intervals: General Method   177
9.3. Normal Distribution   179
    9.3.1. Confidence Intervals for the Mean   180
    9.3.2. Confidence Intervals for the Variance   182
    9.3.3. Confidence Regions for the Mean and Variance   183
9.4. Poisson Distribution   184
9.5. Large Samples   186
9.6. Confidence Intervals Near Boundaries   187
9.7. Bayesian Confidence Intervals   189
    Problems 9   190

10. Hypothesis Testing I: Parameters   193

10.1. Statistical Hypotheses   194
10.2. General Hypotheses: Likelihood Ratios   198
    10.2.1. Simple Hypothesis: One Simple Alternative   198
    10.2.2. Composite Hypotheses   201
10.3. Normal Distribution   204
    10.3.1. Basic Ideas   204
    10.3.2. Specific Tests   206
10.4. Other Distributions   214
10.5. Analysis of Variance   215
    Problems 10   218

11. Hypothesis Testing II: Other Tests   221

11.1. Goodness-of-Fit Tests   221
    11.1.1. Discrete Distributions   222
    11.1.2. Continuous Distributions   225
    11.1.3. Linear Hypotheses   228
11.2. Tests for Independence   231
11.3. Nonparametric Tests   233
    11.3.1. Sign Test   233
    11.3.2. Signed-Rank Test   234
    11.3.3. Rank-Sum Test   236
    11.3.4. Runs Test   237
    11.3.5. Rank Correlation Coefficient   239
    Problems 11   241

Appendix A. Miscellaneous Mathematics   243

A.1. Matrix Algebra   243
A.2. Classical Theory of Minima   247

Appendix B. Optimization of Nonlinear Functions   249

B.1. General Principles   249
B.2. Unconstrained Minimization of Functions of One variable   252
B.3. Unconstrained Minimization of Multivariable Functions   253
    B.3.1. Direct Search Methods   253
    B.3.2. Gradient Methods   254
B.4. Constrained Optimization   255

Appendix C. Statistical Tables 257

C.1. Normal Distribution 257
C.2. Binomial Distribution 259
C.3. Poisson Distribution 266
C.4. Chi-squared Distribution 273
C.5. Student's t Distribution 275
C.6. F Distribution 277
C.7. Signed-Rank Test 283
C.8. Rank-Sum Test 284
C.9. Runs Test 285
C.10. Rank Correlation Coefficient 286

Appendix D. Answers to Odd-Numbered Problems 287

Bibliography 293
Index 295

Preface

Almost all physical scientists — physicists, astronomers, chemists, earth scientists, and others — at some time come into contact with statistics. This is often initially during their undergraduate studies, but rarely is it via a full lecture course. Usually, some statistics lectures are given as part of a general mathematical methods course, or as part of a laboratory course; neither route is entirely satisfactory. The student learns a few techniques, typically unconstrained linear least-squares fitting and analysis of errors, but without necessarily the theoretical background that justifies the methods and allows one to appreciate their limitations. On the other hand, physical scientists, particularly undergraduates, rarely have the time, and possibly the inclination, to study mathematical statistics in detail. What I have tried to do in this book is therefore to steer a path between the extremes of a recipe of methods with a collection of useful formulas, and a detailed account of mathematical statistics, while at the same time developing the subject in a reasonably logical way. I have included proofs of some of the more important results stated in those cases where they are fairly short, but this book is written by a physicist for other physical scientists and there is no pretense to mathematical rigor. The proofs are useful for showing how the definitions of certain statistical quantities and their properties may be used. Nevertheless, a reader uninterested in the proofs can easily skip over these, hopefully to come back to them later. Above all, I have contained the size of the book so that it can be read in its entirety by anyone with a basic exposure to mathematics, principally calculus and matrices, at the level of a first-year undergraduate student of physical science.

Statistics in physical science is principally concerned with the analysis of numerical data, so in Chapter 1 there is a review of what is meant by an experiment, and how the data that it produces are displayed and characterized by a few simple numbers. This leads naturally to a discussion in Chapter 2 of the vexed question of probability — what do we mean by this term and how is it calculated. There then follow two chapters on probability distributions: the first reviews some basic concepts and in the second there is a discussion of the properties of a number of specific theoretical distributions commonly met in the physical sciences. In practice, scientists rarely have access to the whole population of events, but instead have to rely on a sample from which to draw inferences about the population; so in Chapter 5 the basic ideas involved in sampling are discussed. This is followed in Chapter 6 by a review of some sampling distributions associated with the important and ubiquitous normal distribution, the latter more familiar to physical scientists as the Gaussian function. The next two chapters explain how estimates are inferred for individual parameters of a population from sample statistics, using several practical techniques. This is called point estimation. It is generalized in Chapter 9 by considering how to obtain estimates for the interval within which an estimate may lie. In the final two chapters, methods for testing hypotheses about statistical data are discussed. In the first of these the emphasis is on hypotheses about individual parameters, and in the second we discuss a number of other hypotheses, such as whether a sample comes from a given population distribution and goodness-of-fit tests. This chapter also briefly describes tests that can be made in the absence of any information about the underlying population distribution.

All the chapters contain worked examples. Most numerical statistical analyses are of course carried out using computers, and several statistical packages exist to enable this. But the object of the present book is to provide a first introduction to the ideas of statistics, and so the examples are simple and illustrative only, and any numerical calculations that are needed can be carried out easily using a simple spreadsheet. In an introduction to the subject, there is an educational value in doing this, rather than simply entering a few numbers into a computer program. The examples are an integral part of the text, and by working through them the reader's understanding of the material will be reinforced. There is also a short set of problems at the end of each chapter and the answers to the odd-numbered ones are given in Appendix D. The number of problems has been kept small to contain the size of the book, but numerous other problems may be found in the references given in the Bibliography. There are three other appendices: one on some basic mathematics, in case the reader needs to refresh their memory about details; another about the principles of function optimization; and a set of the more useful statistical tables, to complement the topics discussed in the chapters and to make the book reasonably self-contained.

Most books contain some errors, typos, etc., and doubtless this one is no exception. I will maintain a website at www.hep.ucl.ac.uk/~brm/statsbook.html, where any corrections that are brought to my attention will be posted, along with any other comments.

Brian R. Martin
June 2011

Statistics, Experiments, and Data

OUTLINE

1.1 Experiments and Observations	2	1.3.2 Measures of Spread	9
1.2 Displaying Data	4	1.3.3 More than One Variable	12
1.3 Summarizing Data Numerically	7	1.4 Large Samples	15
1.3.1 Measures of Location	8	1.5 Experimental Errors	17

In the founding prospectus of the Statistical Society of London (later to become the Royal Statistical Society), written in 1834, statistics was defined very broadly as 'the ascertaining and bringing together of those facts which are calculated to illustrate the conditions and prospects of society'. In the context of modern physical science, statistics may be defined more narrowly as the branch of scientific method that deals with collecting data from experiments, describing the data (known as *descriptive statistics*), and analyzing them to draw meaningful conclusions (known as *inferential statistics*).[1] Statistics also plays a role in the design of experiments, but for the purposes of this book, this aspect of statistics, which is a specialized subject in its own right, will be excluded. There are many other possible definitions of statistics that differ in their details, but all have the common elements of collections of data, that are the result of experimental measurements, being described in some way and then used to make inferences. The following examples illustrate some of the many applications of statistics.

Consider a situation where several experiments claim to have discovered a new elementary particle by observing its decay modes, but all of them have very few examples to support their claim. Statistics tells us how to test whether the various results are consistent with each

---

[1] The word 'statistics' is used here as the name of the subject; it is a collective noun and hence singular. In Section 1.3 we will introduce another meaning of the word to describe a function (or functions) of the data themselves.

other and if so how to combine them so that we can be more confident about the claims. A second example concerns the efficacy of a medical treatment, such as a drug. A new drug is never licensed on the basis of its effect on a single patient. Regulatory authorities rightly require positive testing on a large number of patients of different types, that is, it is necessary to study its effects on distributions of people. However both time and cost limit how many people can be tested, so in practice samples of patients are used. Statistics specifies how such samples are best chosen to ensure that any inferences drawn are meaningful for the whole population. Another class of situations is where the predictions from a theory, or a law of nature, depend on one or more unknown parameters, such as the electromagnetic force between two charge particles, which depends on the strength of the electric charge. Such parameters can be determined from experiment by fitting data with a function including the unknown parameters. Statistics specifies ways of doing this that lead to precise statements about the best values of the parameters. A related situation is where there are competing theories with different predictions for some phenomenon. Statistical analysis can use experimental data to test the predictions in a way that leads to precise statements about the relative likelihoods of the different theories being correct.

The discussion of statistics will begin in this chapter by considering some aspects of descriptive statistics, starting with what is meant by an experiment.

1.1. EXPERIMENTS AND OBSERVATIONS

An *experiment* is defined formally as a set of reproducible conditions that enables measurements to be made and which produce *outcomes*, or *observations*. The word 'reproducible' is important and implies that in principle *independent* measurements of a given quantity can be made. In other words, the result of a given measurement x_i does not depend on the result of any other measurement. If the outcomes are denoted by x_i, then the set of all possible outcomes $x_i (1 = 1, 2, ..., N)$ is called the *population* and defines a *sample space* S, denoted by $S \equiv \{x_1, x_2, ..., x_N\}$. In principle N could be infinitely large, even if only conceptually. Thus, when measuring the length of a rod, there is no limit to the number of measurements that could in principle be made, and we could conceive of a hypothetical infinite population of measurements. A subset of the population, called a *sample*, defines an *event*, denoted by E. Two simple examples will illustrate these definitions.

EXAMPLE 1.1

An 'experiment' consists of a train traveling from A to B. En route, it must negotiate three sections consisting of a single track, each of which is governed by a set of traffic lights, and the state of these, either red (r) or green (g), is recorded. What is the sample space for the experiment, and write an expression for the event corresponding to the train encountering a red signal at the second traffic light?

The basic data resulting from the experiment are sequences of three signals, each either red or green. The population therefore consists of 2^3 possible outcomes and the sample space is denoted by the eight events

$$S = \{rrr, rrg, rgg, ggg, ggr, grr, rgr, grg\}.$$

The event E defined by the driver encountering a red signal at the second traffic light (and also possibly at other lights) consists of four outcomes. Thus

$$E = (rrr, rrg, grr, grg).$$

In this case the event is called *complex* because it contains a number of *simple* events each containing a single outcome, rrr, etc.

EXAMPLE 1.2

The number of 'heads' obtained by tossing two coins simultaneously can assume the discrete values 0, 1, or 2. If we distinguish between the two coins, what is the sample space for the experiment and show the content of its events?

Denoting $H =$ 'heads' and $T =$ 'tails', there are four events:

$$E_1 = (H,H); \quad E_2 = (H,T); \quad E_3 = (T,H); \quad E_4 = (T,T),$$

and the sample space consists of the four events, $S = \{E_1, E_2, E_3, E_4\}$.

In these examples the observations are non-numeric, but in physical science the outcomes are almost invariably numbers, or sets of numbers, and to relate these definitions to numerical situations, consider firstly the simple experiment of tossing a six-sided die. A simple event would be one of the six numbers on the faces of the die, and the occurrence of the event would be the situation where the number defining the event was observed on the face of the die. The outcome is thus a discrete variable and can take one of the six numbers 1 to 6. If there were two dice, an example of a complex event would be the observation of a 2 on one die and a 3 on the other. Another example is an experiment to measure the heights of all students in a given class. In this case the outcomes are not discrete, but continuous, and in practice an event would be defined by an interval of heights. Then the occurrence of the event would be interpreted as the situation where a measured height fell within a specified range.

In practice, statistics often consists of just the stages of describing and analyzing the data, because they are already available, but if this is not the case, statistics can also play a role in the design of an experiment to ensure that it produces data of a useful form. Although this will not be discussed in detail in this book, there is a general point to be made.

Consider the problem of how to test which of two university staff is more effective in teaching their students. An equal number of students could be assigned to each instructor and they could give each student the same number of lectures and tutorial classes. The outcomes could be the examination pass rates for the two groups, and these could be analyzed to see if there were a significant difference between them. However to interpret the pass rates as a measure of the effectiveness of the teaching skills of the instructors, we have to be sure that as far as possible the experiment had been designed to eliminate bias. Only then would any inferences be meaningful. This is because we would be using only a sample of students to make statistical inferences and not the whole population. The simplest accepted way of ensuring this is to assign students to the two examiners in such a way that all possible choices of members of the groups are equally likely. Alternatively, put another way, that if we had a very large population of students, then every possible sample of a particular size n has an equal chance of being selected. This is called *simple random sampling*. In principle, this condition could be relaxed provided we could calculate the chance of each sample being selected. Samples obtained this way are called *random samples of size n*, and the outcomes are called *random variables*. At first sight this choice might appear counterintuitive. We might think we could do better by assigning students on the basis of their compatibility with the instructor, but any alternative mode of nonrandom selection usually results in one that is inherently biased toward some specific outcome. A random choice removes this bias. Note that the word 'chance' has anticipated the idea of probability that will be discussed in more detail in Chapter 2.

A question that naturally arises is: 'How are random samples chosen?' For small populations this is easy. One could assign a unique number to each member of the population and write it on a ball. The balls could then be thoroughly shaken in a bag and n balls drawn. This is essentially how lottery numbers are decided, with one ball assigned to each integer or set of integers and drawing several balls sequentially. For very large populations, simple methods like this are not practical and more complex methods have to be used. This problem will be considered in more detail when sampling is discussed in Chapter 5.

Finally we should mention that although in practice in physical science 'sampling' almost invariably means simple random sampling, there are other types of sampling that are used in other fields. For example, pollsters often use *systematic sampling* (sometimes mistakenly presented as true random sampling) where every nth member of a population is selected, for example every 100th name in a telephone book. Another method is when the population can be divided into mutually exclusive subpopulations. In this case simple random samples are selected from the subpopulations with a size proportional to the fraction of members that are in that subpopulation. For example, if we know the fractions of men and women that take a degree in physics, we could take simple random samples of sizes proportional to these fractions to make inferences about the populations of all physics students. This is called *stratified sampling* and is very efficient, but not often applicable in physical science.

1.2. DISPLAYING DATA

In physical sciences, experiments almost invariably produce data as a set of numbers, so we will concentrate on numerical outcomes. The measurements could be a set of discrete numbers, i.e., integers, like the numbers on the faces of a die, or a set of real numbers forming a continuous distribution, as in the case of the heights of the students in the example above. We will start by describing how data are displayed.

Experimental results can be presented by simply drawing a vertical line on an axis at the value of every data point, but in practice for both discrete and continuous data it is common to group the measurements into *intervals*, or *bins*, containing all the data for a particular value, or range of values. The binned data can be presented as a *frequency table*, or graphically. For discrete data, binning can be done exactly and the results displayed in the form of a *bar chart*, where a vertical bar is drawn, the height of which represents the number of events of a given type. The total of the heights of the columns is equal to the number of events. The width of the bins is arbitrary and sometimes for clarity a gap is left between one bin and the next. Both are matters of taste. We will also be interested in the frequency with which an outcome takes on values equal to or less than a stated value. This is called the *cumulative frequency*.

EXAMPLE 1.3

The frequency table below shows the results of an experiment where 6 coins are simultaneously tossed 200 times and the number of 'heads' recorded.

Number of heads	0	1	2	3	4	5	6
Frequency	2	19	46	62	47	20	4

Display these data, and the cumulative frequency, as bar charts.

Figure 1.1(a) shows the data displayed as a bar chart. The total of the heights of the columns is 200. Figure 1.1(b) shows a plot of the cumulative frequency of the same data. The numbers on this plot are obtained by cumulatively summing entries on the bar chart of frequencies.

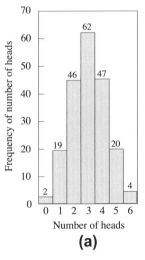
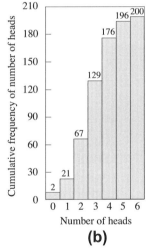

FIGURE 1.1 Bar charts showing (a) the frequency of heads obtained in an experiment where 6 coins were simultaneously tossed 200 times; and (b) the cumulative frequency of heads obtained in the same experiment.

For continuous data, the values of the edges of the bins have to be defined and it is usual to choose bins of equal width, although this is not strictly essential. The raw data are then rounded to a specific accuracy, using the normal rules for rounding real numbers, and assigned to a particular bin. By convention, if a bin has lower and upper values of a and b, respectively, then a data point with value x is assigned to this bin if $a \leq x < b$. There is inevitably some loss of precision in binning data, although it will not be significant if the number of measurements is large. It is the price to be paid for putting the data in a useful form. The resulting plot is called a *histogram*. The only significant difference between this and a bar chart is that the number of events in a histogram is proportional to the area of the bins rather than their heights. The choice of bin width needs some care. If it is too narrow, there will be few events in each bin and fluctuations will be significant. If the bins are too wide, details can be lost by the data being spread over a wide range. About 10 events per bin over most of the range is often taken as a minimum when choosing bin widths, although this could be smaller at the end points.

EXAMPLE 1.4

The table below shows data on the ages of a class of 230 university science students taking a first-year course in mathematical methods. Draw three histograms with bins sizes of 2 yrs, 1yr and ½ yr, respectively, all normalized to a common area of unity. Which bin size is optimal?

Age range	Student numbers
17.0–17.5	2
17.5–18.0	3
18.0–18.5	35
18.5–19.0	27
19.0–19.5	61
19.5–20.0	29
20.0–21.5	28
20.5–21.0	14
21.0–21.5	12
21.5–22.0	8
22.0–22.5	8
22.5–23.0	3

Figure 1.2 shows the three histograms. They have been normalized to a common area of unity by dividing the number of events in a bin by the product of the bin size times the total number of

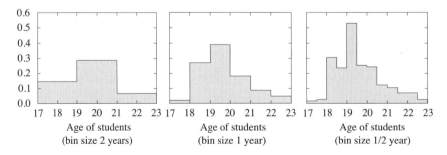

FIGURE 1.2 Normalized histograms of the ages of 230 university science students taking a first-year course on mathematical method, showing the effect of using different bins sizes.

events. For example, with a bin size of 2 years, as shown in the left-hand histogram, the entry in the 19–21 age bin is $(61 + 29 + 28 + 14)/(2 \times 230) = 0.29$. The number of data is probably sufficient to justify a bin size of ½ year because most of the bins contain a reasonable number of events. The other two histograms have lost significant detail because of the larger bin sizes.

Histograms can be extended to three dimensions for data with values that depend on two variables, in which case they are sometimes colloquially called *lego plots*. Two-dimensional data, such as the energies and momenta of particles produced in a nuclear reaction, can also be displayed in *scatter plots*, where points are plotted on a two-dimensional grid. Examples of these types of display are shown in Fig. 1.3. Although there are other ways that data can be displayed, bar charts, histograms, and scatter plots are by far the most common graphical representations of data used in physical sciences.

1.3. SUMMARIZING DATA NUMERICALLY

Although a frequency histogram provides useful information about a set of measurements, it is inadequate for the purposes of making inferences because many histograms can be constructed from the same data set. To make reliable inferences and to test the quality of such inferences, other quantities are needed that summarize the salient features of the data. A quantity constructed from a data sample is called a *statistic*[2] and is conventionally written using the Latin alphabet. The analogous quantity for a population is called a *parameter* and is written using the Greek alphabet. We will look first at statistics and parameters that describe frequency distributions.

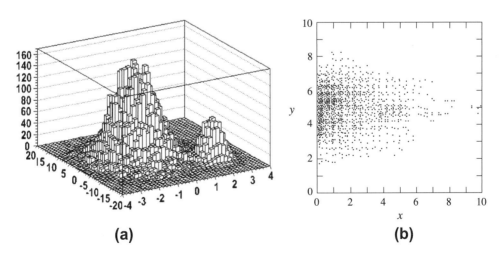

FIGURE 1.3 Examples of displays for two-dimensional data: (a) lego plot; and (b) scatter plot.

[2]This is the second use of this word, as mentioned earlier.

1.3.1. Measures of Location

The first measure of *location*, and the one most commonly used, is the *arithmetic mean*, usually simply called the *mean*. For a finite population $x_i (i = 1, 2, ..., N)$ of size N, the *population mean* is a parameter denoted by μ, and defined by

$$\mu \equiv \frac{1}{N} \sum_{i=1}^{N} x_i. \qquad (1.1)$$

The mean of a sample of size $n < N$ drawn from the population is the statistic, denoted $\bar{x}_n$, or just $\bar{x}$, defined by

$$\bar{x} \equiv \frac{1}{n} \sum_{i=1}^{k} f_i x_i, \qquad (1.2a)$$

where x can take on the values x_i with frequencies f_i ($i = 1, 2, ..., k$) respectively, and

$$\sum_{i=1}^{k} f_i = n, \qquad (1.2b)$$

is the total frequency, or sample size. Alternatively, the sum in (1.2a) can be taken over individual data points. In this case, $k = n$, $f_i = 1$ and

$$\bar{x} = \frac{1}{n} \sum_{i=1}^{n} x_i, \qquad (1.3)$$

so that μ is the limit of the sample mean, when $n \to N$. Note that the values of the mean calculated from (1.2a) and (1.3) will not be exactly the same, although the difference will be small for large samples divided into many bins. (See Problem 1.5.)

Although the mean is the measure of location usually used in physical sciences, there are two other measures that are occasionally used. These are the *mode*, which is the value of the quantity for which the frequency is a maximum; and the *median*, which is the value of the quantity that divides the cumulative frequency into two equal parts. The median is useful in situation where the distribution of events is very asymmetric, because it is less effected by events a long way from the 'center'. An example is the income of a population, where it is common practice for official statistics to quote the median, because this is less influenced by the large incomes of a few very wealthy individuals. In the coin tossing experiment of Example 1.3, the mode is 3, with a frequency 62. Both the 100th and the 101th throws, arranged by order of size, fall in the class '3' and since the quantity in this example can only take on integer values between 0 and 6, the median value is also 3. In the case of a continuous quantity, such as that shown in Fig. 1.2, the mode is the value of the 19–19½-year bin and the median could be estimated by forming the cumulative frequency distribution from the raw data and using the 115th point on the plot to find the median age by interpolation.

The median is an example of a more general measure of location called a *quantile*. This is the value of x below which a specific fraction of the observations must fall. It is thus the inverse of the cumulative frequency. Commonly met quantiles are those that divide a set of observations into 100 equal intervals, ordered from smallest to largest. They are called *percentiles*. Thus, if the percentiles are denoted by $P_p (p = 0.01, 0.02, ..., 1)$, then $100p$ percent of the data are at, or fall below, P_p. The median therefore corresponds to the $P_{0.5}$ percentile. In practice, to find the percentile corresponding to p, we order the data from lowest to highest and calculate $q = np$, where n is the sample size. Then if q is an integer we average the values of the qth and the $(q+1)$th ordered values; if q is not an integer, the median is the kth ordered value, where k is found by rounding up q to the next integer.

EXAMPLE 1.5

Use the sample data below to calculate the mean, the median, and the percentile $P_{0.81}$.

1.6	3.4	9.2	9.6	6.1	7.5	8.0	8.9	11.1	12.3
2.3	4.1	6.8	4.8	12.5	10.0	5.1	8.2	8.5	11.7

The mean is

$$\bar{x} = \frac{1}{20} \sum_{i=1}^{20} x_i = 7.585.$$

To find the median and $P_{0.81}$, we first order the data from lowest to highest:

1.6	2.3	3.4	4.1	4.8	5.1	6.1	6.8	7.5	8.0
8.2	8.5	8.9	9.2	9.6	10.0	11.1	11.7	12.3	12.5

The median is $P_{0.5}$, so $q = np = 20 \times 0.5 = 10$. As q is an integer, the median is the average of the 10th and 11th smallest data values, i.e., 8.1. For $P_{0.81}$, $q = 16.2$, and because q is not an integer, $P_{0.81}$ is the 17th lowest data value, i.e., 11.1.

1.3.2. Measures of Spread

The other most useful quantity to characterize a distribution measures the *spread*, or *dispersion*, of the data about the mean. We might be tempted to use the average of the differences $d_i \equiv x_i - \mu$ from the mean, that is,

$$\bar{d} = \frac{1}{N} \sum_{i=1}^{n} (x_i - \mu),$$

but from the definition of μ, equation (1.1), this quantity is identically zero. So instead we use a quantity called the *variance* that involves the squares of the differences from the means. The *population variance*, denoted by σ^2, is defined by

$$\sigma^2 \equiv \frac{1}{N} \sum_{i=1}^{N} (x_i - \mu)^2, \tag{1.4}$$

and the square root of the variance is called the *standard deviation* σ. The standard deviation is a measure of how spread out the distribution of the data is, with a larger value of σ meaning that the data have a greater spread about the mean. *The sample variance s^2 is defined for a sample of size n by*

$$s^2 \equiv \frac{1}{n-1}\sum_{i=1}^{k} f_i(x_i - \bar{x})^2, \tag{1.5}$$

or, if the sum is over individual data points,

$$\begin{aligned} s^2 &= \frac{1}{n-1}\sum_{i=1}^{n}(x_i - \bar{x})^2 = \frac{1}{n-1}\left[\sum_{i=1}^{n} x_i^2 - \frac{1}{n}\left(\sum_{i=1}^{n} x_i\right)^2\right] \\ &= \frac{1}{n-1}\left[\sum_{i=1}^{n} x_i^2 - n\bar{x}^2\right] = \frac{n}{n-1}\left(\overline{x^2} - \bar{x}^2\right), \end{aligned} \tag{1.6}$$

where equation (1.3) has been used to obtain the second line of (1.6) and an overbar is used to denoted an average. Thus $\overline{x^2}$ is the average value of x^2, defined by analogy with (1.3) as

$$\overline{x^2} \equiv \frac{1}{n}\sum_{i=1}^{n} x_i^2. \tag{1.7}$$

Equation (1.6) is easily proved and is useful when making numerical calculations. Just as for the mean, the sample variance calculated from (1.5) will not be exactly the same as that obtained from (1.6).

Note that the definitions of the sample and population variances differ in their external factors, although for large sample sizes the difference is of little consequence. The reason for the difference is a theoretical one related to the fact that (1.5) contains $\bar{x}$, that has itself been calculated from the data, and we require that for large samples, sample statistics should provide values that on average are close to the equivalent population parameters. In this case, we require the sample variance to provide a 'true', or so-called 'unbiased', estimate of the population variance. This will be discussed in later chapters when we consider sampling in more detail, as will the role of s in determining how well the sample mean is determined.

EXAMPLE 1.6

The price of laboratory consumables from 10 randomly selected suppliers showed the following percentage price increases over a period of one year.

Supplier	1	2	3	4	5	6	7	8	9	10
Price increase (%)	15	14	20	19	18	13	15	16	22	17

Find the average price increase and the sample variance.

To find the average percentage price increase, we calculate the sample mean. This is

$$\bar{p} = \frac{1}{10}\sum_{i=1}^{10} p_i = 16.9.$$

This can then be used to find the sample variance from (1.6),

$$s^2 = \frac{1}{9}\sum_{i=1}^{10}(p_i - \bar{p})^2 = 8.10,$$

and hence the sample standard deviation is $s = \sqrt{8.10} = 2.85$. Thus we could quote the outcome of the observations as an average percentage price increase of $p = 16.9$ percent with a standard deviation of 2.9 percent. In Section 1.3.4 below, an empirical interpretation is given of statements such as this.

The mean and variance involve the first and second powers of x. In general, the nth *moment* of a population about an arbitrary point λ is defined as

$$\mu'_n \equiv \frac{1}{N} \sum_{i=1}^{N} (x_i - \lambda)^n. \tag{1.8}$$

Thus,

$$\mu'_0 = 1, \quad \mu'_1 = \mu - \lambda \equiv d \quad \text{and} \quad \mu'_2 = \sigma^2 + d^2.$$

If λ is taken to be the mean μ, the moments are called *central* and are conventionally written without a prime. For example, $\mu_0 = 1$, $\mu_1 = 0$ and $\mu_2 = \sigma^2$. The general relation between the two sets of moments is

$$\mu_k = \sum_{r=0}^{k} \frac{k!}{(k-r)!r!} \mu'_{k-r} (-\mu'_1)^r, \tag{1.9a}$$

with its inverse

$$\mu'_k = \sum_{r=0}^{k} \frac{k!}{(k-r)!r!} \mu_{k-r} (\mu_1)^r. \tag{1.9b}$$

Moments can also be defined for samples by formulas analogous to those above.

In the case of grouped data, taking the frequencies to be those at the mid-points of the intervals is an approximation and so some error is thereby introduced. This was mentioned previously for the case of the sample mean and sample variance. In many circumstances it is possible to apply corrections for this effect. Thus, if μ' are the true moments, and $\bar{\mu}'$ the moments as calculated from the grouped data with interval width h, then the so-called *Sheppard's corrections* are

$$\mu'_1 = \bar{\mu}'_1; \qquad \mu'_2 = \bar{\mu}'_2 - \frac{1}{12} h^2;$$
$$\mu'_3 = \bar{\mu}'_3 - \frac{1}{4} \bar{\mu}'_1 h^2; \qquad \mu'_4 = \bar{\mu}'_4 - \frac{1}{2} \bar{\mu}'_2 h^2 + \frac{7}{240} h^4; \tag{1.10a}$$

and, in general

$$\mu'_r = \sum_{j=0}^{r} \left\{ \binom{r}{j} (2^{1-j} - 1) B_j h^j \bar{\mu}'_{r-j} \right\}, \tag{1.10b}$$

where B_j is the Bernoulli number of order j.[3]

---

[3] The Bernoulli number B_j is defined as the coefficient of $t^j/j!$ in the expansion of $t/(e^t - 1)$. The first few are: $B_0 = 1$, $B_1 = -1/2$, $B_2 = 1/6$, $B_3 = 0$, $B_4 = -1/30$.

A number of statistics can be defined in terms of low-order moments that measure additional properties, such as *skewness*, the degree to which a distribution of data is asymmetric, although more than one definition of such statistics exists. In practice they are not very useful because using the same data very different distributions can be constructed with similar values of these statistics, and they are seldom used in physical science applications.

Finally, the definition (1.6) can be used to prove a general constraint on how the data points x_i are distributed about the sample mean. If the data are divided into two sets, one denoted S_k, with $N_k (k < n)$ elements with $|x_i - \bar{x}| < ks$ and $s > 0$, and the other containing the rest of the points having $|x_i - \bar{x}| > ks$, then from (1.6)

$$(n-1)s^2 = \sum_{i=1}^{n}(x_i - \bar{x})^2 = \sum_{x_i \in S_k}(x_i - \bar{x})^2 + \sum_{x_i \notin S_k}(x_i - \bar{x})^2 \geq \sum_{x_i \notin S_k}(x_i - \bar{x})^2,$$

where the expression $i \in S_k$ means 'the quantity x_i lies in the set S_k', and the inequality follows from the fact that the terms in the summations are all positive. Using the condition $(x_i - \bar{x})^2 \geq k^2 s^2$ for points not in the set S_k, the right-hand side may be replaced, so that

$$(n-1)s^2 \geq \sum_{x_i \notin S_k} n^2 s^2 = k^2 s^2 (n - N_k).$$

Finally, dividing both sides by $nk^2 s^2$ gives

$$\frac{n-1}{nk^2} \geq 1 - \frac{N_k}{n}. \tag{1.11}$$

This result is called *Chebyshev's inequality* and shows that for any value of k, greater than $100(1 - 1/k^2)$ percent of the data lie within an interval from $\bar{x} - ks$ to $\bar{x} + ks$. For example, if $k = 2$, then 75 percent of the data lies within $2s$ of the sample mean. Although this result is interesting, it is only a weak bound and for the distributions commonly met, the actual percentage of data that lie within the interval $\bar{x} - ks$ to $\bar{x} + ks$ is considerably larger than given by (1.11).

1.3.3. More than One Variable

The mean and variance are the most useful quantities that characterize a set of data, but if the data are defined by more than one variable, then other quantities are needed. The most important of these is the *covariance*. If the data depend on two variables and consist of pairs of numbers $\{(x_1, y_1), (x_2, y_2), \ldots\}$, their population covariance is defined by

$$\text{cov}(x, y) \equiv \frac{1}{N} \sum_{i=1}^{N} (x_i - \mu_x)(y_i - \mu_y), \tag{1.12a}$$

where μ_x and μ_y are the population means of the quantities x and y. The related *sample covariance* is defined by

$$\text{cov}(x,y) \equiv \frac{1}{n-1}\sum_{i=1}^{n}(x_i - \bar{x})(y_i - \bar{y}) \qquad (1.12b)$$
$$= \frac{n}{n-1}(\overline{xy} - \bar{x}\,\bar{y}),$$

where overbars again denote averages. The covariance can be used to test whether the quantity x depends on the quantity y. If small values of x tend to be associated with small values of y, then both terms in the summation will be negative and the sum itself will be positive. Likewise, if large values of x are associated with small values of y, the sum will be negative. If there is no general tendency for values of x to be associated with particular values of y, the sum will be close to zero.

Because the covariance has dimensions, a more convenient quantity is the *correlation coefficient* (also called *Pearson's correlation coefficient*), defined for a sample by

$$r \equiv \frac{\text{cov}(x,y)}{s_x s_y}, \qquad (1.13)$$

which is a dimensionless number between -1 and $+1$. An analogous relation to (1.13) with the sample standard deviation replaced by the population value, defines the *population correlation coefficient* ρ. A positive value of r, i.e., a *positive correlation*, implies that values of x that are larger than the mean $\bar{x}$ tend on average to be associated with values of y that are larger than the mean $\bar{y}$. Likewise, a negative value of r, i.e., a *negative correlation*, implies that that values of x that are larger than $\bar{x}$ tend on average to be associated with values of y that are smaller than $\bar{y}$. If r is $+1$ or -1, then x and y are totally correlated, i.e., knowing one completely determines the other. If $r = 0$, then x and y are said to be *uncorrelated*. Examples of scatter plots for data showing various degrees of correlation are shown in Fig. 1.4. If there are more than two variables present, correlation coefficients for pairs of variables can be defined and form a matrix.[4]

Correlation coefficients must be interpreted with caution, because they measure *association*, which is not necessarily the same as *causation*. For example, although the failure rate of a piece of equipment in a particular month may show an association, i.e., be correlated, with an increase in the number of users of the equipment, this does not necessarily mean that the latter has caused the former. The failures might have occurred because of other

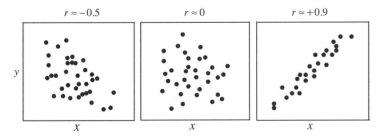

FIGURE 1.4 Scatter plot of two-dimensional data and approximate values of their correlation coefficients r.

---

[4] A brief review of matrices is given in Appendix A.

reasons, such as disruptions in the power supply to the equipment during the month in question. Another example is the ownership of mobile phones over the past decade or so. The number of mobile phones in use correlates positively with a wide range of disparate variables, including the total prison population and the increase in the consumption of 'organic' foods, which common sense would say cannot possibly have caused the increase. The answer lies in realizing that each of the latter quantities has increased with time and it is this that has led to the observed correlations. Thus time is acting as a 'hidden variable' and without knowing this, a misleading conclusion may be drawn from the correlation.

EXAMPLE 1.7

The lengths and electrical resistances (in arbitrary units) of a sample of 10 pieces of copper wire were measured with the results below. Calculate the correlation coefficient for the sample.

Number of piece	1	2	3	4	5	6	7	8	9	10
Length (L)	15	13	10	11	12	11	9	14	12	13
Resistance (R)	21	18	13	15	16	14	10	16	15	12

From these data we can calculate the sample means from (1.3) to be $\bar{L} = 12$ and $\bar{R} = 15$; the sample variances from (1.6) to be $s_L^2 = \frac{30}{9}$ and $s_R^2 = \frac{86}{9}$; and the covariance from (1.12b) to be $\text{cov}(L, R) = \frac{40}{9}$. So, from (1.13), the correlation coefficient of the sample is 0.79, which indicates a strong linear relationship between length and resistance of the pieces of wire, as one might expect.

Just as the mean and variance can be calculated using binned data, so can the correlation coefficient, although the calculations are a little more complicated, as the following example shows.

EXAMPLE 1.8

A class of 100 students has taken examinations in mathematics and physics. The binned marks obtained are shown in the table below. Use them to calculate the correlation coefficient.

		\multicolumn{6}{c}{Mathematics marks}
		40–49	50–59	60–69	70–79	80–89	90–99
Physics marks	40–49	2	5	4			
	50–59	3	7	6	2		
	60–69	2	4	8	5	2	
	70–79	1	1	5	7	8	1
	80–89			2	4	6	5
	99–99				2	4	4

Here we are working with binned data, but for the whole population, i.e., $N = 100$ students. The variances and covariance are easiest to calculate from formulas analogous to (1.6) and (1.12b), but for binned data. For the population, using x for the mathematics marks and y for the physics mark, these are:

$$\sigma_x^2 = \frac{1}{N}\sum_{i=1}^{6} f_i^{(x)} x_i^2 - \frac{1}{N^2}\left(\sum_{i=1}^{6} f_i^{(x)} x_i\right)^2,$$

$$\sigma_y^2 = \frac{1}{N}\sum_{i=1}^{6} f_i^{(y)} y_i^2 - \frac{1}{N^2}\left(\sum_{i=1}^{6} f_i^{(y)} y_i\right)^2$$

and

$$\mathrm{cov}(x,y) = \frac{1}{N}\sum_{i,j=1}^{6} h_{ij}\, x_i y_j - \frac{1}{N^2}\left(\sum_{i=1}^{6} f_i^{(x)} x_i\right)\left(\sum_{i=1}^{6} f_i^{(y)} y_i\right),$$

where

$$f_i^{(x)} = \sum_{j=1}^{6} h_{ij} \quad \text{and} \quad f_j^{(y)} = \sum_{i=1}^{6} h_{ij}$$

and h_{ij} is the frequency in the individual bin corresponding to (x_i, y_j), that is, $f_i^{(x)}(f_j^{(y)})$ is the total frequency of the bin having a central value $x_i(y_j)$. Using the frequencies given in the table gives $\sigma_x^2 = 206.51$, $\sigma_y^2 = 220.91$, and $\mathrm{cov}(x,y) = 153.21$. Hence the correlation coefficient is $\rho = \mathrm{cov}(x,y)/(\sigma_x \sigma_y) \simeq 0.72$.

1.4. LARGE SAMPLES

The total area under a histogram is equal to the total number of entries n multiplied by the bin width Δx. Thus the histogram may be normalized to unit area by dividing each entry by the product of the bin width (assumed for convenience to be all equal) and the total number of entries, as was done for the data shown in Fig. 1.2. As the number of entries increases and the bin widths are reduced, the normalized histogram usually approximates to a smooth curve and in the limit that the bin width tends to zero and the number of events tends to infinity, the resulting continuous function $f(x)$ is called a *probability density function*, abbreviated to *pdf*, or simply a *density function*. This is illustrated in Fig. 1.5, which shows the results of repeated measurements of a quantity, represented by the random variable x. The three normalized histograms $N(x)$ show the effect of increasing the number of measurements and at the same time reducing the bin width. Figure 1.5(d) shows the final normalized histogram, together with the associated density function $f(x)$.

The properties of density functions will be discussed in detail in Chapter 3, but one feature worth noting here is that $f(x)$ is very often of a symmetrical form known as a *Gaussian* or *normal* distribution, like that shown in Fig. 1.5(d), the latter name indicating its importance

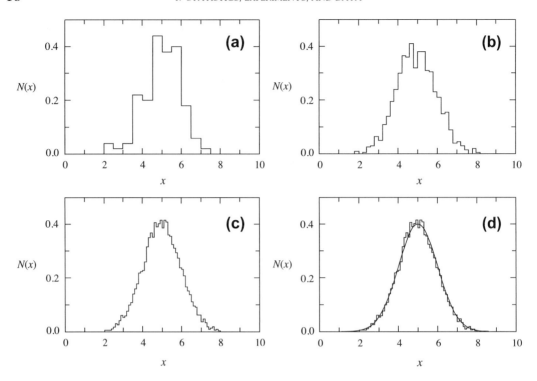

FIGURE 1.5 Normalized histograms $N(x)$ obtained by observations of a random variable x: (a) $n = 100$ observations and bin width $\Delta x = 0.5$; (b) $n = 1000$, $\Delta x = 0.2$; (c) $n = 10000$, $\Delta x = 0.1$; and (d) as for (c), but also showing the density function $f(x)$ as a smooth curve.

in statistical analysis. If the function $f(x)$ is of approximately normal form, an empirical rule is that:

1. approximately 68.3% of observations lie within 1 sample standard deviation of the sample mean;
2. approximately 95.4% of observations lie within 2 sample standard deviations of the sample mean;
3. approximately 99.7% of observations lie within 3 sample standard deviations of the sample mean.

These results could be used in principle to interpret the results of experiments like that in Example 1.4, although in that case with only 10 events the distribution of observations is unlikely to closely approximate a normal distribution.

The question of to what extent a set of measurements can be assumed *a priori* to be normally distributed is an interesting and important one. It is often remarked that physical scientists make this assumption because they believe that mathematicians have proved it, and that mathematicians assume it because they believe that it is experimentally true. In fact there is positive evidence from both mathematics and experiments that the approximation is often very good, but it is not universally true. In later chapters we will discuss the circumstances in which one can be confident about the assumption.

A problem with the standard deviation defined in (1.6) as a measure of spread is that because the terms in the definition of the variance are squared, its value can be strongly influenced by a few points far from the mean. For this reason, another measure sometimes used is related to the form of the probability density function. This is the *full width at half maximum height (FWHM)*, often (rather confusingly) called the *half-width*, which is easily found by measuring the width of the distribution at half the maximum frequency. This quantity depends far less on values a long way from the maximum frequency, that is, data points in the tails of the distribution. For an exact normal distribution, the half-width is 2.35σ. (See Problem 1.8.)

1.5. EXPERIMENTAL ERRORS

In making inferences from a set of data, it is essential for experimenters to be able to assess the reliability of their results. Consider the simple case of an experiment to measure a single parameter, the length of a rod. The rod clearly has a true length, although unknown, so the results of the experiment would be expressed as an average value, obtained from a sample of measurements, together with a precise statement about the relationship between it and the true value, i.e., a statement about the experimental uncertainty, called in statistics the *error*. Without such a statement, the measurement has little or no value. The closeness of the measured value to the true value defines the *accuracy* of the measurement, and an experiment producing a measured value closer to the true value than that of a second experiment is said to be more accurate.

There are several possible contributions to the error. The first is a simple mistake — a reading of 23 from a measuring device may have been recorded incorrectly as 32. These types of errors usually quickly reveal themselves as gross discrepancies with other measurements, particularly if data are continually recorded and checked during the experiment, and can usually be eliminated by repeating the measurement.

Then there are contributions that are inherent to the measuring process. If the length of the rod is measured with a meter rule, the experimenter will have to estimate how far the end of the rod is between calibrations. If it is equally likely that the experimenter will over- or underestimate this distance, the errors are said to be *statistical*, or *random*. Analogous errors are also present in realistic experiments, such as those that involve electronic counting equipment. Mathematical statistics is largely concerned with the analysis of random errors. One general result that will emerge later is that they can be reduced by accumulating larger quantities of data, i.e., taking more readings. The statistical error on a measurement is a measure of its *precision*. Denoting the measurement as x and the statistical error as Δ, the result of the experiment is expressed as $x \pm \Delta$.

An experiment with a smaller statistical error is said to be more *precise* than one with a larger statistical error. In statistics, 'precision' and 'accuracy' are not the same. This is illustrated in Fig. 1.6, which shows a set of measurements made at different values of x of a quantity y that is known to be a linear function of x, as shown by the straight line. The data in Fig. 1.6(a) are more precise than those in Fig. 1.6(b) because they have smaller errors, as shown by the error bars. (These are the vertical lines of length 2Δ drawn vertically through the data points to show the range of values $x \pm \Delta$.) However the data clearly show

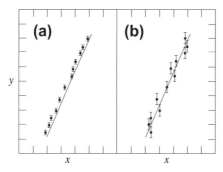

FIGURE 1.6 Illustration of the difference between (a) precision and (b) accuracy.

a systematic deviation from the straight line that gives the known dependence of y on x. The data in graph (b) have larger error bars and so are less precise, but they are scattered about the line and are a better representation of the true relationship between x and y. Thus they are more accurate. Later chapters will show how these statements may be expressed quantitatively.

The deviation of the data in Fig. 1.6(a) from the true values is an indication of the presence of a third type of error, called a *systematic* error. There are many possible sources of these, which may, or may not, be known and they are by no means as obvious as Fig. 1.6 might suggest. If a meter rule is used to measure the rod it may have been wrongly calibrated during its manufacture, so that each scale unit, for example one millimeter, is smaller than it should be. In a more realistic case where an experiment counts the particles emitted by a radioactive isotope, the detectors could also have been wrongly calibrated, but in addition the source might contain other isotopes, or the detectors may be sensitive to particles other than those that identify the decay. In the simple case of measuring the length of a rod, repeating the experiment with another meter rule would reveal the problem and enable it to be eliminated, but in real situations using substantial equipment this may well not be possible and one of the skills of a good experimentalist is to anticipate possible sources of systematic errors and design them out at the planning stage of the experiment. Those that cannot be eliminated must be carefully investigated and taken account in the final estimation of the overall error.

Systematic errors are a potentially serious problem, because you can never be sure that you have taken all of them into account. There is no point in producing a very precise measurement by taking more data to reduce the statistical error if the systematic error is larger. This would only lead to a spurious accuracy. Books on mathematical statistics usually have little to say about systematic errors, or ignore them all together, because in general there is no way a full mathematical treatment of systematic errors can be made. However in the real world of science we do not have the luxury of ignoring this type of error, and a limited analysis is possible in some circumstances, particularly if the systematic effect is the same for all data points, or its dependence on the measuring process is known, as is often the case. We will return to this point in Section 5.4, when we discuss how to combine data from different experiments.

In practice, it is better for clarity to quote the statistical and systematic errors separately, by writing $x \pm \Delta_R \pm \Delta_S$, where the subscripts stand for 'random' and 'systematic', respectively.

PROBLEMS 1

1.1 An experiment consists of tossing a die followed by tossing a coin (both unbiased). The coin is tossed once if the number on the die is odd and twice if it is even. If the order of the outcomes of each toss of the coin is taken into account, list the elements of the sample space.

1.2 Five students, denoted by S_1, S_2, S_3, S_4 and S_5, are divided into pairs for laboratory classes. List the elements of the sample space that define possible pairings.

1.3 The table gives the examination scores out of a maximum of 100 for a sample of 40 students.

22	67	45	76	90	87	27	45	34	36
67	68	97	73	56	59	76	67	63	45
55	59	90	82	74	34	68	56	53	68
28	39	43	66	67	59	38	39	56	61

Cast the data in the form of a frequency histogram with 8 equally spaced bins. What is the frequency of each bin and the numbers in the bins of the cumulative distribution?

1.4 Calculate the median and the percentile $P_{0.67}$ for the unbinned data of Example 1.3.

1.5 Use the data of Problem 1.3 to compute the sample mean $\bar{x}$ and the sample standard deviation s, both for the unbinned and binned data. How would Shepard's corrections change the results? What percentage of the unbinned data falls within $\bar{x} \pm 2s$ and compare this with the predictions that would follow if the data were approximately 'normally' distributed.

1.6 Verify that the second moment of a population about an arbitrary point λ is given by $\mu'_2 = \sigma^2 + d^2$, where $d = \mu - \lambda$ and μ and σ^2 are the mean and variation, respectively, of the population.

1.7 Show that for the normal (Gaussian) density
$$f(x) = \frac{1}{\sigma\sqrt{2\pi}} \exp\left[-\frac{(x-\mu)^2}{2\sigma^2}\right],$$
the half-width is equal to 2.35σ.

1.8 One measure of the skewness of a population is the parameter

$$\gamma = \frac{1}{N\sigma^3} \sum_{i=1}^{N} (x_i - \mu)^3.$$

Show that this may be written as

$$\gamma = \frac{1}{\sigma^3}\left(\overline{x^3} - 3\overline{x}\,\overline{x^2} + 2\overline{x}^3\right),$$

where the overbars denote averages over the population.

1.9 The electrical resistance per meter R was measured for a sample of 12 standard lengths of cable of varying diameters D. The results (in arbitrary units) were:

D	1	3	4	2	7	2	9	9	7	8	5	3
R	10	9	7	8	4	9	3	2	5	4	3	8

Calculate the correlation coefficient for the data.

CHAPTER 2

Probability

OUTLINE

2.1 Axioms of Probability 21
2.2 Calculus of Probabilities 23
2.3 The Meaning of Probability 27
2.3.1 Frequency Interpretation 27
2.3.2 Subjective Interpretation 29

Statistics is intimately connected with the branch of mathematics called *probability theory*, and so before we can meaningfully discuss the subject of statistics and how it is used to analyze data we must say something about probabilities. This chapter therefore starts with a brief review of the axioms of probability and then proceeds to the mathematical rules for their application. The meaning of probability is more problematic and there is no single interpretation. The final section examines two interpretations that are used in physical sciences.

2.1. AXIOMS OF PROBABILITY

Let S denote a sample space consisting of a set of events $E_i (i = 1, 2, ..., n)$, where for specific events subscripts will be avoided by instead using the notation A, B, C, etc. If we have two events in S, denoted by A and B, then the event in which *both* occur (called the *intersection* of A and B) is denoted $A \cap B$, or equivalently $B \cap A$. If $A \cap B = \emptyset$, where the symbol $\emptyset$ denotes a sample space with no elements (called a *null space*), then the events are said to be *mutually exclusive* (also called *disjoint* or *distinct*). The event in which *either A or B, or both*, occurs is called the *union* of A and B and is denoted $A \cup B$. It also follows that if $\overline{A}$ denotes the event 'not A', called the *complement* of A, then $\overline{A} = S - A$. To summarize[1]

---

[1]Readers should be aware that the notations in probability theory are not unique. For example, the complement of A is often written A^c and other differences in notation, particularly for the union, are common.

2. PROBABILITY

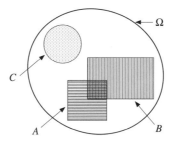

FIGURE 2.1 A Venn diagram (see text for a detailed description).

$A \cap B$ intersection of A and B (both A and B occur),
$A \cup B$ union of A and B (either A or B, or both occur),
$A \cap B = \emptyset$ A and B are mutually exclusive (no comment elements),
$\overline{A}$ complement of A (the event 'not A').

These definitions may be illustrated on a so-called *Venn diagram*, as shown in Fig. 2.1. The sample space S consists of all events within the boundary Ω. A, B, and C are three such events. The doubly shaded area is $A \cap B$ and the sum of both shaded regions A and B is $A \cup B$. The area outside the region occupied by A and B is $\overline{A \cup B}$ and includes the area occupied by the third event C. The latter is disjoint from both A and B, and so $C \cap A = C \cap B = \emptyset$, i.e., A and C, and B and C are pairwise mutually exclusive. An example will illustrate these definitions.

EXAMPLE 2.1

A sample space S consists of all the numbers from 1 to 8 inclusive. Within S there are four events:

$$A = (2,4,7,8),\ B = (1,3,5,7),\ C = (2,3,4,5)\ \text{and}\ D = (1,7,8).$$

Construct the content of the events $\overline{A} \cup C$, $B \cap \overline{C}$, $\overline{S \cap B}$, $(\overline{C \cap D}) \cup B$, $(B \cap \overline{C}) \cup A$, *and* $A \cap C \cap \overline{D}$.
From the above

$$\overline{A} = (1,3,5,6),\ \overline{B} = (2,4,6,8),\ \overline{C} = (1,6,7,8)\ \text{and}\ \overline{D} = (2,3,4,5,6),$$

and so the events are

$$\overline{A} \cup C = (1,2,3,4,5,6),\ B \cap \overline{C} = (1,7),\ \overline{S \cap B} = (1,3,5,7),$$

and

$$(\overline{C \cap D}) \cup B = (1,3,5,7,8),\ (B \cap \overline{C}) \cup A = (1,2,4,7,8),\ A \cap C \cap \overline{D} = (2,4).$$

The axioms of probability may now be stated as follows.

1. Every event E_i in S can be assigned a *probability* $P[E_i]$ that is a real number satisfying

$$0 \leq P[E_i] \leq 1. \tag{2.1a}$$

2. Since S consists of all the events E_i, then

$$\sum_i P[E_i] = P[S] = 1. \qquad (2.1b)$$

3. If the events A and B are such that the occurrence of one excludes the occurrence of the other, i.e., they are mutually exclusive, then

$$P[A \cup B] = P[A] + P[B]. \qquad (2.1c)$$

2.2. CALCULUS OF PROBABILITIES

A number of basic results follow from these axioms and will be used in later chapters. For the case of discrete events A, B, C, etc., these are:

1. If $A \subset B$, i.e., A is a subset of B, then $P[A] \leq P[B]$ and $P[B - A] = P[B] - P[A]$.
2. $P[\overline{A}] = 1 - P[A]$, from which it follows that for any two events, A and B,

$$P[A] = P[A \cap B] + P[A \cap \overline{B}].$$

3. $P[A \cup B] = P[A] + P[B] - P[A \cap B]$, which reduces to

$$P[A \cup B] = P[A] + P[B] \qquad (2.2a)$$

if A and B are mutually exclusive, or in general

$$P[A \cup B \cup C ...] = P[A] + P[B] + P[C]... \qquad (2.2b)$$

This is called the *additive rule* and generalizes (2.1c).

We will also need a number of other definitions involving multiple events. Thus, if the sample space contains two subsets A and B, then provided $P[B] \neq 0$, the probability of the occurrence of A *given that B has occurred* is called the *conditional probability of A*, written $P[A|B]$, and is defined by

$$P[A|B] \equiv \frac{P[A \cap B]}{P[B]}, \quad P[B] \neq 0. \qquad (2.3a)$$

If the occurrence of A does not depend on that fact that B has occurred, i.e.

$$P[A|B] = P[A], \qquad (2.3b)$$

then the event A is said to be *independent* of the event B. (Note that independence is not the same as being distinct.) An important result that follows in a simple way from these definitions is the *multiplicative rule*, which follows from rewriting (2.3a) as

$$P[A \cap B] = P[B] \, P[A|B] = P[A] \, P[B|A], \qquad (2.4a)$$

and reduces to

$$P[A \cap B] = P[A] \, P[B], \qquad (2.4b)$$

if A and B are independent. This may be generalized in a straightforward way. For example, if A, B, and C are three events, then

$$P[A \cap B \cap C] = P[A]P[B|A]P[C|A \cap B]. \tag{2.4c}$$

Finally, if the event A must result in one of the mutually exclusive events B, C, ..., then

$$P[A] = P[B]P[A|B] + P[C]P[A|C] + \ldots$$

or, reverting to index notation,

$$P[A] = \sum_{i=1}^{n} P[E_i]P[A|E_i]. \tag{2.5}$$

The use of these various results is illustrated in the following two examples.

EXAMPLE 2.2

A student takes a multiple-choice exam where each question has $n = 4$ choices from which to select an answer. If $p = 0.5$ is the probability that the student knows the answer, what is the probability that a correct answer indicates that the student really did know the answer and that it was not just a 'lucky guess'?

Let Y be the event where the student answers correctly, and let $+$ and $-$ be the events where the student knows, or does not know, the answer, respectively. Then we need to find $P[+|Y]$, which from (2.3a) is given by

$$P[+|Y] = \frac{P[+ \cap Y]}{P[Y]}.$$

From (2.4a) the numerator is

$$P[+ \cap Y] = P[+]P[Y|+] = p \times 1 = p,$$

and from (2.5), the denominator is

$$P[Y] = P[+]P[Y|+] + P[-]P[Y|-]$$
$$= p + (1-p) \times \frac{1}{n}.$$

So, finally,

$$P[+|Y] = \frac{p}{p + (1-p)/n} = \frac{np}{1 + (n-1)p}.$$

Thus for $n = 4$ and $p = 0.5$, the probability is 0.8 that a correct solution was because the student did really know the answer to the question.

EXAMPLE 2.3

A class of 100 physical science students has a choice of courses in physics (Ph), chemistry (Ch), and mathematics (Ma). As a result, 30 choose to take physics and mathematics, 20 physics and chemistry, 15 chemistry and mathematics, and 10 take all three courses. If the total numbers of students taking each subject are 70 (Ph), 60 (Ma), and 50 (Ch), find the probabilities that a student chosen at random from the group will be found to be taking the specific combinations of courses: (a) physics but not mathematics, (b) chemistry and mathematics, but not physics, and (c) neither physics nor chemistry.

These probabilities may be found using the formulas given above.

(a) $P[Ph \cap \overline{Ma}] = P[Ph] - P[Ph \cap Ma] = 0.70 - 0.30 = 0.40$
(b) $P[Ch \cap Ma \cap \overline{Ph}] = P[Ch \cap Ma] - P[Ch \cap Ma \cap Ph] = 0.15 - 0.10 = 0.05$
(c) $P[\overline{Ph} \cap \overline{Ch}] = P[Ma] - P[Ch \cap Ma] - P[Ph \cap Ma] + P[Ph \cap Ch \cap Ma]$
$= 0.60 - 0.15 - 0.30 + 0.10 = 0.25$

They can also be found by constructing the Venn diagram shown in Fig. 2.2.

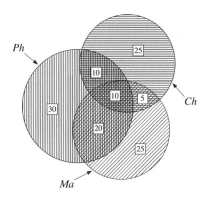

FIGURE 2.2 Venn diagram for Example 2.3.

These ideas can be generalized to the situation where an event can be classified under multiple criteria. Consider, for example, the case of three classifications. If the classifications under the criteria are

$$A_1, A_2, \ldots, A_r;\ B_1, B_2, \ldots, B_s;\ \text{and}\ C_1, C_2, \ldots, C_t;$$

with

$$\sum_{i=1}^{r} P[A_i] = \sum_{i=1}^{s} P[B_i] = \sum_{i=1}^{t} P[C_i] = 1,$$

then a table of the possible values of the three random variables, together with their associated probabilities, defines the *joint probability* of A, B, and C. The *marginal probability* of A_i and C_k is then defined as

$$P[A_i \cap C_k] \equiv \sum_{j=1}^{s} P\left[A_i \cap B_j \cap C_k\right] \tag{2.6a}$$

and likewise the marginal probability of C_k is

$$P[C_k] \equiv \sum_{i=1}^{r}\sum_{j=1}^{s} P[A_i \cap B_j \cap C_k] = \sum_{i=1}^{r} P[A_i \cap C_k] \qquad (2.6b)$$
$$= \sum_{j=1}^{s} P[B_j \cap C_k] = \sum_{j=1}^{s} P[B_j|C_k]\, P[B_j],$$

where (2.3a) has been used in the last expression. This result is known as the *law of total probability* and is a generalization of (2.5).

The final result we shall need is the basic theorem of permutations. The number of ways of *permuting* (i.e., arranging) m objects selected from n distinct objects is

$$_nP_m \equiv \frac{n!}{(n-m)!}, \qquad (2.7a)$$

where $n!$, called *n factorial*, is defined as $n! \equiv n(n-1)(n-2)\ldots 1$, with $0! \equiv 1$. If, on the other hand, the set of n objects consists of k distinct subsets each containing n_k objects, indistinguishable within the subset, with $n = n_1 + n_2 + \ldots + n_k$, then the number of distinct permutations of the objects is

$$_nP_{n_1,\, n_2,\, \ldots,\, n_k} = \frac{n!}{n_1! n_2! \ldots n_k!}. \qquad (2.7b)$$

It follows from (2.7a) that the total number of *combinations* of the m objects without regard to arrangement is

$$_nC_m \equiv \binom{n}{m} = \frac{_nP_m}{m!} = \frac{n!}{m!(n-m)!}, \qquad (2.8)$$

i.e., the coefficient of x^m in the binomial expansion of $(1+x)^n$.

EXAMPLE 2.4

A set of books is arranged on a shelf purely by their subject matter. Four are physics books, three are chemistry books, and two are mathematics books. What is the total number of possible arrangements?

The number of different arrangements N can be found by applying (2.7b) with $n_p = 4$, $n_c = 3$, $n_m = 2$ and $n = n_p + n_c + n_m = 9$. Thus

$$N = {}_nP_{n_p,\, n_c,\, n_m} = \frac{9!}{4!\, 3!\, 2!} = 1260.$$

EXAMPLE 2.5

A committee of 4 people is to be selected at random from a group of 6 physicists and 9 engineers. What is the probability that the committee consists of 2 physicists and 2 engineers?

There is a total of $_{15}C_4$ possible choices of 4 people for the committee and each choice has an equal probability of being chosen. However there are $_6C_2$ possible choices of 2 physicists from the 6 in the group, and $_9C_2$ possible choices of 2 engineers from the 9 in the group. So the required probability is

$$\frac{_6C_2 \times \, _9C_2}{_{15}C_4} = \frac{36}{91} = 0.396.$$

2.3. THE MEANING OF PROBABILITY

The axioms and definitions discussed so far specify the rules that probability satisfies and can be used to calculate the probability of complex events from the probabilities of simple events, but they do not tell us how to assign specific probabilities to actual events. Mathematical statistics proceeds by assigning a *prior* probability to an event on the basis of a given mathematical model, specified by known parameters, about the possible outcomes of the experiment. In physical situations, even if the mathematical form is known, its parameters rarely are and one of the prime objectives of statistical analysis is to obtain values for them when there is access to only incomplete information. Without complete knowledge we cannot make absolutely precise statements about the correct mathematical form and its parameters, but we can make less precise statements in terms of probabilities. So we now turn to examine in more detail what is meant by the word 'probability'.

2.3.1. Frequency Interpretation

We all use the word 'probability' intuitively in everyday language. We say that an unbiased coin when tossed has an equal probability of coming down 'heads' or 'tails'. What we mean by this is that if we were to repeatedly toss such a coin, we would expect the average number of heads and tails to be very close to, but not necessarily exactly equal to, 50%. Thus we adopt a view of probability operationally defined as *the limit of the relative frequency of occurrence*. While this is a common-sense approach to probability, it does have an element of circularity. What do we mean by an unbiased coin? Presumably one that, when tossed a large number of times, tends to give an equal number of heads and tails! We have already used the word 'random' in this context in Chapter 1 when discussing statistical errors, and again when using the words 'equally likely' when discussing the example of an experiment to test lecturers' teaching capabilities.

The *frequency definition of probability* may be stated formally as follows. In a sequence of n trials of an experiment in which the event E of a given type occurs n_E times, the ratio $R[E] = n_E/n$ is called the relative frequency of the event E, and the probability $P[E]$ of the event E is the limit approached by $R[E]$ as n increases indefinitely, it being assumed that this limit exists.

If we return to the sample space of Example (1.2), and denote by X the number of heads obtained in a single throw, then

$$X(E_1) = 2;\ X(E_2) = 1;\ X(E_3) = 1;\ X(E_4) = 0,$$

and we can calculate $P[X]$ using the frequency approach as follows:

$$P[X = 2] = \frac{1}{4};\ P[X = 1] = \frac{1}{2};\ P[X = 0] = \frac{1}{4}.$$

It is worth noting that this definition of probability differs somewhat from the mathematically similar one: that for some arbitrary small quantity ε there exists a large n, say n_L, such that $|R[E] - P[E]| < \varepsilon$ for all $n > n_L$. The frequency definition has an element of uncertainty in it being derived from the fact that in practice only a finite number of trials can ever be made. This way of approaching probability is essentially experimental. A probability, referred to as the *posterior* probability,[2] is assigned to an event on the basis of experimental observation. A typical situation that occurs in practice is when a model of nature is constructed and from the model certain *prior* probabilities concerning the outcomes of an experiment are computed. The experiment is then performed, and from the results obtained posterior probabilities are calculated for the same events using the frequency approach. The correctness of the model is judged by the agreement of these two sets of probabilities and on this basis modifications may be made to the model. These ideas will be put on a more quantitative basis in later chapters when we discuss estimation and the testing of hypotheses.

Most physical scientists would claim that they use the frequency definition of probability, and this is what has been used in the previous sections and examples in this book (such as Examples 2.3 and 2.5), but it is not without its difficulties, as we have seen. It also has to be used in context and the results interpreted with care. A much quoted example that illustrates this is where an insurance company analyzes the death rates of its insured men and finds there is about a 1% probability of them dying at age 40. This does *not* mean that a particular insured man has a 1% probability of dying at this age. For example he may be a member of a group where the risk of dying is increased, such as being a regular participant in a hazardous sport, or having a dangerous occupation. So had an analysis been made of members of those groups, the probability of his death at age 40 may well have been much greater. Another example is where canvassers questioning people on a busy street claim to deduce the 'average' view of the population about a specific topic. Even if the sample of subjects approached is random (a dubious assumption) the outcome would only be representative of the people who frequent that particular street, and may well not represent the views of people using other streets or in other towns.

Crucially, the frequency approach usually ignores prior information. Consider, for example, a situation where two identically made devices are to be tested sequentially. If the probability of a successful test is assessed to be p, then the two tests would be considered as independent and so the combined probability for both tests to be successful is $P[1,2] = P[1]P[2] = p^2$. However, common sense would suggest that the probability $P[2]$

---

[2] The *prior* and *posterior* probabilities were formerly, and sometimes still are, called the *a priori* and *a posteriori* probabilities.

should be decreased (or increased) if the first test was a failure (or a success), so really $P[1,2] = P[2|1]P[1] \neq p^2$.

The frequency approach also assumes the repeatability of experiments, under identical conditions and with the possibility of different outcomes, for example tossing a coin many times. So what are we to make of an everyday statement such as 'It will probably rain tomorrow', when there is only one tomorrow? Critics also argue that quoting the result of the measurement of a physical quantity, such as the mass of a body as 10 ± 1 kg, together with a statement about the probability that the quantity lies within the range specified by the uncertainty of 1kg, is incompatible with the frequency definition. This is because the quantity measured presumably does have a true value and so either it lies within the error bars or it does not, i.e., the probability is either 1 or 0. These various objections are addressed in the next section.

2.3.2. Subjective Interpretation

The calculus of probabilities as outlined above proceeds from the definition of probabilities for simple events to the probabilities of more complex events. In practice, what is required in physical applications is the inverse, that is, given certain experimental observations we would like to deduce something about the parent population and the generating mechanism by which the events were produced. This, in general, is the problem of *statistical inference* alluded to in Chapter 1.

To illustrate how this leads to an alternative interpretation of probability, we return to the definition of conditional probability, which can be written using (2.3a) as

$$P[B \cap A] = P[A]P[B|A]. \tag{2.9a}$$

Since $A \cap B$ is the same as $B \cap A$, we also have

$$P[B \cap A] = P[B]P[A|B], \tag{2.9b}$$

and by equating these two quantities we deduce that

$$P[B|A] = \frac{P[B]P[A|B]}{P[A]}, \tag{2.10}$$

provided $P[A] \neq 0$. Finally, we can generalize to the case of multiple criteria and use the law of total probability (2.6b) to write

$$P[A] = \sum_{j}^{n} P[A \cap B_j] = \sum_{j}^{n} P[B_j] P[A|B_j],$$

so that (2.10) becomes

$$P[B_i|A] = \frac{P[B_i] P[A|B_i]}{\sum_{j}^{n} P[B_j] P[A|B_j]}. \tag{2.11}$$

This result was first published in the 18th century by an English clergyman, Thomas Bayes, and is known as *Bayes' theorem*. It differs from the frequency approach to probability by introducing an element of subjectivity into the definition — hence its name: *subjective probability*.

In this approach, the sample space is interpreted as a set of n mutually exclusive and exhaustive hypotheses (i.e., all possible hypotheses are included in the set). Suppose an event

A can be explained by the mutually exclusive hypotheses represented by $B_1, B_2, \ldots, B_n$. These hypotheses have certain *prior* probabilities $P[B_i]$ of being true. Each of them can give rise to the occurrence of the event A, but with distinct probabilities $P[A|B_i]$, which are the probabilities of observing A *when B_i is known to be satisfied*. The interpretation of (2.11) is then: the probability of the hypothesis B_i, given the observation of A, is the *prior* probability that B_i is true, multiplied by the probability of the observation assuming that B_i is true, divided by the sum of the same product of probabilities for all allowed hypotheses. In this approach, the *prior* probabilities are statements of one's belief that a particular hypothesis is true. Thus in the subjective approach, quoting the measurement of a mass as 10 ± 1 kg is a valid statement that expresses one's current belief about the true value of that quantity.

To illustrate how Bayes' theorem can be used, consider an example where a football team is in a knockout tournament and will play either team A or team B next depending on their performance in earlier rounds. The manager assesses their prior probabilities of winning against A or B as $P[A] = 3/10$ and $P[B] = 5/10$ and their probabilities of winning given that they know their opponents as $P[W|A] = 5/10$ and $P[W|B] = 7/10$. If the team wins their next game, we can calculate from (2.11) the probabilities that their opponents were either A or B as $P[A|W] = 3/10$ and $P[B|W] = 7/10$. So if you had to bet, the odds favor the hypothesis that the opponents were team B.

The result of this simple example is in agreement with common sense, but the following examples illustrate that Bayes' theorem can sometimes lead to results that at first sight are somewhat surprising.

EXAMPLE 2.6

The process of producing microchips at a particular factory is known to result in 0.2% that do not satisfy their specification, i.e., are faulty. A test is developed that has a 99% probability of detecting these chips if they are faulty. There is also a 3% probability that the test will give a false positive, i.e., a positive result even though the chip is not faulty. What is the probability that a chip is faulty if the test gives a positive result?

If we denote the presence of a fault by f and its absence by $\bar{f}$, then $P[f] = 0.002$ and $P[\bar{f}] = 0.998$. The test has a 99% probability of detecting a fault if present, so it follows that the test yields a 'false negative' result in 1% of tests, that is, the probability is 0.01 that the test will be negative even though the chip tested does have a fault. So if we denote a positive test by $+$ and a negative one by $-$, then $P[+|f] = 0.99$ and $P[-|f] = 0.01$. There is also a 3% probability of the test giving a false positive, i.e., a positive result even though the chip does not have a fault, so $P[+|\bar{f}] = 0.03$ and $P[-|\bar{f}] = 0.97$. Then from Bayes' theorem,

$$P[f|+] = \frac{P[+|f]P[f]}{P[+|f]P[f] + P[+|\bar{f}]P[\bar{f}]}$$

$$= \frac{0.99 \times 0.002}{(0.99 \times 0.002) + (0.03 \times 0.998)} = 0.062.$$

So the probability of a chip having a fault given a positive test result is only 6.2%.[3]

---

[3]The same reasoning applied to a medical test for rare conditions shows that a positive test result often means only a low probability for having the condition.

EXAMPLE 2.7

An experiment is set up to detect particles of a particular type A in a beam, using a detector that has a 95% efficiency for their detection. However, the beam also contains 15% of particles of a second type B and the detector has a 10% probability of misrecording these as particles of type A. What is the probability that the signal is due to a particle of type B?

If the observation of a signal is denoted by S, we have

$$P[A] = 0.85, \quad P[B] = 0.15,$$

and

$$P[S|A] = 0.95, \quad P[S|B] = 0.10.$$

Then from Bayes' theorem,

$$P[B|S] = \frac{P[B]\,P[S|B]}{P[A]\,P[S|A] + P[B]\,P[S|B]}$$

$$= \frac{0.15 \times 0.10}{(0.85 \times 0.95) + (0.15 \times 0.10)} = 0.018.$$

Thus there is a probability of only 1.8% that the signal is due a particle of type B, even though 15% of the particles in the beam are of type B.

If we had to choose an hypothesis from the set B_i we would choose that one with the greatest posterior probability. However (2.11) shows that this requires knowledge of the prior probabilities $P[B_i]$ and these are, in general, unknown. *Bayes' postulate* is the hypothesis that, in the absence of any other knowledge, the prior probabilities should all be taken as equal. A simple example will illustrate the use of this postulate.

EXAMPLE 2.8

A container has four balls, which could be either all white (hypothesis 1), or two white and two black (hypothesis 2). If n balls are withdrawn, one at a time, replacing them after each drawing, what are the probabilities of obtaining an event E with n white balls under the two hypotheses? Comment on your answer.

If n balls are withdrawn, one at a time, replacing them after each drawing, the probabilities of obtaining an event E with n white balls under the two hypotheses are

$$P[E|H_1] = 1 \quad \text{and} \quad P[E|H_2] = 2^{-n}.$$

Now from Bayes' postulate,

$$P[H_1] = P[H_2] = \frac{1}{2}$$

and so from (2.11),

$$P[H_1|E] = \frac{2^n}{1 + 2^n} \quad \text{and} \quad P[H_2|E] = \frac{1}{1 + 2^n}.$$

Providing no black ball appears, the first hypothesis should be accepted because it has the greater posterior probability.

While Bayes' postulate might seem reasonable, it is the subject of controversy and can lead to erroneous conclusions. In the frequency theory of probability it would imply that events corresponding to the various B_i are distributed with equal frequency in some population from which the actual B_i has arisen. Many statisticians reject this as unreasonable.[4]

Later in this book we will examine some of the many other suggested alternatives to Bayes' postulate, including the principle of least squares and minimum chi-squared. This discussion will be anticipated by briefly mentioning here one principle of general application, that of *maximum likelihood*.

From (2.11) we see that

$$P[B_i|A] \propto P[B_i]\, L, \tag{2.12}$$

where $L = P[A|B_i]$ is called the *likelihood*. *The principle of maximum likelihood* states that when confronted with a set of hypotheses B_i, we choose the one that maximizes L, if one exists, that is, the one that gives the greatest probability to the observed event. Note that this is *not* the same as choosing the hypothesis with the greatest probability. It is not at all self-evident why one should adopt this particular choice as a principle of statistical inference, and we will return to this point in Chapter 7. For the simple case above, the maximum likelihood method clearly gives the same result as Bayes' postulate.

There are other ways of defining probabilities and statisticians do not agree among themselves on the 'best' definition, but in this book we will not dwell too much on the differences between them,[5] except to note that the frequency definition is usually used, although the subjective approach will be important when discussing some aspects of interval estimation and hypothesis testing.

PROBLEMS 2

2.1 The diagram shows an electrical circuit with four switches S ($S = 1, 4$) that when closed allow a current to flow. If the switches act independently and have a probability p for being closed, what is the probability for a current to flow from I to O? Check your calculation by calculating the probability for no current to flow.

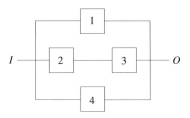

FIGURE 2.3 Circuit diagram.

---

[4]Bayes himself seems to have had some doubts about it and it was not published until after his death.

[5]To quote a remark attributed to the eminent statistician Sir Maurice Kendall: "In statistics it is a mark of immaturity to argue over much about the fundamentals of probability theory."

PROBLEMS 2

2.2 In the sample space $S = \{1,2,3, \ldots, 9\}$, the events A, B, C, and D are defined by $A = (2,4,8)$, $B = (2,3,5,9)$, $C = (1,2,4)$, and $D = (6,8,9)$. List the structure of the events

(a) $\bar{A} \cap D$, (b) $(B \cap \bar{C}) \cup A$, (c) $(A \cap \bar{B} \cap C)$, (d) $A \cap (\bar{B} \cup \bar{D})$.

2.3 A technician is 70% convinced that a circuit is failing due to a fault in a particular component. Later it emerges that the component was part of a batch supplied from a source where it is known that 15% of batches are faulty. How should this evidence alter the technician's view?

2.4 Over several seasons, two teams A and B have met 10 times. Team A has won 5 times, team B has won 3 times, and 2 matches have been drawn. What is the probability that in their next two encounters (a) team B will win both games and (b) team A will win at least one game?

2.5 Five physics books, 4 maths books, and 3 chemistry books are to be placed on a shelf so that all books on a given subject are together. How many arrangements are possible?

2.6 a. One card is drawn at random from a standard deck of 52 cards. What is the probability that the card is nine (9) or a club (C)?

b. If four cards are drawn, what is the probability that at least three will be of the same suite?

2.7 A lie detector test is used to detect people who have committed a crime. It is known to be 95% reliable when testing people who have actually committed a crime and 99% reliable when they are innocent. The test is given to a suspect chosen at random from a population of which 3% have been convicted of a crime, and the result is positive, i.e., indicates guilt. What is the probability the suspect has not actually committed a crime?

2.8 Box A contains 4 red and 2 blue balls, and box B contains 2 red and 6 blue balls. One of the boxes is selected by tossing a die and then a ball is chosen at random from the selected box. If the ball selected is red, what is the probability that it came from box A?

2.9 Three balls are drawn at random successively from a bag containing 12 balls, of which 3 are red, 4 are white, and 5 are blue. In case (a), each ball is not replaced after it has been drawn, and in case (b) they are replaced. In both cases all three balls are found to be in the order red, white, and blue. What are the probabilities for this in the two cases?

CHAPTER 3

Probability Distributions I: Basic Concepts

OUTLINE

3.1 Random Variables	35	3.3.1 Joint Probability Distributions	45
3.2 Single Variates	36	3.3.2 Marginal and Conditional Distributions	45
3.2.1 Probability Distributions	36	3.3.3 Moments and Expectation Values	49
3.2.2 Expectation Values	40		
3.2.3 Moment Generating, and Characteristic Functions	42	3.4 Functions of a Random Variable	51
3.3 Several Variates	45		

In the present chapter we will define random variables and discuss their probability distributions. This will be followed in the next chapter by a discussion of the properties of some of the population distributions that are commonly encountered in physical science and that govern the outcome of experiments.

3.1. RANDOM VARIABLES

The events discussed in Chapters 1 and 2 could be arbitrary quantities, heads, tails, etc., or numerical values. It is useful to associate a set of real numbers with the outcomes of an experiment even if the basic data are non-numeric. This association can be expressed by a real-valued function that transforms the points in S to points on the x axis. The function is called a *random variable* and to distinguish between random variables and other variables, the former will also be called *variates*.[1] Returning to the coining tossing experiment of Example 1.2, the four events were

$$E_1 = (H,H); \quad E_2 = (H,T); \quad E_3 = (T,H); \quad E_4 = (T,T),$$

and so if we let x be a random variable[2] that can assume the values given by the number of 'heads', then

$$x(E_1) = 2; \quad x(E_2) = 1; \quad x(E_3) = 1; \quad x(E_4) = 0.$$

And, using the frequency definition of probability,

$$P[x=0] = P[E_4] = 1/4; \quad P[x=1] = P[E_2 \cup E_3] = 1/2; \quad P[x=2] = P[E_1] = 1/4,$$

where we have assumed an unbiased coin, so that throwing a 'head' or a 'tail' is equally likely. From this example we see that a random variable can assume an ensemble of numerical values in accord with the underlying probability distribution. These definitions can be extended to continuous variates and to situations involving multivariates, as we will see below. In general, it is the quantities $P[x]$ that are the objects of interest and it is to these that we now turn.

3.2. SINGLE VARIATES

In this section we will examine the case of a single random variable. The ideas discussed here will be extended to the multivariate case in Section 3.3.

3.2.1. Probability Distributions

First we will need some definitions that extend those given in Chapter 2 in the discussion of the axioms of probability, starting with the case of a single discrete random variable. If x is a discrete random variable that can take the values $x_k (k = 1, 2, \ldots)$ with probabilities $P[x_k]$, then we can define a probability distribution $f(x)$ by

$$P[x] \equiv f(x). \tag{3.1a}$$

Thus,

$$P[x_k] = f(x_k) \quad \text{for} \quad x = x_k, \quad \text{otherwise } f(x) = 0. \tag{3.1b}$$

To distinguish between the cases of discrete and continuous variables, the probability distribution for the former is often called the *probability mass function* (or

---

[1] Some authors use the word 'variate' to mean any variable, random or otherwise, that can take on a numerical value.

[2] A convention that is often used is to denote random variables by upper case letters and the values they can take by the corresponding lower case letter. As there is usually no ambiguity, lower case letters will be used for both.

simply a *mass function*) sometimes abbreviated to *pmf*. A pmf satisfies the following two conditions:

1. $f(x)$ is a single-valued non-negative real number for all real values of x, i.e., $f(x) \geq 0$;
2. $f(x)$ summed over all values of x is unity:

$$\sum_x f(x) = 1. \tag{3.1c}$$

We saw in Chapter 1 that we are also interested in the probability that x is less than or equal to a given value. This was called the *cumulative distribution function* (or simply the *distribution function*), sometimes abbreviated to *cdf*, and is given by

$$F(x) = \sum_{x_k \leq x} f(x_k). \tag{3.2a}$$

So, if x takes on the values $x_k (k = 1, 2, \ldots n)$, the cumulative distribution function is

$$F(x) = \begin{cases} 0 & -\infty < x < x_1 \\ f(x_1) & x_1 \leq x < x_2 \\ f(x_1) + f(x_2) & x_2 \leq x < x_3 \\ \vdots & \vdots \\ f(x_1) + \cdots + f(x_n) & x_n \leq x < \infty \end{cases} \tag{3.2b}$$

$F(x)$ is a nondecreasing function with limits 0 and 1 as $x \to -\infty$ and $x \to +\infty$, respectively. The quantile x_α of order α, defined in Chapter 1, is thus the value of x such that $F(x_\alpha) = \alpha$, with $0 \leq \alpha \leq 1$, and so $x_\alpha = F^{-1}(\alpha)$, where F^{-1} is the inverse function of F. For example, the median is $x_{0.5}$.

As sample sizes become larger, frequency plots tend to approximate smooth curves and if the area of the histogram is normalized to unity, as in Fig. 1.5, the resulting function $f(x)$ is a continuous *probability density function* (or simply a *density function*) abbreviated to *pdf*, introduced in Chapter 1. The definitions above may be extended to continuous random variables with the appropriate changes. Thus, for a continuous random variable x, with a pdf $f(x)$, (3.2a) becomes

$$F(x) = \int_{-\infty}^{x} f(x') dx', \quad (-\infty < x < \infty). \tag{3.3}$$

It follows from (3.3) that if a member of a population is chosen at *random*, that is, by a method that makes it equally likely that each member will be chosen, then $F(x)$ is the probability that the member will have a value $\leq x$. While all this is clearly consistent with earlier definitions, once again we should note the element of circularity in the concept of randomness defined in terms of probability. In mathematical statistics it is usual to start from the cumulative distribution and define the density function as its derivative. For the mathematically well-behaved distributions usually met in physical science the two approaches are equivalent.

The density function $f(x)$ has the following properties analogous to those for discrete variables.

1. $f(x)$ is a single-valued non-negative real number for all real values of x.

In the frequency interpretation of probability, $f(x)dx$ is the probability of observing the quantity x in the range $(x, x+dx)$. Thus, the second condition is:

2. $f(x)$ is normalized to unity:

$$\int_{-\infty}^{+\infty} f(x)dx = 1.$$

It follows from property 2 that the probability of x lying between any two real values a and b for which $a < b$ is given by

$$P[a \leq x \leq b] = \int_a^b f(x)dx, \tag{3.4}$$

and so, unlike a discrete random variable, the probability of a continuous random variable assuming *exactly* any of its values is zero. This result may seem rather paradoxical at first until you consider that between any two values a and b there is an infinite number of other values and so the probability of selecting an exact value from this infinitude of possibilities must be zero. The density function cannot therefore be given in a tabular form like that of a discrete random variable.

EXAMPLE 3.1

A family has 5 children. Assuming that the birth of a girl or boy is equally likely, construct a frequency table of possible outcomes and plot the resulting probability mass function $f(g)$ and the associated cumulative distribution function $F(g)$.

The probability of a sequence of births containing g girls (and hence $b = 5 - g$ boys) is $\left(\frac{1}{2}\right)^g \left(\frac{1}{2}\right)^b = \left(\frac{1}{2}\right)^5$. However there are ${}_5C_g$ such sequences, and so the probability of having g girls is $P[g] = {}_5C_g/32$. The probability mass function $f(g)$ is thus as given in the table below.

g	0	1	2	3	4	5
$f(g)$	1/32	5/32	10/32	10/32	5/32	1/32

From this table we can find the cumulative distribution function using (3.2b), and $f(g)$ and $F(g)$ are plotted in Figs 3.1(a) and (b), respectively, below.

FIGURE 3.1 Plots of the probability mass function $f(g)$ and the cumulative distribution function $F(g)$.

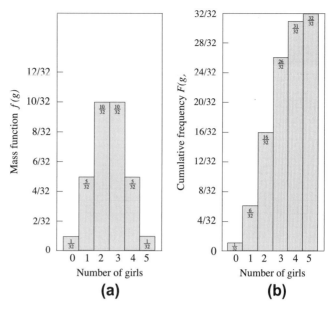

EXAMPLE 3.2

Find the value of N in the continuous density function:

$$f(x) = \begin{cases} Ne^{-x}x^2/2 & x \geq 0 \\ 0 & x < 0 \end{cases},$$

and find its associated distribution function $F(x)$. Plot $f(x)$ and $F(x)$.

Because $f(x)$ has to be correctly normalized, to find N we evaluate the integral:

$$\frac{N}{2}\int_0^\infty e^{-x}x^2 dx = 1.$$

Integrating by parts, gives

$$\frac{1}{N} = \frac{1}{2}\int_0^\infty e^{-x}x^2 dx = -\frac{1}{2}[e^{-x}(x^2 + 2x + 2)]_0^\infty = 1,$$

so that $N = 1$. The resulting density function is plotted in Fig. 3.2(a). The associated distribution function is

$$F(x) = \frac{1}{2}\int_0^x e^{-u}u^2 du = -\frac{1}{2}[e^{-u}(u^2 + 2u + 2)]_0^x = -\frac{1}{2}e^{-x}(x^2 + 2x + 2) + 1,$$

and is shown in Fig. 3.2(b).

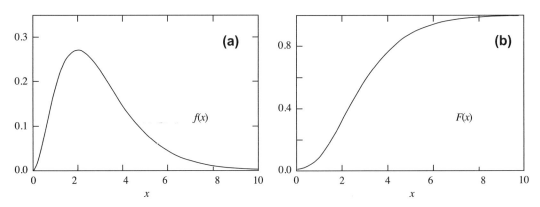

FIGURE 3.2 Probability density function $f(x) = e^{-x}x^2/2 (x \geq 0)$ and the corresponding cumulative distribution function $F(x)$.

Some of the earlier definitions of Chapter 1 may now be rewritten in terms of these formal definitions. Thus, the general moments about an arbitrary point λ are, for a continuous variate,

$$\mu'_n = \int_{-\infty}^{+\infty} f(x)(x - \lambda)^n, \tag{3.5}$$

so that the mean and variance, also with respect to the point λ, are

$$\mu_\lambda = \int_{-\infty}^{+\infty} f(x)(x - \lambda)\,dx \quad \text{and} \quad \sigma^2 = \int_{-\infty}^{+\infty} f(x)(x - \mu_\lambda)^2\,dx, \tag{3.6}$$

respectively. The integrals in (3.5) may not converge for all n, and some distributions possess only the trivial zero-order moment. For convenience, usually $\lambda = 0$ will be used in what follows.

3.2.2. Expectation Values

The *expectation value*, also called the *expected value*, of a random variable is obtained by finding the average value of the variate over all its possible values weighted by the probability of their occurrence. Thus, if x is a discrete random variable with the possible values $x_1, x_2, \ldots, x_n$, then the expectation value of x is defined as

$$E[x] \equiv \sum_{i=1}^{n} x_i P[x_i] = \sum_x x f(x), \tag{3.7}$$

where the second sum is over all relevant values of x and $f(x)$ is their probability mass distribution. The analogous quantity for a continuous variate with density function $f(x)$ is

$$E[x] = \int_{-\infty}^{+\infty} x f(x)\,dx. \tag{3.8a}$$

3.2. SINGLE VARIATES

We can see from this definition that the *n*th moment of a distribution about any point λ is

$$\mu'_n = E[(x-\lambda)^n]. \tag{3.8b}$$

In particular, the *n*th *central moment* is

$$\mu_n = E[(x - E[x])^n] = \int_{-\infty}^{+\infty} (x-\mu)^n f(x) dx, \tag{3.8c}$$

and for $\lambda = 0$ the *n*th *algebraic moment* is

$$\mu'_n = E[x^n] = \int_{-\infty}^{+\infty} x^n f(x) dx \tag{3.8d}$$

Thus, the mean is the first algebraic moment and the variance is the second central moment. It follows from (3.8) that if c is a constant, then

$$E[cx] = cE[x], \tag{3.9a}$$

and for a set of random variables A, B, C, etc.:

$$E[A + B + C + \cdots] = E[A] + E[B] + E[C] + \cdots \tag{3.9b}$$

In addition, if the random variables A, B, C, etc. are independent, then

$$E[ABC...] = E[A]E[B]E[C]... \tag{3.9c}$$

EXAMPLE 3.3

Three 'fair' dice are thrown and yield face values a, b, and c. What is the expectation value for the sum of their face values?

From (3.7),

$$E[a] = \sum_{i=1}^{6} i(1/6) = 7/2,$$

and since $E[a] = E[b] = E[c]$, then from (3.9b) $E[a+b+c] = 21/2$.

EXAMPLE 3.4

Find the mean of the continuous distribution of Example 3.2.
Using (3.8d), the mean is

$$\mu = \frac{1}{2} \int_0^\infty x^3 e^{-x} dx.$$

Integrating by parts gives

$$\mu = -\frac{e^{-x}}{2}[x^3 + 3x^2 + 6x + 6]_0^\infty = 3.$$

3.2.3. Moment Generating, and Characteristic Functions

The usefulness of moments partly stems from the fact that knowledge of them determines the form of the density function. Formally, if the moments μ'_n of a random variable x exist and the series

$$\sum_{n=1}^{\infty} \frac{\mu'_n}{n!} r^n \tag{3.10}$$

converges absolutely for some $r > 0$, then the set of moments μ'_n uniquely determines the density function. There are exceptions to this statement, but fortunately it is true for all the distributions commonly met in physical science. In practice, knowledge of the first few moments essentially determines the general characteristics of the distribution and so it is worthwhile to construct a method that gives a representation of all the moments. Such a function is called a *moment generating function (mgf)* and is defined by

$$M_x(t) \equiv E[e^{xt}]. \tag{3.11}$$

For a discrete random variable x, this is

$$M_x(t) = \sum e^{xt} f(x), \tag{3.12a}$$

and for a continuous variable,

$$M_x(t) = \int_{-\infty}^{+\infty} e^{xt} f(x) \, dx. \tag{3.12b}$$

The moments may be generated from (3.11) by first expanding the exponential,

$$M_x(t) = E\left[1 + xt + \frac{1}{2!}(xt)^2 + \cdots\right] = \sum_{n=0}^{\infty} \frac{1}{n!} \mu'_n t^n,$$

then differentiating n times and setting $t = 0$, that is:

$$\mu'_n = \left.\frac{\partial^n M_x(t)}{\partial t^n}\right|_{t=0}. \tag{3.13}$$

For example, setting $n = 0$ and $n = 1$, gives $\mu'_0 = 1$ and $\mu'_1 = \mu$. Also, since the mgf about any point λ is

$$M_\lambda(t) = E[\exp\{(x - \lambda)t\}],$$

then if $\lambda = \mu$,

$$M_\mu(t) = e^{-\mu t} M_x(t). \tag{3.14}$$

An important use of the mgf is to compare two density functions $f(x)$ and $g(x)$. If two random variables possess mgfs that are equal for some interval symmetric about the origin, then $f(x)$ and $g(x)$ are identical density functions. It is also straightforward to show that the mgf of a sum of independent random variables is equal to the product of their individual mgfs.

3.2. SINGLE VARIATES

It is sometimes convenient to consider, instead of the mgf, its logarithm. The Taylor expansion[3] for this quantity is

$$\ln M_x(t) = \kappa_1 t + \kappa_2 \frac{t^2}{2} + \cdots,$$

where κ_n is the *cumulant* of order n, and

$$\kappa_n = \left.\frac{\partial^n \ln M_x(t)}{\partial t^n}\right|_{t=0}.$$

Cumulants are simply related to the central moments of the distribution, the first few relations being

$$\kappa_i = \mu_i \ (i = 1, 2, 3), \quad \kappa_4 = \mu_4 - 3\mu_2^2.$$

For some distributions the integral defining the mgf may not exist and in these circumstances the Fourier transform of the density function, defined as

$$\phi_x(t) \equiv E[e^{itx}] = \int_{-\infty}^{+\infty} e^{itx} f(x) dx = M_x(it), \qquad (3.15)$$

may be used. In statistics, $\phi_x(t)$ is called the *characteristic function (cf)*. The density function is then obtainable by the Fourier transform theorem (known in this context as the *inversion theorem*):

$$f(x) = \frac{1}{2\pi} \int_{-\infty}^{+\infty} e^{-itx} \phi_x(t) dt. \qquad (3.16)$$

The cf obeys theorems analogous to those obeyed by the mgf, that is: (a) if two random variables possess cfs that are equal for some interval symmetric about the origin then they have identical density functions; and (b) the cf of a sum of independent random variables is equal to the product of their individual cfs. The converse of (b) is however untrue.

EXAMPLE 3.5

Find the moment generating function of the density function used in Example 3.2 and calculate the three moments μ'_1, μ'_2, and μ'_3.

Using definition (3.12b),

$$M_x(t) = \int_0^\infty e^{xt} f(x) dx = \frac{1}{2} \int_0^\infty e^{xt} x^2 e^{-x} dx = \frac{1}{2} \int_0^\infty e^{-x(1-t)} x^2 dx,$$

---

[3]Some essential mathematics is reviewed briefly in Appendix A.

which integrating by parts gives:

$$M_x(t) = \left\{ -\frac{e^{-x(1-t)}}{2(1-t)^3}[(1-t)^2 x^2 + 2(1-t)x + 2] \right\}_0^\infty = \frac{1}{(1-t)^3}.$$

Then, using (3.13), the first three moments of the distribution are found to be

$$\mu_1' = 3, \quad \mu_2' = 12, \quad \mu_3' = 60.$$

EXAMPLE 3.6

(a) Find the characteristic function of the density function:

$$f(x) = \begin{cases} 2x/a^2 & a \leq x < 0 \\ 0 & \text{otherwise} \end{cases},$$

and (b) the density function corresponding to a characteristic function $e^{-|t|}$.

(a) From (3.15),

$$\phi_x(t) = E[e^{itx}] = \frac{2}{a^2} \int_0^a e^{itx} x \, dx.$$

Again, integration by parts gives

$$\phi_x(t) = \frac{2}{a^2} \left[\frac{e^{itx}}{(it)^2}(itx - 1) \right]_0^a = -\frac{2}{a^2 t^2} \left[e^{ita}(ita - 1) + 1 \right].$$

(b) From the inversion theorem,

$$f(x) = \frac{1}{2\pi} \int_{-\infty}^\infty e^{-|t|} e^{-itx} dx = \frac{1}{\pi} \int_0^\infty e^{-t} \cos(tx) dx,$$

where the symmetry of the circular functions has been used. The second integral may be evaluated by parts to give

$$\pi f(x) = \left[-e^{-t}\cos(tx) \right]_0^\infty - x \int_0^\infty e^{-t}\sin(tx) dt$$

$$= 1 - x\left\{ \left[-e^{-t}\sin(tx) \right]_0^\infty + x \int_0^\infty e^{-t}\cos(tx) dt \right\} = 1 - \pi x^2 f(x).$$

Thus,

$$f(x) = \frac{1}{\pi(1+x^2)}, \quad -\infty \leq x \leq \infty.$$

This is the density of the Cauchy distribution that we will meet again in Section 4.5.

3.3. SEVERAL VARIATES

All the results of the previous sections may be extended to multivariate distributions. We will concentrate on continuous variates, but the formulas may be transcribed in a straightforward way to describe discrete variates.

3.3.1. Joint Probability Distributions

The *multivariate joint density function* $f(x_1, x_2, \ldots, x_n)$ of the n continuous random variables $x_1, x_2, \ldots, x_n$ is a single-valued non-negative real number for all real values of $x_1, x_2, \ldots, x_n$, normalized so that

$$\int_{-\infty}^{+\infty} \cdots \int_{-\infty}^{+\infty} f(x_1, x_2, \ldots, x_n) \prod_{i=1}^{n} dx_i = 1, \tag{3.17}$$

and the probability that x_1 falls between any two numbers a_1 and b_1, x_2 falls between any two numbers a_2 and b_2, ..., and x_n falls between any two numbers a_n and b_n, *simultaneously*, is defined by

$$P[a_1 \leq x_1 \leq b_1; \ldots; a_n \leq x_n \leq b_n] \equiv \int_{a_n}^{b_n} \cdots \int_{a_1}^{b_1} f(x_1, x_2, \ldots, x_n) \prod_{i=1}^{n} dx_i. \tag{3.18}$$

Similarly, the *multivariate joint distribution function* $F(x_1, x_2, \ldots, x_n)$ of the n random variables $x_1, x_2, \ldots, x_m$ is

$$F(x_1, x_2, \ldots, x_n) \equiv \int_{-\infty}^{x_n} \cdots \int_{-\infty}^{x_1} f(t_1, t_2, \ldots, t_n) \prod_{i=1}^{n} dt_i. \tag{3.19}$$

For simplicity, consider the case of just two random variables x and y. These could correspond to the energy and angle of emission of a particle emitted in a nuclear scattering reaction. If an event A corresponds to the variable x being observed in the range $(x, x + dx)$ and the event B corresponds to the variable y being observed in the range $(y, y + dy)$, then

$$P[A \cap B] = \text{probability of } x \text{ being in } (x, x+dx) \text{ and } y \text{ being in } (y, y+dy)$$
$$= f(x, y) dx\, dy.$$

As noted in Chapter 1, the joint density function corresponds to the density of points on a scatter plot of x and y in the limit of an infinite number of points. This is illustrated in Fig. 3.3, using the data shown on the scatter plot of Fig. 1.3(b).

3.3.2. Marginal and Conditional Distributions

We may also be interested in the density function of a subset of variables. This is called the *marginal density function* f^M, and in general for a subset x_1 ($i = 1, 2, \ldots, m < n$) of the variables is given by integrating the joint density function over all the variables other than $x_1, x_2, \ldots, x_m$. Thus,

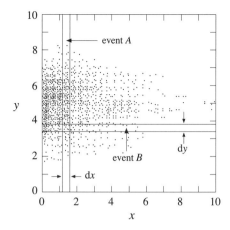

FIGURE 3.3 A scatter plot of 1000 events that are functions of two random variables x and y showing two infinitesimal bands dx and dy. The area of intersection of the bands is $dx\,dy$ and $f(x,y)dx\,dy$ is the probability of finding x in the interval $(x, x+dx)$ and y in the interval $(y, y+dy)$.

$$f^M(x_1, x_2, \ldots, x_m) = \int_{-\infty}^{+\infty} \cdots \int_{-\infty}^{+\infty} f(x_1, x_2, \ldots, x_m, x_{m+1}, \ldots, x_n) \prod_{i=m+1}^{n} dx_i. \quad (3.20a)$$

In the case of two variables, we may be interested in the density function of x regardless of the value of y, or the density function of y regardless of x. For example, the failure rate of a resistor may be a function of its operating temperature and the voltage across it, but in some circumstances we might be interested in just the dependence on the former. In these cases (3.20a) becomes

$$f^M(x) = \int_{-\infty}^{+\infty} f(x,y)dy \quad \text{and} \quad f^M(y) = \int_{-\infty}^{+\infty} f(x,y)dx. \quad (3.20b)$$

These density functions correspond to the normalized histograms obtained by projecting a scatter plot of x and y onto one of the axes. This is illustrated in Fig. 3.4, again using the data of Fig. 1.3(b).

We can also define the *multivariate conditional density function* of the random variables $x_1 (i = 1, 2, \ldots, m < n)$ by

$$f^C(x_1, x_2, \ldots, x_m | x_{m+1}, x_{m+2}, \ldots, x_n) \equiv \frac{f(x_1, x_2, \ldots, x_n)}{f(x_{m+1}, x_{m+2}, \ldots, x_n)}. \quad (3.21)$$

Again, if we consider the case of two variables x and y, the probability for y to be in the interval $(y, y+dy)$ with any x (event B), given that x is in the interval $(x, x+dx)$ with any y (event A), is

$$P[B|A] = \frac{P[A \cap B]}{P[A]} = \frac{f(x,y)dx\,dy}{f^M(x)dx},$$

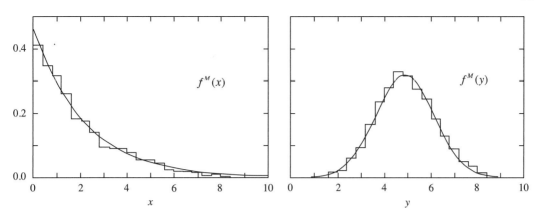

FIGURE 3.4 Normalized histograms obtained by projecting the data of Fig. 1.3(b) onto the x and y axes, together with the corresponding marginal probability density functions $f^M(x)$ and $f^M(y)$.

where $f^M(x)$ is the marginal density function for x. The conditional density function for y given x, is thus

$$f^C(y|x) = \frac{f(x,y)}{f^M(x)} = \frac{f(x,y)}{\int f(x,y')dy'}. \tag{3.22a}$$

This is the density function of the single random variable y where x is treated as a constant. It corresponds to projecting the events in a band dx centered at some value x onto the y axis and renormalizing the resulting density so that it is unity when integrated over y. The form of $f^C(y|x)$ will therefore vary as different values of x are chosen.

The conditional density function for x given y is obtained from (3.22a) by interchanging x and y, so that

$$f^C(x|y) = \frac{f(x,y)}{f^M(y)} = \frac{f(x,y)}{\int f(x',y)dx'}, \tag{3.22b}$$

and combining these two equations gives

$$f^C(x|y) = \frac{f^C(y|x)f^M(x)}{f^M(y)}, \tag{3.22c}$$

which is Bayes' theorem for continuous variables.

We can use these definitions to generalize the law of total probability (2.6b) to the case of continuous variables. Using conditional and marginal density functions we have

$$f(x,y) = f^C(y|x)\,f^M(x) = f^C(x|y)\,f^M(y), \tag{3.23}$$

so the marginal density functions may be written as

$$f^M(y) = \int_{-\infty}^{+\infty} f^C(y|x)\,f^M(x)dx$$

and

$$f^M(x) = \int_{-\infty}^{+\infty} f^C(x|y) f^M(y) dy.$$

With more than one random variable we also have to consider the question of statistical independence (by analogy with the work of Chapter 2). If the random variables may be split into groups such that their joint density function is expressible as a product of marginal density functions of the form

$$f(x_1, x_2, \ldots, x_n) = f_1^M(x_1, x_2, \ldots, x_i) f_2^M(x_{i+1}, x_{i+2}, \ldots, x_k) \ldots f_n^M(x_{l+1}, x_{l+2}, \ldots, x_n),$$

then the sets of variables

$$(x_1, x_2, \ldots, x_i); (x_{i+1}, x_{i+2}, \ldots, x_k); \ldots ; (x_{l+1}, x_{l+2}, \ldots, x_n),$$

are said to be *statistically independent*, or *independently distributed*. So two random variables x and y are independently distributed if

$$f(x,y) = f^M(x) f^M(y). \tag{3.24}$$

It follows from (3.22) that in this case the conditional density function of one variate does not depend on knowledge about the other variate.

EXAMPLE 3.7

The joint mass function for two discrete variables x and y is given by

$$f(x,y) = \begin{cases} k(2x + 3y) & 0 \le x \le 3, \ 0 \le y \le 2 \\ 0 & \text{otherwise} \end{cases},$$

where k is a constant. Find: (a) the value of k, (b) $P[x \ge 2, y \le 1]$, and (c) the marginal density of x. The mass function may be tabulated as below.

x	y = 0	1	2	total
0	0	3k	6k	9k
1	2k	5k	8k	15k
2	4k	7k	10k	21k
3	6k	9k	12k	27k
total	12k	24k	36k	72k

(a) The normalization condition is $\sum_{x,y} f(x,y) = 1$, so $k = 1/72$.

(b) $P[x \ge 2, y \le 1] = P[x = 2, y = 1] + P[x = 2, y = 0] + P[x = 3, y = 1] + P[x = 3, y = 0]$
$= 7k + 4k + 9k + 6k = 26k = 13/36$.

(c) The marginal probability of x is

$$P[x] = \sum_y P[x,y] = \begin{cases} 9k = 3/24 & x = 0 \\ 15k = 5/24 & x = 1 \\ 21k = 7/24 & x = 2 \\ 27k = 9/24 & x = 3 \end{cases}$$

EXAMPLE 3.8

If $f(x,y)$ is the joint density function of two continuous random variables x and y, defined by

$$f(x,y) = \begin{cases} e^{-(x+y)} & x, y \geq 0 \\ 0 & otherwise \end{cases},$$

find their conditional distribution.

From (3.22b),

$$f^C(x|y) = \frac{f(x,y)}{f^M(y)},$$

where the marginal density of y is given from (3.20b) as

$$f^M(y) = \int_0^\infty f(x,y)dx = e^{-y}[-e^{-x}]_0^\infty = e^{-y}.$$

Thus

$$f^C(x|y) = \frac{e^{-(x+y)}}{e^{-y}} = e^{-x}.$$

3.3.3. Moments and Expectation Values

The definition of moments and expectation values can be generalized to the multivariable case. Thus the rth algebraic moment of the random variable x_i is given by

$$E[x_i^r] = \int_{-\infty}^{+\infty} \cdots \int_{-\infty}^{+\infty} x_i^r f(x_1, x_2, \ldots, x_n) \prod_{j=1}^n dx_j, \qquad (3.25)$$

from which we obtain the results

$$\mu_i = \int_{-\infty}^{+\infty} \cdots \int_{-\infty}^{+\infty} x_i f(x_1, x_2, \ldots, x_n) \prod_{j=1}^n dx_j \qquad (3.26a)$$

and

$$\sigma_i^2 = \int_{-\infty}^{+\infty} \cdots \int_{-\infty}^{+\infty} (x_i - \mu_i)^2 f(x_1, x_2, \ldots, x_n) \prod_{j=1}^n dx_j, \qquad (3.26b)$$

for the mean and variance. In addition to the individual moments of (3.25) we can also define *joint moments*. In general these are given by

$$E\left[x_a^i x_b^j \ldots x_c^k\right] \equiv \int_{-\infty}^{+\infty} \ldots \int_{-\infty}^{+\infty} (x_a^i x_b^j \ldots x_c^k) f(x_1, x_2, \ldots, x_n) \prod_{m=1}^{n} dx_m. \quad (3.27)$$

The most important of these is the *covariance*, introduced in Chapter 1 for two random variables, and now defined more generally for any pair of variates as

$$\operatorname{cov}(x_i, x_j) \equiv \sigma_{ij} = \int_{-\infty}^{+\infty} \ldots \int_{-\infty}^{+\infty} (x_i - \mu_i)(x_j - \mu_j) f(x_1, x_2, \ldots, x_n) \prod_{m=1}^{n} dx_m, \quad (3.28)$$

where the means are given by (3.26a). In Chapter 1, the *correlation coefficient* $\rho(x_i, x_j)$ was defined by

$$\rho(x_i, x_j) \equiv \frac{\operatorname{cov}(x_i, x_j)}{\sigma(x_i)\sigma(x_j)}. \quad (3.29)$$

This is a number lying between -1 and $+1$. It is a necessary condition for statistical independence that $\rho(x_i, x_j)$ is zero. However, this is *not* a sufficient condition and $\rho(x_i, x_j) = 0$ does *not* always imply that x_i and x_j are independently distributed.

EXAMPLE 3.9

Find the means, variances, and covariance for the density of Example 3.8.
The mean μ_x (which is equal to μ_y by symmetry) is, from (3.26a),

$$\mu_x = \int_{-\infty}^{+\infty} \int_{-\infty}^{+\infty} x f(x, y) dx\, dy = \int_0^\infty x e^{-x} dx \int_0^\infty e^{-y} dy = 1.$$

The variance follows from (3.26b), and is

$$\sigma_x^2 = \int_{-\infty}^{+\infty} \int_{-\infty}^{+\infty} (x-1)^2 f(x, y) dx\, dy = \int_0^\infty (x - \mu_x)^2 e^{-x} dx \int_0^\infty e^{-y} dy = 1,$$

with $\sigma_x^2 = \sigma_y^2$ by symmetry. Finally, from (3.28)

$$\sigma_{xy} = \int_{-\infty}^{+\infty} \int_{-\infty}^{+\infty} (x - \mu_x)(y - \mu_y) f(x, y) dx\, dy = \int_0^\infty (x-1)e^{-x} dx \int_0^\infty (y-1)e^{-y} dy = 0.$$

EXAMPLE 3.10

If x has a density function that is symmetric about the mean, find the covariance of the two random variables $x_1 = x$ and $x_2 = y = x^2$. Comment on your answer.

The covariance is

$$\text{cov}(x, y) = E[xy] - E[x]E[y] = E[x^3] - E[x]E[x^2].$$

However because x has a density function that is symmetric about the mean, all the odd-order moments vanish and, in particular,

$$E[x] = E[x^3] = 0.$$

Thus $\text{cov}(x, y) = 0$ and hence $\rho(x, y) = 0$, even though x and y are not independent. Thus $\text{cov}(x, y) = 0$ is a necessary but not sufficient condition for statistical independence.

3.4. FUNCTIONS OF A RANDOM VARIABLE

In practice, it is common to have to consider a function of a random variable, for example $y(x)$. This is also a random variable, and the question arises: what is the density function of y, given that we know the density function of x? If y is monotonic (strictly increasing or decreasing) then the solution is simply

$$f(y\{x\}) = f(x\{y\}) \left| \frac{dx}{dy} \right|, \tag{3.30}$$

the absolute value being necessary to ensure that probabilities are always non-negative. If instead y has a continuous nonzero derivative at all but a finite number of points, the range must be split into a finite number of sections in each of which $y(x)$ is a strictly monotonic increasing or decreasing function of x with a continuous derivative, and then (3.30) applied to each section separately. Thus, at all points where (i) $dy/dx \neq 0$ *and* (ii) $y = y(x)$ has a real finite solution for $x = x(y)$, the required density function is

$$g(y\{x\}) = \prod_{\text{all } x} f(x\{y\}) \left| \frac{dy}{dx} \right|^{-1}. \tag{3.31}$$

If the above conditions are violated, then $g(y\{x\}) = 0$ at that point.

The method may be extended to multivariate distributions. Consider n random variables $x_i (i = 1, 2, \ldots, n)$ with a joint probability density $f(x_1, x_2, \ldots, x_n)$, and suppose we wish to find the joint probability density $g(y_1, y_2, \ldots, y_n)$ of a new set of variates y_i, which are themselves a function of the n variables $x_i (i = 1, 2, \ldots, n)$, defined by $y_i = y_i(x_1, x_2, \ldots, x_n)$. To do this we impose the probability condition:

$$|f(x_1, x_2, \ldots, x_n) \, dx_1 \, dx_2 \cdots dx_n| = |g(y_1, y_2, \ldots, y_n) \, dy_1 \, dy_2 \cdots dy_n|.$$

It follows that

$$g(y_1, y_2, \ldots, y_n) = f(x_1, x_2, \ldots, x_n)|J|, \qquad (3.32a)$$

where $|J|$ is the modulus of the determinant of the Jacobian of x_i with respect to y_i, i.e., the matrix

$$J = \frac{\partial(x_1, x_2, \ldots, x_n)}{\partial(y_1, y_2, \ldots, y_n)} = \begin{pmatrix} \frac{\partial x_1}{\partial y_1} & \cdots & \frac{\partial x_n}{\partial y_1} \\ \vdots & \ddots & \vdots \\ \frac{\partial x_1}{\partial y_n} & \cdots & \frac{\partial x_n}{\partial y_n} \end{pmatrix}, \qquad (3.32b)$$

where again the absolute value is necessary to ensure that probabilities are always non-negative. If the partial derivatives are not continuous, or there is not a unique solution for x_i in terms of the y_i, then the range of the variables can always be split into sections as for the single variable case and (3.32) applied to each section. The marginal density of one of the random variables can then be found by integrating the joint density over all the other variables. For several random variables it is usually too difficult in practice to carry out the above program analytically and numerical methods have to be used.

EXAMPLE 3.11

A random variable x has a density function:

$$f(x) = \frac{1}{\sqrt{2\pi}} \exp\left(\frac{-x^2}{2}\right).$$

What is the density function of $y = x^2$?

Now

$$x = \pm\sqrt{y} \quad \text{and} \quad \frac{dy}{dx} = 2x = \pm 2\sqrt{y}.$$

Thus, for $y < 0$, x is not real and so $g(y\{x\}) = 0$. For $y = 0$, $dx/dy = 0$, so again $g(y\{x\}) = 0$. Finally, for $y > 0$, we may split the range into two parts, $x > 0$ and $x < 0$. Then, applying (3.30) gives

$$g(y) = \frac{1}{2\sqrt{y}}[f(x = -\sqrt{y}) + f(x = +\sqrt{y})] = \frac{1}{\sqrt{(2\pi y)}} \exp\left(\frac{-y}{2}\right).$$

EXAMPLE 3.12

The single-variable density function of Example 3.11 may be generalized to two variables, i.e.,

$$f(x_1, x_2) = \frac{1}{2\pi} \exp\left\{-\frac{1}{2}\left(x_1^2 + x_2^2\right)\right\}.$$

Find the joint density function $g(y_1, y_2)$ of the variables $y_1 = x_1/x_2$ and $y_2 = x_1$.
The Jacobian of the transformation is

$$J = \det \begin{pmatrix} x_2 & -x_2^2/x_1 \\ 1 & 0 \end{pmatrix} = \frac{x_2^2}{x_1} = \frac{y_2}{y_1^2}.$$

Thus, applying (3.32), (provided $y_1 \neq 0$) gives

$$g(y_1, y_2) = \frac{1}{2\pi} \frac{|y_2|}{y_1^2} \exp\left\{ -\frac{1}{2}\left(y_2^2 + \frac{y_2^2}{y_1^2} \right) \right\} \quad (y_1 \neq 0)$$

A particular example of interest is the density of the sum of two random variables x and y. If we set $u = x + y$ and $v = x$, where the second choice is arbitrary, the Jacobian of the transformation is

$$J = \begin{vmatrix} \dfrac{\partial x}{\partial u} & \dfrac{\partial y}{\partial u} \\ \dfrac{\partial x}{\partial v} & \dfrac{\partial y}{\partial v} \end{vmatrix} = \begin{vmatrix} 0 & 1 \\ 1 & -1 \end{vmatrix} = -1.$$

Thus the joint density of u and v is

$$g(u, v) = f(x, y) = f(v, u - v).$$

The density function of u, denoted $h(u)$, is then

$$h(u) = g^M(u) = \int_{-\infty}^{\infty} f(x, u - x)dx, \tag{3.33a}$$

and in the special case where x and y are independent, so that $f(x,y) = f_1(x)f_2(y)$, (3.33a) reduces to

$$h(u) = \int_{-\infty}^{\infty} f_1(x) f_2(u - x) dx, \tag{3.33b}$$

which is called the *convolution* of f_1 and f_2 and is denoted $f_1 * f_2$.

Convolutions obeys the commutative, associative, and distributive laws of algebra, i.e.,

(a) $f_1 * f_2 = f_2 * f_1$,   (b) $f_1 * (f_2 * f_3) = (f_1 * f_2) * f_3$,

and

(c) $f_1 * (f_2 + f_3) = (f_1 * f_2) + (f_1 * f_3)$.

They occur frequently in physical science applications. An example is the problem of determining a physical quantity represented by a random variable x with a density $f_1(x)$ from measurements having experimental errors y distributed like the normal distribution of Problem 1.6 with zero mean and variance σ^2. The measurements yield values of the sum $u = x + y$. Then, using the form of the normal distribution given in Problem 1.6, equation (3.33b) gives

$$f(u) = \frac{1}{\sqrt{2\pi}\sigma} \int_{-\infty}^{\infty} f_1(x) \exp\left[-\frac{(u-x)^2}{2\sigma^2}\right] dx,$$

and we wish to find $f_1(x)$ from the experimental values of $f(u)$ and σ^2. In general this is a difficult problem unless $f(u)$ turns out to be particularly simple. More usually, the form of $f_1(x)$ is assumed, but is allowed to depend on one or more parameters. The integral is then evaluated and compared with the experimental values of $f(u)$ and the parameters varied until a match is obtained. Even so, exact evaluation of the integral is rarely possible and numerical methods have to be used. One of these is the so-called Monte Carlo method that we will discuss briefly in Section 5.1.1.

EXAMPLE 3.13

If two random variables x and y have probability densities of the form

$$f(x) = \frac{1}{\sigma_x \sqrt{2\pi}} \exp\left[-\frac{x^2}{2\sigma_x^2}\right],$$

and similarly for y, find the density function $h(u)$ of the random variable $u = x + y$.

From (3.33b), the density $h(u)$ is given by

$$h(u) = \int_{-\infty}^{\infty} f_1(x) f_2(u-x) dx = \frac{1}{2\pi \sigma_x \sigma_y} \int_{-\infty}^{\infty} \exp\left[-\frac{x^2}{2\sigma_x^2} - \frac{(u-x)^2}{2\sigma_y^2}\right] dx.$$

Completing the square for the exponent gives

$$-\frac{\sigma^2}{2\sigma_x^2 \sigma_y^2}\left[\left(x - \frac{\sigma_x^2}{\sigma^2} u\right)^2 - \frac{\sigma_x^4}{\sigma^4} u^2 + \frac{\sigma_x^2}{\sigma^2} u^2\right],$$

where $\sigma^2 = \sigma_x^2 + \sigma_y^2$. Then changing variables to $v = (\sigma/\sigma_x \sigma_y)(x - \sigma_x^2 u/\sigma^2)$ in the integral and simplifying yields,

$$h(u) = \frac{1}{2\pi\sigma} \exp\left[\left\{\frac{(\sigma_x^2 - \sigma^2)\sigma_x^2}{2\sigma^2 \sigma_x^2 \sigma_y^2}\right\} u^2\right] \int_{-\infty}^{\infty} \exp\left(-\frac{v^2}{2}\right) dv = \frac{1}{\sigma\sqrt{2\pi}} \exp\left(-\frac{u^2}{2\sigma^2}\right).$$

Expectation values may also be found for functions of x. If $h(x)$ is a function of x, then its expectation value is

$$E[h(x)] = \sum_{i=1}^{n} h(x_i) P[x_i] = \sum_x h(x) f(x), \tag{3.34a}$$

if the variate is discrete, and

$$E[h(x)] = \int_{-\infty}^{+\infty} h(x) f(x) dx. \tag{3.34b}$$

if it is continuous. Note that in general $E[h(x)] \neq h(E[x])$.

Using these definitions, the following results are easily proved, where h_1 and h_2 are two functions of x and c is a constant:

$$E[c] = c, \tag{3.35a}$$

$$E[ch(x)] = cE[h(x)], \tag{3.35b}$$

$$E[h_1(x) + h_2(x)] = E[h_1(x)] + E[h_2(x)], \tag{3.35c}$$

and if x_1 and x_2 are independent variates,

$$E[h_1(x_1)h_2(x_2)] = E[h_1(x_1)]E[h_2(x_2)]. \tag{3.35d}$$

PROBLEMS 3

3.1 Use the method of characteristic functions to find the first two moments of the distribution whose pdf is

$$f(x) = \frac{a^\gamma}{\Gamma(\gamma)} e^{-ax} x^{\gamma-1}, \quad 0 \le x \le \infty; \quad a > 0, \gamma > 0,$$

where $\Gamma(\gamma)$ is the gamma function, defined by

$$\Gamma(\gamma) \equiv \int_0^\infty e^{-x} x^{\gamma-1} dx, \quad 0 < \gamma < \infty.$$

3.2 A disgruntled employee types n letters and n envelopes, but assigns the letters randomly to the envelopes. What is the expected number of letters that will arrive at their correct destination?

3.3 An incompetent purchasing clerk repeatedly forgets to specify the magnitude of capacitors when ordering them from a manufacturer. If the manufacturer makes capacitors in 10 different sizes and sends one at random, what is the expected number of different capacitor values received after 5 orders are placed?

3.4 Find the probability distribution of the discrete random variable whose characteristic function is $\cos \omega$.

3.5 Two random variables x and y have a joint density function:

$$f(x,y) = \begin{cases} 3e^{-x}e^{-3y} & 0 < x, y < \infty \\ 0 & \text{otherwise} \end{cases}.$$

Find (a) $P[x < y]$ and (b) $P[x > 1, y < 2]$.

3.6 If $h = ax + by$, where x and y are random variables and a and b are constants, find the variance of h in terms of the variances of x and y and their covariance.

3.7 Two random variables x and y have a joint density function:

$$f(x,y) = \begin{cases} cx^2(1+x-y) & 0 < x, y < 1 \\ 0 & \text{otherwise} \end{cases},$$

where c is a constant. Find the conditional density of x given y.

3.8 The table shows the joint probability mass function of two discrete random variables x and y defined in the ranges $1 \leq x \leq 3$ and $1 \leq y \leq 4$. (Note that the probabilities are correctly normalized.)

		x		
y	1	2	3	Row totals
1	3/100	4/25	1/20	6/25
2	3/25	7/50	1/20	31/100
3	1/10	9/100	3/50	1/4
4	1/20	1/20	1/10	1/5
Column totals	3/10	11/25	13/50	1

Construct the following marginal probabilities:

(a) $P[x \leq 2, y = 1]$, (b) $P[x > 2, y \leq 2]$ and (c) $P[x+y = 5]$.

3.9 Find the probability density of the random variable $u = x + y$ where x and y are two independent random variables distributed with densities of the form

$$f_x(x) = \begin{cases} 1 & 0 \leq x < 1 \\ 0 & \text{otherwise} \end{cases},$$

and similarly for $f_y(y)$.

CHAPTER 4

Probability Distributions II: Examples

OUTLINE

4.1 Uniform	57	4.5 Cauchy	68
4.2 Univariate Normal (Gaussian)	59	4.6 Binomial	69
4.3 Multivariate Normal	63	4.7 Multinomial	74
4.3.1 Bivariate Normal	65	4.8 Poisson	75
4.4 Exponential	66		

In this chapter we will examine the properties of some of the theoretical distributions commonly met in physical sciences, for both discrete and continuous variates, including the all-important and ubiquitous so-called 'normal' distribution that we have discussed briefly in earlier chapters.

4.1. UNIFORM

The *uniform distribution* for a continuous random variable x has a density function:

$$f(x) \equiv u(x; a, b) = \begin{cases} \dfrac{1}{b-a} & a \leq x \leq b \\ 0 & \text{otherwise} \end{cases}, \tag{4.1}$$

where a and b are constants[1]. The distribution function from (4.1) is

$$F(x) = \begin{cases} 0 & x < a \\ \dfrac{x-a}{b-a} & a \leq x < b \\ 1 & x \geq b \end{cases} \tag{4.2}$$

---

[1]Here, and in the distributions that follow, we use the convention of separating the random variable from any constants by a semicolon.

An example of the uniform distribution is the distribution of rounding errors made in arithmetical calculations. Using equations (3.5) and (3.6) we can easily show that the mean and variance are given by

$$\mu = \frac{a+b}{2}; \qquad \sigma^2 = \frac{(b-a)^2}{12}. \qquad (4.3)$$

The value of a random variable uniformly distributed in the interval (0,1) is called a *(uniform) random number*. These random numbers are useful because they enable various probabilities and expectation values to be evaluated empirically. The theoretical importance of the uniform distribution is enhanced by the fact that any density function $f(x)$ of a continuous random variable x may be transformed to the uniform density function

$$g(u) = 1, \qquad 0 \le u \le 1$$

by the transformation $u = F(x)$, where $F(x)$ is the distribution function of x. This follows from the fact that

$$\frac{du}{dx} = \frac{d}{dx} \int_{-\infty}^{x} f(x') dx' = f(x),$$

and hence by changing variables,

$$g(u) = f(x) \left| \frac{du}{dx} \right|^{-1} = 1, \qquad 0 \le u \le 1$$

This property is useful in generating random numbers from an arbitrary distribution by transforming a set of uniformly distributed random numbers. It enables many properties of continuous distributions to be exhibited, by proving them for the particular case of the uniform distribution. It also follows that there is at least one transformation that transforms any continuous distribution to any other; it is simply the product of the transformations that take each distribution into the uniform distribution.

EXAMPLE 4.1

Trains to a given destination depart on the hour and at 30 minutes past the hour. A passenger arrives at the station at a time that is uniformly distributed in the interval from one hour to the next. What is the probability that they will have to wait at least 10 minutes for a train?

Let t denote the time in minutes past the hour that the passenger arrives at the station. Because t is a random variable uniformly distributed in the interval (0,60), it follows that the passenger will have to wait at least 10 minutes if they arrive up to 20 minutes past the hour, or between 30 and 50 minutes past the hour. Thus the required probability is

$$P[0 < t < 20] + P[30 < t < 50] = \frac{20}{60} + \frac{20}{60} = \frac{2}{3}.$$

4.2. UNIVARIATE NORMAL (GAUSSIAN)

This distribution is by far the most important in statistics because many distributions encountered in practice are believed to be of approximately this form, a point that was mentioned in Section 1.4 and that will be discussed in more detail in Chapter 5. The name is perhaps unfortunate, because it might imply that all other distributions are somehow 'abnormal', which of course they are not. In physical sciences the normal distribution is more usually known as a *Gaussian distribution*, although several people in addition to Gauss have claims to have studied this function. In this book, the name used in statistics has been adopted. We start with the case of a single variate.

The *normal density function* for a single continuous random variable x is defined as

$$f(x) \equiv n(x; \mu, \sigma) = \frac{1}{\sqrt{2\pi}\sigma} \exp\left[-\frac{1}{2}\left(\frac{x-\mu}{\sigma}\right)^2\right], \qquad (\sigma > 0) \tag{4.4}$$

and its distribution function is

$$F(x) \equiv N(x; \mu, \sigma) = \frac{1}{\sqrt{2\pi}\sigma} \int_{-\infty}^{x} \exp\left[-\frac{1}{2}\left(\frac{t-\mu}{\sigma}\right)^2\right] dt. \tag{4.5}$$

Graphs of $f(x)$ and $F(x)$ are shown in Fig. 4.1 for $\mu = 0$ and $\sigma = 0.5$, 1.0 and 2.0. Keeping the value of σ fixed, but changing the value of the parameter μ simply moves the curves along the x axis.

As this is the first nontrivial distribution we have encountered it will be useful to implement some of our previous definitions. First, it is clear from (4.4) that $f(x)$ is a single-valued non-negative real number for all values of x. Furthermore, by the substitution

$$t^2 = \frac{1}{2}\left(\frac{x-\mu}{\sigma}\right)^2,$$

we can write

$$\int_{-\infty}^{+\infty} f(x) dx = \frac{1}{\sqrt{\pi}} \int_{-\infty}^{+\infty} e^{-t^2} dt.$$

Since the integral on the right-hand side has the well-known value of $\sqrt{\pi}$, we see that $f(x)$ is normalized to unity and is thus a valid density function.

To find the moments of the normal distribution we first find the mgf. From equation (3.11),

$$M_x(t) = E[\exp(tx)] = \exp(t\mu) E[\exp\{t(x-\mu)\}]$$

$$= \frac{\exp(t\mu)\exp(\sigma^2 t^2/2)}{(2\pi)^{1/2}\sigma} \int_{-\infty}^{\infty} \exp\left[\frac{-(x-\mu-\sigma^2 t)^2}{2\sigma^2}\right] dx.$$

The integral is related to the area under a normal curve with mean $(\mu + \sigma^2 t)$ and variance σ^2. Thus

$$M_x(t) = \exp(t\mu + \sigma^2 t^2/2). \tag{4.6}$$

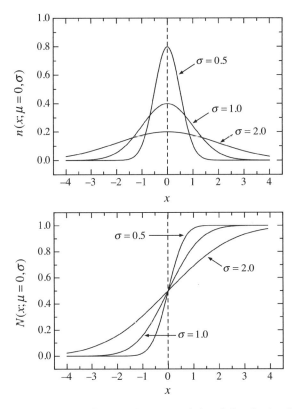

FIGURE 4.1 Normal (Gaussian) density function (upper graphs) and distribution function (lower graphs) for $\mu = 0$ and $\sigma = 0.5$, 1.0 and 2.0.

On differentiating (4.6) twice and setting $t = 0$, we have

$$\mu'_1 = \mu, \qquad \mu'_2 = \sigma^2 + \mu^2$$

and

$$\text{var}(x) = \mu'_2 - (\mu'_1)^2 = \sigma^2.$$

The mean and variance of the normal distribution are therefore μ and σ^2, respectively. The same technique for moments about the mean gives

$$\mu_{2n} = \frac{(2n)!}{n! 2^n} \sigma^{2n} \quad \text{and} \quad \mu_{2n+1} = 0, \quad n \geq 1. \tag{4.7}$$

The odd order moments are zero by virtue of the symmetry of the distribution. Using (4.7) we can calculate quantities that are sometimes used to measure skewness and the degree of peaking in a distribution. These are denoted β_1 and β_2, respectively (β_2 is also called the *kurtosis*) and are

$$\beta_1 \equiv \mu_3^2/\mu_2^3 = 0 \quad \text{and} \quad \beta_2 \equiv \mu_4/\mu_2^2 = 3 \tag{4.8}$$

This value of β_2 is taken as a standard against which the kurtosis of other distributions may be compared.

Using essentially the same technique that was used to derive the mgf we can show that the cf of the normal distribution is

$$\phi(t) = \exp[it\mu - t^2\sigma^2/2], \qquad (4.9)$$

which agrees with (3.15) and which may be confirmed by applying the result of the inversion theorem.

Any normal distribution may be transformed to a normal distribution in z with $\mu = 0$ and $\sigma^2 = 1$ by setting $z = (x - \mu)/\sigma$. Then, from (4.4) and (4.5),

$$n(z; 0, 1) = \frac{1}{(2\pi)^{1/2}} \exp(-z^2/2) \qquad (4.10)$$

and

$$N(z; 0, 1) = \frac{1}{(2\pi)^{1/2}} \int_{-\infty}^{z} \exp(-u^2/2) du. \qquad (4.11)$$

These forms are called the *standard normal density function* and the *standard normal distribution function*, respectively, and will usually be denoted by $n(z)$ and $N(z)$, omitting the constants. Values of $N(z)$ are given in Appendix C, Table C.1. If these functions are required for negative values of z, they may be found from the relations

$$n(-z) = n(z), \qquad (4.12)$$

and

$$N(-z) = 1 - N(z), \qquad (4.13)$$

which follow from the symmetry of the distribution. Another useful relation that follows from (4.13) is

$$2 \int_{0}^{z} n(u) du = \int_{-z}^{z} n(u) du = 2N(z) - 1. \qquad (4.14)$$

Using (4.12)–(4.14) and Table C.1, the following results may be deduced:

1. The proportion of standard normal variates contained within 1, 2, and 3 standard deviations from the mean is 68.3%, 95.4%, and 99.7%, respectively;
2. If t_α denotes that value of the standard normal distribution for which

$$\int_{t_\alpha}^{\infty} n(t; 0, 1) dt = \alpha, \qquad (4.15)$$

then $(\mu \pm t_\alpha \sigma)$ defines a $100(1 - 2\alpha)\%$ symmetric interval centered on μ.

The first of these results was mentioned in Section 1.4, when discussing the behavior of experimental frequency plots for cases where the sample size becomes large. They are stronger conditions than the constraints implied by the Chebyshev inequality (1.11).

The reason why such plots tend to the normal distribution is embedded in the so-called Law of Large Numbers and will be discussed in the next chapter. The usefulness of the second result will be evident when confidence intervals are discussed in Chapter 9.

We will see in Chapter 5 when we consider sampling in more detail that one can often assume that the measurement errors on a quantity x are distributed according to a normal distribution $n(x)$ with mean zero, so that the probability of obtaining a value between x and $x + dx$ is $n(x)dx$. The dispersion σ of the distribution is called the *standard error* and (4.15) gives the probability for the true value being within an interval of plus or minus one, two, or three standard errors about the measured value, given a single measurement of x.

One final useful result is that the distribution of a linear sum

$$T = \sum_i a_i x_i$$

of n independent random variables x_i, having normal distributions $N(x_i; \mu_i, \sigma_i^2)$, is distributed as $N(T; \mu, \sigma^2)$, i.e., is also normally distributed. To show this, we can use the characteristic function. Because the x_i are independent, this may be written

$$\phi_T(t) = \prod_{i=1}^{n} E[\exp(i t a_i x_i)] = \prod_{i=1}^{n} \phi_i(t),$$

where $\phi_i(t)$ is the cf of the random variable $a_i x_i$. We have previously shown in (4.9) that

$$\phi_i(t) = \exp[i t a_i \mu_i - t \sigma_i^2 a_i^2 / 2]$$

and so

$$\phi_T(t) = \exp\left\{ \sum_{i=1}^{n} (i t a_i \mu_i - t^2 a_i^2 \sigma_i^2 / 2) \right\} = \exp(i t \mu - t^2 \sigma / 2),$$

where

$$\mu = \sum_{i=1}^{n} a_i \mu_i \quad \text{and} \quad \sigma^2 = \sum_{i=1}^{n} a_i^2 \sigma_i^2.$$

However this is the cf of a normal variate whose mean is μ and whose variance is σ^2. Thus, by the inversion theorem, T is distributed as $N(T; \mu, \sigma^2)$.

EXAMPLE 4.2

If x is a random variable normally distributed with mean μ and variance σ^2, show that for any constants a and b, with $b \neq 0$, $y = a + bx$ is a random variable with mean $(a + b\mu)$ and variance $b^2 \sigma^2$.

Let $F_x(x)$ and $F_y(y)$ be the distribution functions of x and y, respectively. Then, for $b > 0$,

$$F_y(y) = P[(a + bx) \leq y] = P[x \leq (y - a)/b] = F_x(\{y - a\}/b).$$

Similarly, for $b < 0$,

$$F_y(y) = P[(a + bx) \leq y] = P[x \geq (y - a)/b] = 1 - F_x(\{y - a\}/b).$$

The associated density functions $f_x(x)$ and $f_y(y)$ are obtained by differentiating the distribution functions, so

$$f_y(y) = \begin{cases} \dfrac{1}{b} f_x\left(\dfrac{y-a}{b}\right), & b > 0 \\ -\dfrac{1}{b} f_x\left(\dfrac{y-a}{b}\right), & b < 0 \end{cases},$$

which combined is

$$f_y(y) = \dfrac{1}{|b|} f_x\left(\dfrac{y-a}{b}\right) = \dfrac{1}{\sigma|b|\sqrt{2\pi}} \exp\left[-\dfrac{1}{2\sigma^2}\left(\dfrac{y-a}{b} - \mu\right)^2\right]$$

$$= \dfrac{1}{\sigma|b|\sqrt{2\pi}} \exp\left[-\dfrac{1}{2}\left(\dfrac{y-a-b\mu}{b\sigma}\right)^2\right].$$

This is a normal distribution with mean $(a + b\mu)$ and variance $b^2\sigma^2$.

EXAMPLE 4.3

A company makes electrical components with a mean life of 800 days and a standard deviation of 20 days. If the distribution of lifetimes is normal, what is the probability that a component chosen at random will last between 780 and 850 days? Also, what is the minimum lifetime of the longest lived 12% of the components?

First convert to standard form by using the transformation $z = (x - \mu)/\sigma$, so that $z_1 = -1.0$ and $z_2 = 2.5$. Then

$$P[780 < x < 850] = P[-1.0 < z < 2.5] = P[z < 2.5] - P[z < -1.0],$$

and using Table C.1, the right-hand side is $0.9938 - (1 - 0.8413) = 0.8351$. To answer the second question, we can use the 'inverse' transformation to find the value of z, say z_0, such that $P[z < z_0] = 0.88$. From Table C.1 this is $z_0 = 1.175$. Thus from the inverse transformation, we have

$$x_0 = z_0 \sigma + \mu = (1.175 \times 20) + 800 = 823.5 \text{ days}$$

4.3. MULTIVARIATE NORMAL

If $x_1, x_2, \ldots, x_n \equiv \mathbf{x}$ are n random variables, then the *multivariate normal density function*, of order n, is defined as

$$f(\mathbf{x}) \equiv \dfrac{1}{(2\pi)^{n/2}|\mathbf{V}|^{1/2}} \exp\left[-\dfrac{1}{2}(\mathbf{x} - \boldsymbol{\mu})^T \mathbf{V}^{-1}(\mathbf{x} - \boldsymbol{\mu})\right], \tag{4.16}$$

where the constant vector $\boldsymbol{\mu}$ is the mean of the distribution, and $\mathbf{V}$ is a symmetric positive-definite matrix, which is the variance matrix of the vector $\mathbf{x}$. The quantity

$$Q = (\mathbf{x} - \boldsymbol{\mu})^T \mathbf{V}^{-1}(\mathbf{x} - \boldsymbol{\mu}), \tag{4.17}$$

is called the *quadratic form* of the multivariate normal distribution. The distribution possesses a number of important properties, and three are discussed below.

The first concerns the form of the joint marginal distribution of a subset of the n variables. If the n random variables $x_1, x_2, \ldots, x_n$ are distributed as an n-variate normal distribution, then the joint marginal distribution of any set $x_i (i = 1, 2, \ldots m < n)$ is the m-variate normal. This result can be proved in a straightforward way by constructing the joint marginal distribution from (4.16) using the definition (3.20a). It follows that the distribution of any single random variable in the set x_i (this is the case $m = 1$) is distributed as the univariate normal.

This result can be used to derive the second property: the necessary condition under which the variables of the distribution are independent. If we set $\text{cov}(x_i, x_j) = 0$ for $i \neq j$, then this implies that $\mathbf{V}$ is diagonal, so the quadratic form becomes

$$(\mathbf{x} - \boldsymbol{\mu})^T \mathbf{V}^{-1}(\mathbf{x} - \boldsymbol{\mu}) = \sum_{i=1}^{n}(x_i - \mu_i)V_{ii}^{-1},$$

and the density function may be written as

$$f(\mathbf{x}) = \prod_{i=1}^{n} f_i(x_i),$$

where

$$f_i(x_i) = \frac{1}{(2\pi)^{1/2}} \frac{1}{V_{ii}^{1/2}} \exp\left[-\frac{(x_i - \mu_i)^2}{2V_{ii}}\right]. \quad (4.18)$$

Equation (4.18) is the density function for a univariate normal distribution and so, by virtue of the earlier result on the marginal distribution, and the definition of statistical independence, equation (3.24), the variables x_i are independently distributed. Thus a necessary condition for the components of $\mathbf{x}$ to be jointly independent is if $\text{cov}(x_i, x_j) = 0$ for all $i \neq j$. In the case of the multivariate normal distribution, this is also a sufficient condition. It is straightforward, by an analogous argument, to establish the inverse, i.e., that if x_i are jointly independent then $\mathbf{V}$ is diagonal.

The third, and final, property concerns the distribution of linear combinations:

$$S = \sum_{i=1}^{n} a_i x_i = \mathbf{x}^T \mathbf{A},$$

of random variables $\mathbf{x} = x_i (i = 1, 2, \ldots, n)$, each of which has a univariate normal distribution, where a_i are constants and $\mathbf{A} = a_i(i = 1, 2, \ldots, n)$. The moment generating function of S is

$$M_S(t) = E[\exp(St)] = E\left[\exp\{(\mathbf{x}^T \mathbf{A})t\}\right] = \exp\left[(\boldsymbol{\mu}^T \mathbf{A})t\right]\exp\left[(\mathbf{x} - \boldsymbol{\mu})^T \mathbf{A} t\right].$$

Now, if $\mathbf{x}$ has a multivariate normal distribution with mean $\boldsymbol{\mu}$ and variance matrix $\mathbf{V}$, then

$$\exp[(\mathbf{x} - \boldsymbol{\mu})^T \mathbf{A} t] = \exp[(\mathbf{A}^T \mathbf{V} \mathbf{A})t^2/2],$$

and thus

$$M_S(t) = \exp\left[(\boldsymbol{\mu}^T \mathbf{A})t + (\mathbf{A}^T \mathbf{V} \mathbf{A})t^2/2\right]$$

However from (4.6), this is the mgf of a normal variate with mean

$$\mu = \boldsymbol{\mu}^T \mathbf{A} = \sum_{i=1}^{n} a_i \mu_i$$

and variance

$$\sigma^2 = \mathbf{A}^T \mathbf{V} \mathbf{A} = \sum_{i=1}^{n} \sum_{j=1}^{n} a_i a_j V_{ij},$$

and so S is distributed as $N(S; \mu, \sigma^2)$. This result is a generalization of the result obtained at the end of Section 4.2.

4.3.1. Bivariate Normal

An important example of a multivariate normal distribution is the bivariate case, which occurs frequently in practice. Its density function is

$$n(x, y; \mu_x, \mu_y, \sigma_x, \sigma_y, \rho) \equiv n(x, y) = \frac{1}{2\pi\sigma_x\sigma_y(1-\rho^2)^{1/2}} \exp\left[\frac{-R}{2(1-\rho^2)}\right], \quad (4.20)$$

where

$$R \equiv \left(\frac{x-\mu_x}{\sigma_x}\right)^2 - 2\rho\left(\frac{x-\mu_x}{\sigma_x}\right)\left(\frac{y-\mu_y}{\sigma_y}\right) + \left(\frac{y-\mu_y}{\sigma_y}\right)^2, \quad (4.21)$$

and ρ is the correlation coefficient, defined in (3.29). If the exponent in (4.20) is a constant $(-K)$, i.e.,

$$R = 2(1-\rho^2)K,$$

then the points (x,y) lie on an ellipse with center (μ_x, μ_y). The density function (4.20) is a bell-shaped surface, and any plane parallel to the xy plane that cuts this surface will intersect it in an elliptical curve. Any plane perpendicular to the xy plane will cut the surface in a curve of the normal form.

Just as for the univariate normal distribution, we can define a *standard bivariate normal density function*

$$n(u,v) = \frac{1}{2\pi(1-\rho^2)^{1/2}} \exp\left[-\frac{(u^2 - 2\rho uv + v^2)}{2(1-\rho^2)}\right]. \quad (4.22)$$

where

$$u = \frac{x-\mu_x}{\sigma_x}; \quad v = \frac{y-\mu_y}{\sigma_y}.$$

A feature of this distribution is that for $\rho = 0$

$$n(u,v) = n(u)n(v), \quad (4.23)$$

which implies that u and v are independently distributed, a result that is *not* generally true for all bivariate distributions.

Finally, the joint moment generating function may be obtained from the definition

$$M_{xy}(t_1,t_2) = E[\exp(t_1 x + t_2 y)] = \int_{-\infty}^{\infty} \int_{-\infty}^{\infty} \exp(t_1 x + t_2 y) f(x,y) dx\, dy, \quad (4.24)$$

After changing variables to u and v, this becomes

$$M_{xy}(t_1,t_2) = \frac{\exp(t_1 \mu_x + t_2 \mu_y)}{2\pi(1-\rho^2)^{1/2}} \iint e^{(t_1 \sigma_x u + t_2 \sigma_y v)} \exp\left[\frac{u^2 - 2\rho u v + v^2}{-2(1-\rho^2)}\right] du\, dv,$$

which, after some algebra, gives

$$M_{xy}(t_1,t_2) = \exp\left[t_1 \mu_x + t_2 \mu_y + \frac{1}{2}\left(t_1^2 \sigma_x^2 + 2\rho t_1 t_2 \sigma_x \sigma_y + t_2^2 \sigma_y^2\right)\right]. \quad (4.25)$$

The moments may be obtained in the usual way by evaluating the derivatives of (4.25) at $t_1 = t_2 = 0$. For example,

$$E[x^2] = \left.\frac{\partial^2 M_{xy}(t_1,t_2)}{\partial t_1^2}\right|_{t_1=t_2=0} = \sigma_x^2 + \mu_x^2.$$

4.4. EXPONENTIAL

The *exponential* density function for a continuous random variable x is

$$f(x) \equiv f(x;\lambda) = \begin{cases} \lambda e^{-\lambda x} & \lambda > 0,\ x \geq 0 \\ 0 & \text{otherwise}. \end{cases} \quad (4.26)$$

It is an example of a more general class of *gamma distributions* of the form

$$f(x;\alpha,\lambda) = \frac{\lambda^\alpha x^{\alpha-1} e^{-\lambda x}}{\Gamma(\alpha)}, \quad \alpha, \lambda > 0,\ x > 0, \quad (4.27)$$

for $\alpha = 1$, where the gamma function $\Gamma(\alpha)$ was defined in Problem 3.1 and $\Gamma(1) = 1$. From (4.26) the distribution function is

$$F(x) = \int_0^x f(x') dx' = 1 - e^{-\lambda x}.$$

The mgf $M_x(t) = E[e^{tx}]$ may be found from (4.26) and is

$$M_x(t) = \lambda \int_0^\infty e^{tx} e^{-\lambda x} dx = \frac{\lambda}{\lambda - t}, \quad t < \lambda. \quad (4.28)$$

Differentiating as usual gives the mean μ and variance σ^2 as

$$\mu = 1/\lambda \quad \text{and} \quad \sigma^2 = 1/\lambda^2. \quad (4.29)$$

The exponential density is used to model probabilities where there is an interval of time before an event occurs. Examples are the lifetimes of electronic components. In this context the parameter λ is the *(failure) rate*, or *inverse lifetime*, of the component. An interesting property of exponential random variables is that they are *memoryless*, that is, for example, the probability that a component will function for at least an interval $(t+s)$, having already operated for at least an interval t, is the same as the probability that a new component would operate for at least the interval s, if it were activated at time zero.

The proof of this key property, which is unique to exponentially distributed random variables, follows from the definition of conditional probability:

$$P[x > t+s | x > t] = \frac{P[x > t+s]}{P[x > t]}. \tag{4.30}$$

The probabilities on the right-hand side may be calculated from (4.26):

$$P[x > t+s] = \lambda \int_{t+s}^{\infty} e^{-\lambda x'} dx' = e^{-\lambda(t+s)},$$

and

$$P[x > t] = \lambda \int_{t}^{\infty} e^{-\lambda x'} dx' = e^{-\lambda t}.$$

Thus,

$$P[x > t+s | x > t] = e^{-\lambda s} = P[x > s].$$

EXAMPLE 4.4

A system has a critical component whose average lifetime is exponentially distributed with a mean value of 2000 hours. What is the probability that the system will not fail after 1500 hours?

From the memoryless property of the exponential distribution, the distribution of the remaining lifetime of the component is exponential with parameter $\lambda = 1/2000$. Then,

$$P[\text{remaining lifetime} > 1500] = 1 - F[1500]$$

$$= \exp\left[-\frac{1500}{2000}\right] = e^{-3/4} \approx 0.47.$$

If there are several independent random variables $x_1, x_2, \ldots, x_n$, each exponentially distributed with parameters $\lambda_1, \lambda_2, \ldots, \lambda_n$, respectively, then because the smallest value of a set of numbers is greater than some value x if, and only if, all values are greater than x,

$$P[\min(x_1, x_2, \ldots, x_n) > x] = P[x_1 > x, x_2 > x, \ldots, x_n > x].$$

However, because the variables are independently distributed,

$$P[\min(x_1, x_2, \ldots, x_n) > x] = \prod_{i=1}^{n} P[x_i > x] = \exp\left[-\sum_{i=1}^{n} \lambda_i x\right].$$

This result may be used to model the lifetime of a complex system of several independent components, all of which must be working for the system to function.

In the exponential distribution, the quantity λ is a constant, but there are many situations where it is more appropriate to assume that λ is not constant. An example is when calculating the failure rate with time of aging components. In this case we could assume that $\lambda(t) = \alpha\beta t^{\beta-1} (t > 0)$, where α and β are positive constants, so that $\lambda(t)$ increases or decreases when $\beta > 1$ or $\beta < 1$, respectively. It was for precisely this situation, where the components were light bulbs, that the *Weibull distribution* was devised, with a density function

$$f(x;\alpha,\beta) = \alpha\beta x^{\beta-1}\exp(-\alpha x^\beta), \quad x > 0$$

which reduces to the exponential distribution when $\beta = 1$. It is a useful distribution for representing a situation where a probability rises from small values of x to a maximum and then falls again at large values of x.

4.5. CAUCHY

The density function of the *Cauchy* distribution is

$$f(x;\theta) = \frac{1}{\pi}\frac{1}{1+(x-\theta)^2}, \quad -\infty < x < \infty.$$

The parameter θ can be interpreted as the mean μ of the distribution only if the definition is extended as follows:

$$\mu = \lim_{N\to\infty}\int_{-N}^{N} f(x;\theta)x\,dx.$$

This is somewhat questionable and we will, in general, set $\theta = 0$. Then the distribution function becomes

$$F(x) = \frac{1}{2} + \frac{1}{\pi}\arctan(x).$$

The moment about the mean (taken to be zero) of order $2n$ is

$$\mu_{2n} = \frac{1}{\pi}\int_{-\infty}^{\infty}\frac{x^{2n}}{1+x^2}\,dx, \qquad (4.31)$$

but the integral converges only for $n = 0$, so only the trivial moment $\mu_0 = 1$ exists. Likewise, the mgf does not exist, although the cf does and is given by (see Example 3.6b)

$$\phi(t) = e^{-|t|}.$$

It can be shown that the ratio of two standardized normal variates has a Cauchy density function (see Example 4.5 below), which is one reason why it is encountered in practice. The Cauchy distribution is also met frequently in physical science because it describes the line shape seen in the decay of an excited quantum state. In this context it is usually called the *Lorentz distribution* or *Breit–Wigner formula* and is written as

$$f(E; E_0, \Gamma) = \frac{1}{\pi} \frac{\Gamma/2}{(E - E_0)^2 + \Gamma^2/4},$$

where E is the energy of the system and the parameters Γ and E_0 are interpreted as the 'width' of the state, i.e., the full width at half maximum height of the line shape, and its energy, respectively. This distribution must be treated with care because of the non-convergence of the moment integrals, which is due to the long tails of the Cauchy density compared with those of the normal density. In these regions the distribution is not necessarily a good approximation to the physical system.

EXAMPLE 4.5

Two random variables x and y are each distributed with standardized normal density functions. Show that the ratio x/y has a Cauchy probability density.

Define new variables $r = x/y$ and $s = y$. Then if $h(r,s)$ is the joint probability density of r and s, this may be found from probability conservation, that is,

$$|h(r,s) dr\, ds| = |n(x)n(y) dx\, dy|,$$

where n is the standard normal density. Changing variables on the right-hand side to r and s gives

$$h(r,s) dr\, ds = n(rs)n(s) J dr\, ds,$$

where

$$J = \begin{vmatrix} \dfrac{\partial x}{\partial r} & \dfrac{\partial y}{\partial r} \\ \dfrac{\partial x}{\partial s} & \dfrac{\partial y}{\partial s} \end{vmatrix} = \begin{vmatrix} y & 0 \\ 0 & 1 \end{vmatrix} = y = s$$

is the Jacobian of the transformation, as discussed in Section 3.4. Using the symmetry of the normal density about zero, the probability density of r is given by

$$f(r) = 2 \int_0^\infty n(rs)n(s) s\, ds = \frac{1}{\pi} \int_0^\infty \exp\left[-\frac{1}{2} s^2 (1+r^2)\right] s\, ds$$

$$= \frac{1}{\pi} \left[-\frac{\exp\left[-\frac{1}{2} s^2 (1+r^2)\right]}{(1+r^2)} \right]_0^\infty = \frac{1}{\pi} \frac{1}{1+r^2},$$

which is a Cauchy density.

4.6. BINOMIAL

The binomial distribution concerns a population of members each of which either possesses a certain attribute P, or does not possess this attribute, which we will denote by Q. If the proportion of members possessing P is p and that possessing Q is q, then clearly

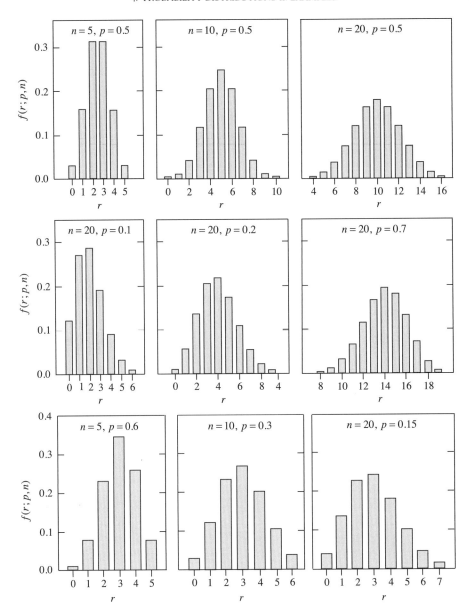

FIGURE 4.2 Plots of the binomial probability function. The top row shows $f(r;p,n)$ for $p = 0.5$ and various values of n; the middle row shows values for $n = 20$ and various values of p; and the lower row shows $f(r;p,n)$ for a fixed value of $np = 3$.

$(p+q) = 1$. An experiment involving such a population is called a *Bernoulli trial*, i.e., one with only two possible outcomes. A simple example is tossing a coin, where the two outcomes are 'heads' and 'tails', with $p = q = 0.5$ if the coin is unbiased and thin, so that the probability of landing on its edge can be neglected. Suppose now we wish to choose sets from the population, each of which contains n members. From the work of Chapter 2, the proportion of cases containing rPs and $(n-r)$Qs is

$$_nC_r p^r q^{n-r} = \binom{n}{r} p^r q^{n-r}, \qquad (4.32)$$

i.e., the rth term of the binomial expansion of

$$f(p,q) = (q+p)^n, \qquad (4.33)$$

hence the name of the distribution. Expressed in another way, if p is the chance of an event happening in a single trial, then for n independent trials the terms in the expansion

$$f(q,p) = q^n + nq^{n-1}p + \cdots + p^n,$$

give the chances of 0, 1, 2, ..., n events happening. Thus, we are led to the following definition. The probability function of the *binomial distribution* is defined as

$$f(r;p,n) \equiv \binom{n}{r} p^r q^{n-r}, \qquad (4.34)$$

and gives the probability of obtaining $r = 0, 1, 2, \ldots, n$ successes, i.e., events having the attribute P, in an experiment consisting of n Bernoulli trials. Note that f is not a probability *density*, but gives the actual probability. Tables of the cumulative binomial distribution are given in Appendix C, Table C.2, and plots of the probability function for some values of its parameters are shown in Fig. 4.2.

EXAMPLE 4.6

If a machine making components has a failure rate of 2%, i.e., 2% are rejected as being defective, what is the probability that less than 3 components will be defective in a random sample of size 100?

Using the binomial distribution, the probability that less than 3 components will be found to be defective is, with $p = 0.02$,

$$P[r < 3] = P[r = 0] + P[r = 1] + P[r = 2]$$

$$= \sum_{r=0}^{2} \binom{100}{r} (0.02)^r (0.98)^{(100-r)} = 0.6767.$$

EXAMPLE 4.7

A device consists of n components, each of which will function independently with a probability p and operates successfully if at least 50% of its components are fully functioning. A researcher can buy 4 components with $p = 0.5$ or, for the same price, 3 components with a higher value of p. For what higher value of p would it be better to use a 3-component system rather than one with 4 components?

The probability that a 4-component system will function is

$$P_4 = \binom{4}{2}p^2(1-p)^2 + \binom{4}{3}p^3(1-p) + \binom{4}{4}p^4(1-p)^0,$$

which for $p = 0.5$ is $11/16$. The probability that a 3-component system will function is

$$P_3 = \binom{3}{2}p^2(1-p) + \binom{3}{3}p^3(1-p)^0 = 3p^2 - 2p^3.$$

A 3-component system is more likely to function than a 4-component one if $P_3 > P_4$, i.e., if

$$3p^2 - 2p^3 - \frac{11}{16} > 0,$$

which is true for (approximately) $p > 0.63$.

The moment distribution function may be found directly from (4.34) and the definition (3.12a) and is

$$M_r(t) = \sum_{r=0}^{n} f(r; p, n)e^{tr} = \sum_{r=0}^{n} \binom{n}{r} p^r q^{n-r} e^{tr} = (pe^t + q)^n, \quad (4.35)$$

from which

$$\mu_1' = \mu = np, \quad \mu_2' = np + n(n-1)p^2, \quad (4.36)$$

and

$$\sigma^2 = \mu_2' - (\mu_1')^2 = npq. \quad (4.37)$$

The mgf for moments about the mean is

$$M_\mu(t) = e^{-\mu t} M(t) \quad (4.38)$$

and gives

$$\mu_3 = npq(q-p), \quad \mu_4 = npq[1 + 3(n-2)pq]. \quad (4.39)$$

So, using the definitions given in (4.8), we have

$$\beta_1 = (q-p)^2/(npq) \quad \text{and} \quad \beta_2 = 3 + (1 - 6pq)/npq, \quad (4.40)$$

which tend to the values for a normal distribution as $n \to \infty$.

The plots shown in Fig. 4.2 suggest that the limiting form of the binomial distribution is indeed the normal. This may be proved using the characteristic function, although it requires several stages. From the relation (3.15) and the form of the mgf (4.35), the cf is

$$\phi_r(t) = (q + pe^{it})^n, \quad (4.41)$$

and the binomial distribution may be expressed in standard measure (i.e., with $\mu = 0$ and $\sigma^2 = 1$) by the transformation:
$$x = (r - \mu)/\sigma. \tag{4.42}$$
This can be considered as the sum of two independent random variables r/σ and $-\mu/\sigma$, even though the second term is actually a constant. From the work of Chapter 3, the characteristic function of x is the product of the characteristic functions of these two variates. Setting $\mu = np$ and using (4.41), we have
$$\phi_x(t) = \exp\left[\frac{-itnp}{\sigma}\right]\left\{q + p\exp\left[\frac{it}{\sigma}\right]\right\}^n.$$
Taking logarithms and using $p = 1 - q$, gives
$$\ln \phi_x(t) = \frac{-itnp}{\sigma} + n\ln\left\{1 + p\left[\exp\left(\frac{it}{\sigma}\right) - 1\right]\right\},$$
and because $t/\sigma = t/\sqrt{npq} \to 0$ as $n \to \infty$, the exponential may be expanded giving
$$\ln \phi_x(t) = \frac{-itnp}{\sigma} + n\ln\left\{1 + p\left[\frac{it}{\sigma} - \frac{1}{2}\left(\frac{t}{\sigma}\right)^2 + \cdots\right]\right\}.$$
Next we expand the logarithm on the right-hand side using
$$\ln(1 + \varepsilon) = \varepsilon - \varepsilon^2/2 + \varepsilon^3/3 - \cdots,$$
with the result
$$\ln \phi_x(t) = \frac{-itnp}{\sigma} + n\left\{\frac{itp}{\sigma} - \frac{1}{2}\left(\frac{t}{\sigma}\right)^2(p - p^2) + O(n^{-1/2})\right\}$$
Finally, letting $n \to \infty$ and keeping t finite, gives
$$\ln \phi_x(t) = -t^2/2 + O(n^{-1/2}),$$
where we have used $\sigma^2 = npq = np(1 - p)$. So, for any finite t,
$$\phi(t) \to \exp(-t^2/2).$$
This is the form of the cf of a standardized normal distribution and so by the inversion theorem, the associated density function is
$$f(x) = \frac{1}{(2\pi)^{1/2}}\exp\left(-\frac{x^2}{2}\right), \tag{4.43}$$
which is the standard form of the normal distribution.

The normal approximation to the binomial is excellent for large values of n and is still good for small values provided p is reasonably close to ½. A working criterion is that the approximation is good if np and nq are both greater than 5. This is confirmed by the plots in Fig. 4.2.

EXAMPLE 4.8

If the probability of a success in a single Bernoulli trial is $p = 0.4$, compare the exact probability of obtaining $r = 5$ successes in 20 trials with the normal approximation.

The binomial probability is

$$P_B = \binom{20}{5}(0.4)^5(0.6)^{15} = 0.075,$$

to three significant figures. In the normal approximation this corresponds to the area under a normal curve in standard form between the points corresponding to $r_1 = 4.5$ and $r_2 = 5.5$. Using $\mu = np = 8$ and $\sigma = \sqrt{npq} = 2.19$, the corresponding standardized variables are $z_1 = -1.60$ and $z_2 = -1.14$. Thus we need to find

$$P_N = P[z < -1.14] - P[z < -1.60] = F(-1.14) - F(-1.60),$$

where F is the standard normal distribution function. Using $F(-z) = 1 - F(z)$ and Table C.1 gives $P_N = 0.072$, so the approximation is good.

4.7. MULTINOMIAL

The multinomial distribution is the generalization of the binomial distribution to the case of n repeated trials where there are more than two possible outcomes to each. It is defined as follows. If an event may occur with k possible outcomes, each with a probability $p_i (i = 1, 2, ..., k)$, with

$$\sum_{i=1}^{k} p_i = 1, \tag{4.44}$$

and if r_i is the number of times the outcome associated with p_i occurs, then the random variables $r_i (i = 1, 2, ..., k-1)$ have a *multinomial* probability defined as

$$f(r_1, r_2, ..., r_{k-1}) \equiv n! \prod_{i=1}^{k} p_i^{r_i} \bigg/ \prod_{i=1}^{k} r_i!, \quad r_i = 0, 1, 2, ..., n. \tag{4.45}$$

Note that each of the r_i may range from 0 to n inclusive, and that only $(k-1)$ variables are involved because of the linear constraint:

$$\sum_{i=1}^{k} r_i = n.$$

Just as the binomial distribution tends to the univariate normal, so does the multinomial distribution tend in the limit to the multivariate normal distribution.

With suitable generalizations the results of Section 4.6 may be extended to the multinomial. For example the mean and variance of the random variables r_i are np_i and $np_i(1 - np_i)$, respectively. Multiple variables mean that we also have a covariance matrix, given by

$$V_{ij} = E[\{r_i - E[r_i]\}\{r_i - E[r_i]\}].$$

It is straightforward to show that

$$V_{ij} = \begin{cases} np_i(1-p_i) & i=j \\ -np_ip_j & \text{otherwise.} \end{cases}$$

An example of a multinomial distribution is if we were to construct a histogram of k bins from n independent observations on a random variable, with r_i entries in bin i. The negative sign in the off-diagonal elements of the covariance matrix shows that if bin i contains a greater than average number of events, then the probability is increased that a different bin j will contain a smaller than average number, as expected.

EXAMPLE 4.9

A bag contains 5 white balls, 4 red balls, 3 blue balls, and 2 yellow balls. A ball is drawn at random from the bag and then replaced. If 10 balls are drawn and replaced, what is the probability of obtaining 3 white, 3 red, 2 blue, and 2 yellow?

The probability of obtaining a given number of balls of a specified color after n drawings is given by the multinomial probability. We know that in a single drawing

$$P[w] = \frac{5}{14}, \quad P[r] = \frac{4}{14}, \quad P[b] = \frac{3}{14} \text{ and } P[y] = \frac{2}{14}.$$

Thus if 10 balls are drawn and replaced, the required probability is, using (4.45),

$$\frac{10!}{3!3!2!2!}\left(\frac{5}{14}\right)^3\left(\frac{4}{14}\right)^3\left(\frac{3}{14}\right)^2\left(\frac{2}{14}\right)^2 = 0.0251.$$

4.8. POISSON

The Poisson distribution is an important distribution occurring frequently in practice and that is derived from the binomial distribution by a special limiting process. Consider the binomial distribution for the case when p, the probability of achieving the outcome P, is very small, but n, the number of members of a given sample, is large such that

$$\lim_{p \to 0} (np) = \lambda, \qquad (4.46)$$

where λ is a finite positive constant, i.e., where $n \gg np \gg p$. The kth term in the binomial distribution then becomes

$$\left[\binom{n}{k} p^k q^{n-k}\right] = \frac{n!}{k!(n-k)!}\left(\frac{\lambda}{n}\right)^k \frac{(1-\lambda/n)^n}{(1-\lambda/n)^k}$$

$$= \frac{\lambda^k}{k!}\left[\frac{n(n-1)(n-2)\cdots(n-k+1)}{n^k}\right]\frac{(1-\lambda/n)^n}{(1-\lambda/n)^k}$$

$$= \frac{\lambda^k}{k!}\left(1-\frac{\lambda}{n}\right)^n\left[\frac{(1-1/n)(1-2/n)\cdots(1-(k-1)/n)}{(1-\lambda/n)^k}\right].$$

Now as $n \to \infty$,

$$\lim_{n \to \infty} \left(1 - \frac{\lambda}{n}\right)^n = e^{-\lambda}$$

and all the terms in the square bracket tend to unity, so in the limit that $n \to \infty$ and $p \to 0$ but $np \to \lambda$,

$$\lim_{np \to \lambda} \left[\binom{n}{k} p^k q^{n-k}\right] = f(k; \lambda) = \frac{\lambda^k}{k!} \exp(-\lambda), \quad \lambda > 0, \; k = 0, 1 \ldots. \quad (4.47)$$

This is the probability function of the *Poisson distribution* and gives the probability for different events when the chance of an event is small, but the total number of trials is large.

Although in principle the number of values of k is infinite, the rapid convergence of successive terms in (4.47) means that in practice the distribution function is accurately given by the first few terms. Some examples of the Poisson probability are shown in Fig. 4.3 and tables of the cumulative distribution are given in Appendix C, Table C.3.

Although in deriving the Poisson distribution we have taken the limit as $n \to \infty$, the approximation works well for modest values of n, provided p is small. This is illustrated in Table 4.1, which shows probability values of the binomial distribution for various values of n and p such that $np = 3$ (see also the plots in Fig. 4.2) compared with the probabilities of the Poisson distribution for $\lambda = 3$.

The moment generating function for the Poisson distribution is

$$M_k(t) = E[e^{kt}] = e^{-\lambda} \sum_{k=0}^{\infty} \frac{(\lambda e^t)^k}{k!} = e^{-\lambda} \exp(\lambda e^t). \quad (4.48)$$

Differentiating (4.48) and setting $t = 0$ gives

$$\begin{aligned} \mu_1' &= \lambda, & \mu_2' &= \lambda(\lambda + 1), \\ \mu_3' &= \lambda[(\lambda + 1)^2 + \lambda], & \mu_4' &= \lambda[\lambda^3 + 6\lambda^2 + 7\lambda + 1], \end{aligned} \quad (4.49)$$

and from (1.11a)

$$\mu_2 = \lambda, \quad \mu_3 = \lambda, \quad \mu_4 = \lambda(3\lambda + 1). \quad (4.50)$$

Thus,

$$\mu = \sigma^2 = \lambda, \quad (4.51)$$

a simple result which is very useful in practice. Also from (4.50) and (4.8), we have

$$\beta_1 = \frac{1}{\lambda}, \quad \beta_2 = 3 + \frac{1}{\lambda} \quad (4.52)$$

From these results, and the fact that the Poisson distribution is derived from the binomial, one might suspect that as $\lambda \to \infty$ the Poisson distribution tends to the standard form of the

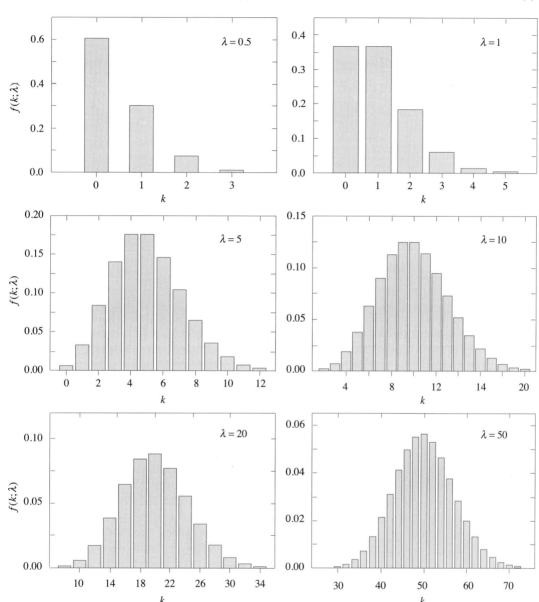

FIGURE 4.3 Plots of the Poisson probability function $f(k; \lambda) = \lambda^k \exp(-\lambda)/k!$ for various values of λ.

TABLE 4.1 Comparison of the Binomial and Poisson probability functions for $np = \lambda = 3$

k	Binomial $p = 0.5$ $n = 6$	$p = 0.2$ $n = 15$	$p = 0.1$ $n = 30$	$p = 0.05$ $n = 60$	$p = 0.02$ $n = 150$	$p = 0.01$ $n = 300$	Poisson $\lambda = 3$
0	0.0156	0.0352	0.0424	0.0461	0.0483	0.0490	0.0490
1	0.0937	0.1319	0.1413	0.1455	0.1478	0.1486	0.1486
2	0.2344	0.2309	0.2276	0.2259	0.2248	0.2244	0.2244
3	0.3125	0.2501	0.2361	0.2298	0.2263	0.2252	0.2252
4	0.2344	0.1876	0.1771	0.1724	0.1697	0.1689	0.1689
5	0.0937	0.1032	0.1023	0.1016	0.1011	0.1011	0.1010
6	0.0156	0.0430	0.0474	0.0490	0.0499	0.0501	0.0501
7	0.0000	0.0138	0.0180	0.0199	0.0209	0.0213	0.0213
8		0.0035	0.0058	0.0069	0.0076	0.0079	0.0079
9		0.0007	0.0016	0.0021	0.0025	0.0026	0.0026
10		0.0001	0.0004	0.0006	0.0007	0.0008	0.0008
11		0.0000	0.0001	0.0001	0.0002	0.0002	0.0002
12			0.0000	0.0000	0.0000	0.0000	0.0000

normal, and indeed this is the case. It can be proved by again using the characteristic function, which using (3.15) and (4.48) is

$$\phi_k(t) = e^{-\lambda} \exp(\lambda e^{it}).$$

Transforming the distribution to standard measure by the relation

$$z = (k - \mu)/\sigma,$$

gives

$$\phi_z(t) = \int_{-\infty}^{\infty} f(z) \exp[it(\sigma z + \mu)] dz = e^{it\mu} \phi_z(\sigma t).$$

However from (4.51), $\mu = \sigma^2 = \lambda$, and so

$$\phi_k(t) = \exp(-it\lambda^{1/2}) e^{-\lambda} \exp(\lambda e^{it\lambda^{-1/2}}).$$

and

$$\ln \phi_z(t) = -it\lambda^{1/2} - \lambda + \lambda \exp(it\lambda^{-1/2}).$$

If we now let $\lambda \to \infty$, keeping t finite, and expand the exponential, then

$$\ln \phi_z(t) = -t^2/2 + O(\lambda^{-1/2}).$$

Thus, for any finite t,

$$\phi(t) \to \exp(-t^2/2),$$

which is the form of the c.f. of a standardized normal distribution and so, by the inversion theorem, the associated density function is the standardized form of the normal distribution. The rate of convergence to normality is the same as for the binomial distribution and so, in particular, the normal approximation to the Poisson distribution is quite adequate for values of $\lambda \geq 10$ and some authors suggest even lower values.

As for the normal distribution, the characteristic function may also be used in a straightforward way to show that the sum of quantities independently distributed as Poisson variates is itself a Poisson variate.

An example of a Poisson distribution is the probability of decay of a radioactive material. A macroscopic amount of the material contains a vast number of atoms, each of which could in principle decay, but the probability of any individual atom decaying in a given time interval is a random event with a very small probability. In this case the quantity $1/\lambda$ is the lifetime of the unstable atom or nucleus. If decays occur randomly in time, with an average of λ events per unit time, then from the Poisson distribution, the probability of N events occurring in an interval t is

$$P[N] = \frac{1}{N!}(\lambda t)^N e^{-\lambda t},$$

and so the probability of no events in time t is an exponential distribution and the probability that the time interval t between events (e.g., the time interval between the detection of the decay particles in a detector) is greater than a specified value x is

$$P[t > x] = e^{-\lambda x}.$$

The memoryless property of the exponential distribution implies that if no events have occurred up to a time y, the probability of no events occurring in a subsequent period x is independent of y.

Finally, it can be shown that if each of a Poisson number of events having mean λ is independently classified as being of one of the types $1, 2, \ldots, r$, with probabilities $p_1, p_2, \ldots, p_r$ respectively, where $\sum_{i=1}^{r} p_r = 1$, then the numbers of events of types $1, 2, \ldots, r$ are independent Poisson random variables with means $\lambda p_1, \lambda p_2, \ldots, \lambda p_r$, respectively.

EXAMPLE 4.10

If the probability of an adverse reaction to a single exposure to a very low dosage of radiation is 0.1% and 10000 people are exposed in an accident, use the Poisson distribution to find the probability that less than 3 will be adversely affected? Why is the use of this distribution justified?

The probability of an adverse reaction is an example of a Bernoulli trial, i.e., there is either a reaction or no reaction. However, if the radiation dose is very low, the probability of an adverse reaction is very small, so in practice the Poisson distribution may be used to predict how many people will suffer an adverse reaction in a large sample. Then,

$$P[k < 3] = P[k = 0] + P[k = 1] + P[k = 2].$$

Using the Poisson distribution with $\lambda = np = 10000 \times 0.001 = 10$, this is

$$P[k < 3] = e^{-10}\left(\frac{10^0}{0!} + \frac{10^1}{1!} + \frac{10^2}{2!}\right) = 61e^{-10} = 0.0028.$$

EXAMPLE 4.11

Use the data of Example 4.10 to investigate the normal approximation to the Poisson for calculating the probability that exactly 5 people will have an adverse reaction.

The Poisson probability that exactly 5 people will have an adverse reaction is

$$P[k = 5] = \frac{e^{-10}10^5}{5!} = 0.0378.$$

In the normal approximation this corresponds to the area under the normal density curve between the points 4.5 and 5.5 and in standard form these points are, using $\mu = \lambda = 10$ and $\sigma = \sqrt{\lambda} = 3.16$, $z_1 = -1.74$ and $z_2 = -1.42$. Then using Table C.1, as in Example 4.8, we find a probability of 0.0384. So the normal approximation is good.

PROBLEMS 4

4.1 The probability of recovering from a certain illness without medical intervention is 50%. A new drug is developed and tested on 20 people with the illness. Fourteen rapidly recover. Is the drug effective?

4.2 In a system designed to destroy incoming missiles, defensive weapons are arranged in layers, each having an efficiency of 95%. To be sure of totally destroying a missile, 'hits' from weapons in at least two defensive layers are required. How many layers would be needed to ensure a probability of at least 99.9% of destroying an incoming missile?

4.3 A biased coin has a probability of 0.48 to fall 'heads' and 0.49 to fall 'tails'. If the coin is thick, so that there is also a probability of it landing on its edge, what is the probability of obtaining 4 heads and 4 tails if it is tossed 10 times?

4.4 Find the coefficient of the term $x^6y^4z^6$ in the expansion of $(2x^2 - 3xy^2 + z^3)^6$.

4.5 A beam of particles is incident on a target with sufficient energy to penetrate it. The particles are mostly absorbed, but there is a small probability p of 5% that this is accompanied by the emission of a new particle from the target. If 100 particles per second are incident on the target, what is the probability that at least 5 particles per second are emitted? Compare your result using the Poisson distribution.

4.6 The average number of car accidents at a dangerous road junction is 5 per month. What is the probability that there will be more than 3 accidents next month?

4.7 A nuclear physics experiment uses 80 detectors. They are checked between data runs and any that have failed are replaced. It is found that the detectors have a 1% probability of failing between checks. If a run can be successfully completed provided no more than 3 detectors fail during the run, find the probability that a data run will be spoiled because of detector failure.

4.8 Resistors are manufactured with a mean value of $R = 50$ ohms and values less than 48.0 ohms, or greater than 51.5 ohms, are considered defective. If the values of R are assumed to be normally distributed with a standard deviation of 1 ohm, what percentage of resistors would be expected to be defective?

4.9 A supply voltage V is assumed to be a normal random variable with mean 100 volts and a variance of 5 volts. It is applied to a resistor $R = 50$ ohms and the power $W = RV^2$ measured. What is the probability that $W > 6 \times 10^5$ watts?

CHAPTER 5

Sampling and Estimation

OUTLINE

5.1 Random Samples and Estimators	83	5.3 Laws of Large Numbers and the Central Limit Theorem		93
5.1.1 Sampling Distributions	84			
5.1.2 Properties of Point Estimators	86	5.4 Experimental Errors		97
5.2 Estimators for the Mean, Variance, and Covariance	90	5.4.1 Propagation of Errors		99

The previous chapters have been concerned almost exclusively with descriptive statistics. The main properties of statistical distributions have been described, and some of the general principles associated with them established. We are now going to consider how to use these ideas to make inferences about a population, given that in practice we usually only have access to a sample of the whole population. This raises several problems, including how to ensure that any sample is random, what is the distribution of the function of the sample data chosen to make statistical inferences, and how to define the desirable properties of such functions so that reliable estimates may be made about the corresponding population parameters. One aspect that will emerge from this discussion is the explanation of why the normal distribution is so important in statistical applications in physical sciences. Finally the formal link is made between theoretical statistics and experimental data.

5.1. RANDOM SAMPLES AND ESTIMATORS

In this section we will consider how random samples are selected, what are their probability distributions, and what are the desirable properties of the functions of random variables that are used to make inferences about the underlying population.

5.1.1. Sampling Distributions

A random sample of size n selected from a population was defined in Section 1.1 as resulting from a situation where every sample of this size had an equal chance of being selected, while noting the intrinsic circularity of this definition. A more formal definition is as follows. If x is a random variable distributed according to a density $f(x)$, then the population is the totality of possible values of x. A sample of size n, i.e., $\mathbf{x} = x_1, x_2, \ldots, x_n$, defines a sample space and is a random sample if (a) all samples are taken from the same distribution as the population and (b) the samples are independently selected. The latter implies that the joint probability density is given by

$$f(\mathbf{x}) = f(x_1)f(x_2)\ldots f(x_n). \tag{5.1}$$

If $f(x)$ is known, random samples can be obtained in principle from (5.1), but if $f(x)$ is unknown, it is often difficult to ensure that the conditions for randomness are strictly met, particularly the first one. It is possible to test whether a given sample is random, but since this is formally testing an hypothesis about the nature of the sample, we will defer discussion of this until hypothesis testing in general is discussed in Chapters 10 and 11. For the present, we will assume that samples have been selected randomly.

We are very often interested in a function y of the sample $x_1, x_2, \ldots, x_n$. Any such function is called a *statistic*, a term introduced in Chapter 1, and is itself a random variable. Because of this, the values of y will vary with different samples of the same size and will be distributed according to a new density function. The formal solution for finding the latter is via construction of the distribution function of y using (5.1), i.e.

$$F(y) = \int \ldots \int \prod_{i=1}^{n} f(x_i) dx_i, \tag{5.2}$$

where the integral is taken over the region such that $y \geq y(x_1, x_2, \ldots, x_n)$. In practice it is often convenient to let $y(x_1, x_2, \ldots, x_n)$ be a new variable and then choose $n-1$ other variables (functions of x_i) such that the n-dimensional integrand in (5.2) takes a simple form. An example will illustrate this.

EXAMPLE 5.1

Find the sampling distribution of the means $\bar{x}_n$ of samples of size n drawn from the Cauchy distribution

$$f(x) = \frac{1}{\pi} \frac{1}{1+x^2}, \quad -\infty \leq x \leq \infty.$$

If we choose new variables $u_i = x_i (i = 1, 2, \ldots, n-1)$ and $u_n = \bar{x}_n$, then the Jacobian of the transformation is

$$J = \frac{\partial(x_1, x_2, \ldots, x_n)}{\partial(u_1, u_2, \ldots, u_{n-1}, \bar{x}_n)},$$

and the distribution function of the means becomes

$$F(\bar{x}_n) = \int \cdots \int f(x_1)f(x_2)\ldots f(x_n) \, J \, d\bar{x}'_n \prod_{i=1}^{n-1} dx_i,$$

where the density functions are expressed in terms of the new set of variables and the integrals are taken over the region such that

$$\bar{x}'_n \geq \frac{1}{n}\sum x_i.$$

Thus

$$F(\bar{x}_n) = \int_{-\infty}^{\bar{x}_n} du_n \int_{-\infty}^{\infty} \cdots \int_{-\infty}^{\infty} \left(\frac{n}{\pi^n}\right) \prod_{j=1}^{n-1}\left\{(1+u_j^2)\left[1+\left(nu_n - \sum_{i=1}^{n-1} u_i\right)^2\right]\right\}^{-1} du_j,$$

and the density function of $\bar{x}_n$ is given by the $(n-1)$-fold integration in $u_j (j = 1, 2, \ldots, n-1)$. The integral can be evaluated, but the algebra is rather lengthy. The result is the probability density

$$f(\bar{x}_n) = \frac{1}{\pi}\frac{1}{1+\bar{x}_n'}$$

which is the same form as the population density for any value of n. We will see later in this chapter that this result is unusual, and the sampling distribution of the sample mean for most distributions commonly met in physical science is a normal distribution for large sample size n.

Even in the simple case given in Example 5.1, the integral is complicated and in practice it is rarely possible to evaluate the required multidimensional integrals analytically. Instead, numerical evaluation is used, but even then conventional techniques are usually far too time consuming and a statistical method, known as the *Monte Carlo method*, mentioned in Section 3.4, is used. The Monte Carlo technique is of rather general application and a technical subject outside the scope of this book, but very briefly the principle of the method in the current context is to use a sequence of random numbers to calculate probabilities and related quantities. Random numbers u uniformly distributed in the interval $0 \leq u \leq 1$ are readily available from computer programs called *random number generators* and may be used to generate a new sequence of numbers distributed with any probability density $f(x)$ that is being studied by using the transformation property of probability distributions referred to in Section 4.1. The new values of x can be viewed as simulated measurements and used to build up an integral of $f(x)$, i.e., the distribution function $F(x)$, such as given in (5.2).

EXAMPLE 5.2

Find an expression for random variables distributed with an exponential density in terms of random variables uniformly distributed with the density

$$p(u) = \begin{cases} 1 & 0 \leq u \leq 1 \\ 0 & \text{otherwise.} \end{cases}$$

The exponential density is, from Equation (4.26),

$$f(x) = \begin{cases} \lambda e^{-\lambda x} & \lambda > 0, \ x \geq 0 \\ 0 & \text{otherwise.} \end{cases}$$

Conservation of probability requires that

$$\int_{-\infty}^{u} p(u') \, du' = \int_{-\infty}^{x} f(x') \, dx'$$

and so $u = 1 - e^{-\lambda x}$, and $x = -\ln(1-u)/\lambda$. As u is distributed uniformly, then so is $(1-u)$. Thus to select random numbers distributed with an exponential density, we generate a sequence u_i and from them generate a new sequence x_i using $x_i = -\ln(u_i)/\lambda$.

Another useful method for finding sampling distributions is to construct the moment generating function, or the characteristic function, for the statistic. If either of these is recognized as that of a known pdf, then this is the pdf of the sampling distribution. This technique is very practical and we shall have occasion to use it later. Alternatively, the inversion theorem may be used to identify the density function. For example, we have shown in Section 4.8 that the cf for a Poisson distribution with a general term $e^{-\lambda}\lambda^k/k!$ is

$$\phi(t) = e^{-\lambda} \exp(\lambda e^{it}),$$

where λ, the Poisson parameter, is also the mean of the distribution. From this we can form the cf for a sample of size n as the product of terms of the form $\phi(t)$ and hence show that the cf for the mean is $\exp\{n\lambda(e^{it} - 1)\}$. Finally, the inversion theorem may be used to show that the sampling distribution of the mean is also a Poisson distribution, but whose general term is $e^{-n\lambda}(n\lambda)^k/k!$

A common situation is where the exact form of $f(x)$ is unknown, but one has a model (or hypothesis) for $f(x)$ that depends on an unknown parameter θ. The central problem is then to construct a function of the observations $x_1, x_2, \ldots, x_n$ that contains no unknown parameters to *estimate* θ, i.e., to give a value to θ, and hence determine $f(x)$. This situation is an example of *parametric statistics*. In these circumstances the statistic is referred to as a *point estimator* of θ and is written as $\hat{\theta}$. The word 'point' will be omitted when it is obvious that we are talking about the estimation of the value of a parameter by a single number. In general there could be several unknown parameters $\boldsymbol{\theta} = \theta_1, \theta_2, \ldots, \theta_m$ and associated estimators. Since the estimator is a function of the random variables $\mathbf{x} = x_1, x_2, \ldots, x_n$, it is itself a random variable and its value will therefore vary with different samples of the same size and be distributed according to a new density function $g(\hat{\theta}; \theta)$. The merit of an estimator is judged by the properties of this distribution and not by the values of a particular estimate. So we will now turn to consider the properties of 'good' estimators.

5.1.2. Properties of Point Estimators

An intuitively obvious desirable property of an estimator is that, as the sample size increases, the estimate tends to the value of the population parameter. Any other result would be inconvenient and even possibly misleading. This property is called *consistency*.

5.1. RANDOM SAMPLES AND ESTIMATORS

Formally, an estimator $\hat{\theta}_n$ computed from a sample of size n, is said to be a *consistent* estimator of a population parameter θ if, for any positive ε, arbitrarily small,

$$\lim_{n \to \infty} P[|\hat{\theta}_n - \theta| > \varepsilon] = 0. \tag{5.3}$$

In these circumstances, $\hat{\theta}_n$ is said to *converge in probability* to θ. Thus, $\hat{\theta}_n$ is a consistent estimator of θ if it converges in probability to θ.

The property of consistency tells us the asymptotic ($n \to \infty$) behavior of a suitable estimator, although the approach to consistency does not have to be monotonic as n increases. Having found such an estimator we may generate an infinite number of other consistent estimators:

$$\hat{\theta}'_n = p(n) \, \hat{\theta}_n, \tag{5.4}$$

provided

$$\lim_{n \to \infty} p(n) = 1. \tag{5.5}$$

However, we may further restrict the possible estimators by requiring that for *all* n the expected value of $\hat{\theta}_n$ is θ, i.e., $E[\hat{\theta}_n] = \theta$, or in full, using (5.1),

$$E[\hat{\theta}(\mathbf{x})] = \int \hat{\theta}(\mathbf{x}) g(\hat{\theta};\theta) \, d\hat{\theta} = \int \cdots \int \hat{\theta}(\mathbf{x}) f(x_1) f(x_2) \ldots f(x_n) \, dx_1 dx_2 \ldots dx_n = \theta. \tag{5.6}$$

Estimators with this property are called *unbiased*, with the *bias* b defined by

$$b \equiv E[\hat{\theta}_n] - \theta. \tag{5.7}$$

Estimators for which $b \to 0$ as $n \to \infty$ are said to be *asymptotically unbiased*. Despite the name, the fact that an estimator is biased is not often a serious problem, because there usually exists a simple factor that converts such an estimator to an unbiased one. Unbiased estimators are just more convenient to use in practice, although as the following example shows, they are not unique.

EXAMPLE 5.3

If $\hat{\theta}_i$ ($i = 1, 2, \ldots, n$) is a set of n unbiased estimators for the parameter θ, show that any linear combination

$$\hat{\theta} = \sum_{i=1}^{m} \lambda_i \hat{\theta}_i, \qquad m \leq n$$

where λ_i are constants, is also an unbiased estimator for θ, provided

$$\sum_{i=1}^{m} \lambda_i = 1.$$

The expectation of $\hat{\theta}$ is

$$E[\hat{\theta}] = E\left[\sum_{i=1}^{m} \lambda_i \hat{\theta}_i\right] = \sum_{i=1}^{m} \lambda_i E[\hat{\theta}_i] = \sum_{i=1}^{m} \lambda_i \theta = \theta.$$

Hence from (5.7), $\hat{\theta}$ is an unbiased estimator for the parameter θ.

The requirements of consistency and lack of bias alone do not produce unique estimators. One can easily show that the sample mean is a consistent and unbiased estimator of the mean of a normal population with known variance. However the same is true of the sample median. Further restrictions must be imposed if uniqueness is required. One of these is the *efficiency* of an estimator. An unbiased estimator with a small variance will produce estimates more closely grouped around the population value θ than one with a larger variance. If two estimators $\hat{\theta}_1$ and $\hat{\theta}_2$, both calculated from samples of size n, have variances such that $\text{var}\hat{\theta}_1 < \text{var}\hat{\theta}_2$, then $\hat{\theta}_1$ is said to be more *efficient* than $\hat{\theta}_2$ for samples of size n. For the normal distribution,

$$\text{var}(\text{mean}) = \sigma^2/n,$$

for any n (this result is proved in Section 5.2 below). But for large n,

$$\text{var}(\text{median}) = \pi\sigma^2/2n > \sigma^2/n.$$

Thus the mean is the more efficient estimator for large n. (In fact this is true for all n.) Consistent estimators whose sampling variance for large samples is less than that of any other such estimators are called *most efficient*. Such estimators serve to define a scale of efficiency. Thus if $\hat{\theta}_2$ has variance v_2 and $\hat{\theta}_1$, the most efficient estimator, has variance v_1, then the efficiency of $\hat{\theta}_2$ is defined as

$$E_2 = v_1/v_2. \tag{5.8}$$

It may still be that there exist several consistent estimators $\hat{\theta}$ for a population parameter θ. Can one choose a 'best' estimator from among them? The criterion of efficiency alone is not enough, since it is possible that for a given finite n, one estimator $\hat{\theta}_n$, which is biased, is consistently closer to θ than an unbiased estimator $\hat{\theta}'_n$. In this case the quantity to consider is not the variance but the second moment of $\hat{\theta}_n$ about θ, which is

$$E\left[(\hat{\theta}_n - \theta)^2\right].$$

Using (5.7) gives (see Problem 5.1)

$$E\left[(\hat{\theta}_n - \theta)^2\right] = \text{var}(\hat{\theta}_n) + b^2. \tag{5.9}$$

This quantity is called the *mean squared error* and we define $\hat{\theta}_n$ to be a *best, or optimal, estimator* of the parameter θ if

$$E\left[(\hat{\theta}_n - \theta)^2\right] \leq E\left[(\hat{\theta}'_n - \theta)^2\right],$$

where $\hat{\theta}'_n$ is any other estimator of θ. Thus an optimal unbiased estimator $\hat{\theta}_n$ is one with minimum variance. We will discuss how to obtain minimum variance estimators in more detail in Chapter 7, Section 7.4.

EXAMPLE 5.4

If $\hat{\theta}_i$ ($i = 1, 2$) are two independent and unbiased estimators for a parameter θ with variances σ_i^2, what value of the constant λ in the linear combination

$$\hat{\theta} = \lambda \hat{\theta}_1 + (1 - \lambda) \hat{\theta}_2$$

ensures that $\hat{\theta}$ is the optimal estimator for θ?

Because $\hat{\theta}_1$ and $\hat{\theta}_2$ are both unbiased, so is $\hat{\theta}$ (see Example 5.3). Thus the optimal estimator of θ is one that minimizes the mean squared error

$$\text{var}(\hat{\theta}) = E\left[(\hat{\theta} - \theta)^2\right].$$

Now because $\hat{\theta}_1$ and $\hat{\theta}_2$ are independent,

$$\text{var}(\hat{\theta}) = \lambda^2 \text{ var}(\hat{\theta}_1) + (1 - \lambda)^2 \text{ var}(\hat{\theta}_2)$$
$$= \lambda^2 \sigma_1^2 + (1 - \lambda)^2 \sigma_2^2,$$

and the minimum of $\text{var}(\hat{\theta})$ is found from $d \text{ var}(\hat{\theta})/d\lambda = 0$, i.e.

$$\lambda = \frac{1/\sigma_1^2}{1/\sigma_1^2 + 1/\sigma_2^2}.$$

The discussion above gives an idea of the desirable properties of estimators, but there is a more general criterion that can be used. Consider the case of estimating a parameter θ and let

$$f(\hat{\theta}_1, \hat{\theta}_2, \ldots, \hat{\theta}_r; \theta)$$

be the joint density function of r independent estimators $\hat{\theta}_i (i = 1, 2, \ldots, r)$. Then, from the definition of the multivariate conditional density, we have (cf. equation (3.23))

$$f(\hat{\theta}_1, \hat{\theta}_2, \ldots, \hat{\theta}_r; \theta) = f^M(\hat{\theta}_1; \theta) f^C(\hat{\theta}_2, \hat{\theta}_3, \ldots, \hat{\theta}_r; \theta | \hat{\theta}_1), \quad (5.10)$$

where $f^M(\hat{\theta}_1; \theta)$ is the marginal density of $\hat{\theta}_1$ and $f^C(\hat{\theta}_2, \hat{\theta}_3, \ldots, \hat{\theta}_r; \theta | \hat{\theta}_1)$ is the conditional density of all the other $\hat{\theta}_i$ given $\hat{\theta}_1$. Now if f^C is independent of θ, then clearly once $\hat{\theta}_1$ is specified the other estimators contribute nothing to the problem of estimating θ, i.e., $\hat{\theta}_1$ contains *all* the information about θ. In these circumstances $\hat{\theta}_1$ is called a *sufficient* statistic for θ. It is more convenient in practice to write (5.10) as a condition on the likelihood function introduced in Chapter 2.

Let $f(x; \theta)$ denote the density function of a random variable x, where the form of f is known, but not the value of θ, which is to be estimated. Then let $x_1, x_2, \ldots, x_n$ be a random sample of size n drawn from $f(x;\theta)$. The joint density function $f(x_1, x_2, \ldots, x_n; \theta)$ of the independent random variables $x_1, x_2, \ldots, x_n$ is given by

$$f(x_1, x_2, \ldots, x_n; \theta) = \prod_{i=1}^{n} f(x_i; \theta), \tag{5.11}$$

where $f(x_i, \theta)$ is the density function for the ith random variable. The function $f(x_1, x_2, \ldots, x_n; \theta)$ is the likelihood function of θ and is written as $L(x_1, x_2, \ldots, x_n; \theta)$. If L is expressible in the form

$$L(x_1, x_2, \ldots, x_n; \theta) = L_1(\hat{\theta}; \theta) \, L_2(x_1, x_2, \ldots, x_n), \tag{5.12}$$

where L_1 does not contain the x's other than in the form θ, and L_2 is independent of θ, then $\hat{\theta}$ is a *sufficient* statistic for the estimation of θ.

EXAMPLE 5.5

Find a sufficient estimator for estimating the variance of a normal distribution with zero mean.
The probability density is

$$f(x) = \frac{1}{\sqrt{2\pi}} \frac{1}{\sigma} \exp\left(-\frac{x^2}{2\sigma^2}\right),$$

and the likelihood function is therefore

$$L(x_1, x_2, \ldots, x_n; \sigma^2) = \left(\frac{1}{\sigma\sqrt{2\pi}}\right)^n \exp\left(-\frac{1}{2\sigma^2} \sum_{i=1}^{n} x_i^2\right).$$

If we let $L_2 = 1$ in (5.12), we have $L_1 = L$ and L_1 is a function of the sample x_i only in terms of $\sum x_i^2$. Thus, $\sum x_i^2$ is a sufficient estimator for σ^2. We will show in Section 5.2 that this estimator is biased.

5.2. ESTIMATORS FOR THE MEAN, VARIANCE, AND COVARIANCE

Estimators for the mean, variance, and covariance are of central importance in statistical analysis, so we consider them in more detail. Let S denote a sample of n observations $x_i (i = 1, 2, \ldots, n)$ selected at random. The sample S is called a *random sample with replacement* (or a *simple random sample*) if, in general, the observation x_{n-1} is returned to the population before x_n is selected. If x_{n-1} is not returned, then S is called a *random sample without replacement*. Sampling with replacement implies, of course, that it is indeed possible to return the 'observation' to the population, as is the case when drawing cards from a deck. In most practical situations this is usually not possible and the sampling is without replacement. Sampling from an infinite population is equivalent to sampling with replacement.

For any continuous population, finite or infinite, the sample mean $\bar{x}$ is an estimator for the population mean μ and since this is true for all possible samples of size n, the sample mean is an unbiased estimator. This result follows simply from the definition of the sample mean, equation (1.3). Thus

$$E[\bar{x}] = E\left[\frac{1}{n}\sum_{i=1}^{n} x_i\right] = \frac{1}{n}\sum_{i=1}^{n} E[x_i].$$

But using (5.1),

$$E[x_i] = \int \ldots \int x_i f(x_1)\ldots f(x_n)\, dx_1\ldots dx_n = \mu$$

and so

$$E[\bar{x}] = \frac{1}{n}\sum_{i=1}^{n} \mu = \mu. \tag{5.13}$$

We can also find the expectation of the sample variance s^2. This is

$$E\left[\frac{1}{n-1}\sum_{i=1}^{n}(x_i - \bar{x})^2\right] = \frac{1}{n-1}E\left[\sum_{i=1}^{n}\left(x_i - \frac{1}{n}\sum_{j=1}^{n} x_j\right)^2\right]$$

$$= \frac{1}{n-1}E\left[\frac{n-1}{n}\sum_{i=1}^{n}(x_i)^2 - \frac{1}{n}\sum_{i\neq j} x_i x_j\right] \tag{5.14}$$

$$= \mu'_2 - (\mu'_1)^2 = \sigma^2.$$

Thus the presence of the factor $1/(n-1)$ in the definition of the sample variance, as we noted in Chapter 1 differs from the analogous definition for the population variance to ensure that s^2 is an unbiased estimator of σ^2. Similarly, the sample covariance defined in (1.12b) is an unbiased estimator for the population covariance of equation (1.12a).

Given any estimator $\hat{\theta}$, one can calculate its variance. For example, the variance of the sample mean drawn from an infinite population, or a finite population with replacement, is by definition

$$\sigma^2_{\bar{x}} \equiv \operatorname{var}(\bar{x}) = E\left[(\bar{x} - E[\bar{x}])^2\right] = E\left[(\bar{x} - \mu)^2\right], \tag{5.15}$$

which may be written as

$$\operatorname{var}(\bar{x}) = \frac{1}{n^2} E\left[\left(\sum_{i=1}^{n}(x_i - \mu)\right)^2\right].$$

If we expand the square bracket on the right-hand side and again use (5.1), there are n terms containing the form $(x_i - \mu)^2$, each of which gives a contribution

$$\int \ldots \int (x_i - \mu)^2 f(x_i)\ldots f(x_n)\, dx_i\ldots dx_n = \sigma^2.$$

The remaining terms are integrals over the forms $(x_i - \mu)(x_j - \mu)$ with $i < j$, each of which, using the definition of μ, is zero. Thus

$$\text{var}(\bar{x}) = \frac{1}{n^2} \sum_{i=1}^{n} \sigma^2 = \frac{\sigma^2}{n}$$

$$= \frac{1}{n(n-1)} \sum_{i=1}^{n} (x_i - \bar{x})^2, \qquad (5.16a)$$

where the second line follows if σ^2 is replaced by its estimator s^2. If the sample is drawn from a finite population without replacement, then this result is modified to

$$\text{var}(\bar{x}) = \frac{\sigma^2}{n}\left(\frac{N-n}{N-1}\right). \qquad (5.16b)$$

The square root of $\sigma_{\bar{x}}^2$, that is, the standard deviation $\sigma_{\bar{x}}$, is called the *(standard) error of the mean* and was introduced briefly when we discussed the normal distribution in Section 4.2. It is worth emphasizing the difference between the standard deviation σ and the standard error on the mean $\sigma_{\bar{x}}$. The former describes the extent to which a single observation is liable to vary from the population mean μ; the latter measures the extent that an estimate of the mean obtained from a sample of size n is liable to differ from the true mean.

The result (5.16a) is of considerable importance, because it shows that as the sample size n increases, the variance of the sample mean decreases, and hence the statistical error on a set of measurements decreases (like $1/\sqrt{n}$ in the case of (5.16a)) and the probability that the sample mean is a good estimation of the population mean increases, a result that was referred to in Chapter 1. Results (5.16) assume that the measurements are random samples and uncorrelated. If this is not the case, then this must be taken into account. They also assume that the samples are obtained by simple random sampling from a single population. Better estimates can be obtained if we have additional information about the sample. One example is stratified sampling, mentioned briefly in Section 1.1. This technique requires that the population can be divided into a number of mutually exclusive subpopulations, with *known* fractions of the whole population in each. Then simple random samples using, for example, sample sizes proportional to these fractions lead to smaller estimates for $\text{var}(\bar{x})$ with the same total sample size. However, as this situation is not usually met in physical science, we will continue to consider only simple random sampling from a single homogeneous population.

We can go further, by using the general results for expectation values, and find the estimator of the variance of s^2 and hence the estimator of the standard deviation σ_s. The latter is not the square root of the former, but, anticipating equation (5.45), is given by

$$\text{var}(s^2) = \left(\frac{ds^2}{ds}\right)^2 \text{var}(s).$$

For a normal distribution, the result is simple:

$$\sigma_s = \frac{\sigma}{\sqrt{2(n-1)}}. \qquad (5.17)$$

Just as in (5.16a), to use this result one would usually have to insert an estimate for σ obtained from the data. Providing n is large, there is little loss in precision in doing this, but for small n

an alternative approach would have to be adopted. This is discussed in Section 6.2. Alternatively, (5.17) can be used to predict how many events would be needed to measure σ to a given precision under different assumptions about its value. This could be useful at the planning stages of an experiment.

EXAMPLE 5.6

A random sample $x_i (i = 1, 2, \ldots, n)$ is drawn from a population with mean μ and variance σ^2. Two unbiased estimators for μ are

$$\hat{\mu}_1 = \frac{1}{2}(x_1 + x_2) \quad \text{and} \quad \hat{\mu}_2 = \bar{x}_n.$$

What is the relative efficiency of $\hat{\mu}_1$ to $\hat{\mu}_2$?

From (5.16a), var$(\hat{\mu}_2)$ = var$(\bar{x}_n) = \sigma^2/n$. If the same steps to derive this result are used for $\hat{\mu}_1$, then there is one term containing the form $(x_1 - \mu)^2$, and one containing the form $(x_2 - \mu)^2$, each of which gives a contribution σ^2 and a single term containing $(x_1 - \mu)(x_2 - \mu)$ which contributes zero. Thus var$(\hat{\mu}_1) = \sigma^2/2$ and so

$$\text{relative efficiency} = \frac{\text{var}(\hat{\mu}_1)}{\text{var}(\hat{\mu}_2)} = \frac{n}{2}.$$

5.3. LAWS OF LARGE NUMBERS AND THE CENTRAL LIMIT THEOREM

The results of Section 5.2 may be stated formally as follows. Let x_i be a population of independent random variables with mean μ and finite variance and let $\bar{x}_n$ be the mean of a sample of size n. Then, given any $\varepsilon > 0$ and δ in the range $0 < \delta < 1$, there exists an integer n such that for all $m \geq n$

$$P[|\bar{x}_m - \mu| \leq \varepsilon] \geq 1 - \delta. \tag{5.18}$$

This is the *weak law of large numbers*. It tells us that $|\bar{x}_n - \mu|$ will ultimately be very small, but does not exclude the possibility that for some finite n it could be large. Since, in practice, we can only have access to finite samples, this possibility could be of some importance. Fortunately there exists the so-called *strong law of large numbers*, which, in effect, states that the probability of such an occurrence is extremely small. It is the laws of large numbers that ensure that the frequency definition of probability adopted in Chapter 2 concurs in practice with the axiomatic one.

The weak law of large numbers may be proved from Chebyshev's inequality that we discussed in Chapter 1, equation (1.11), provided the population distribution has a finite variance. Chebyshev's inequality may be written as

$$P\left[|\bar{x}_n - \mu| \geq \frac{k\sigma}{n^{1/2}}\right] \leq \frac{1}{k^2}, \tag{5.19}$$

so if we choose $k = \delta^{-1/2}$ and $n > \sigma^2/\delta\varepsilon^2$ and substitute in (5.19), then (5.18) results.

The bound given by (5.19) is usually weak, but if we restrict ourselves to the sampling distribution of the mean then we can derive the most important theorem in statistics, the *central limit theorem*, which may be stated as follows. Let the independent random variables x_i of unknown density be identically distributed with mean μ and variance σ^2, both of which are finite. Then the distribution of the sample mean $\bar{x}_n$ tends to the normal distribution with mean μ and variance σ^2/n when n becomes large. Thus, if $u(t)$ is the standard form of the normal density function, then for arbitrary t_1 and t_2,

$$\lim_{n \to \infty} P\left[t_1 \leq \frac{\bar{x}_n - \mu}{\sigma/n^{1/2}} \leq t_2\right] = \int_{t_1}^{t_2} u(t)\, dt. \tag{5.20}$$

The proof of this theorem illustrates the use of several earlier results and definitions and so is worth giving.

By applying the results on expected values given in Chapter 3 to moment-generating functions, it follows immediately that if the components of the sample are independent, then the mean and variance of their sum

$$S = \sum_{i=1}^{n} x_i$$

are given by

$$\mu_S = n\mu \quad \text{and} \quad \sigma_S^2 = n\sigma^2.$$

Now consider the variate

$$u = \frac{S - \mu_S}{\sigma_S} = \frac{1}{\sqrt{n}\sigma} \sum_{i=1}^{n} (x_i - \mu), \tag{5.21}$$

with characteristic function $\phi_u(t)$. If $\phi_i(t)$ is the cf of $(x_i - \mu)$, then

$$\phi_u(t) = \prod_{i=1}^{n} \phi_i(t)\left(\frac{t}{\sqrt{n}\sigma}\right).$$

But all the $(x_i - \mu)$ values have the same distribution and so

$$\phi_u(t) = \left[\phi_i(t)\left(\frac{t}{\sqrt{n}\sigma}\right)\right]^n. \tag{5.22}$$

Just as the mgf can be expanded in an infinite series of moments, we can expand the cf,

$$\phi(t) = 1 + \sum_{r=1}^{\infty} \mu'_r \frac{(it)^r}{r!}, \tag{5.23}$$

and since the first two moments of $(x_i - \mu)$ are zero and σ^2, respectively, we have from (5.22) and (5.23)

$$\phi_u(t) = \left[1 - \frac{t^2}{2n} + O\left(\frac{1}{n}\right)\right]^n.$$

Expanding the square bracket, and then letting $n \to \infty$ but keeping fixed t, gives

$$\phi_u(t) \to e^{-t^2/2}, \tag{5.24}$$

which is the cf of a standardized normal distribution. So by the inversion theorem, S is distributed as the normal distribution $n(S; \mu_S, \sigma_S^2)$, and hence $\bar{x}_n$ is distributed as $n(\bar{x}_n; \mu, \sigma^2/n)$.

In practice, the normal approximation is good for $n \geq 30$ regardless of the shape of the population. For values of n less than about 30, the approximation is good only if the population distribution does not differ much from a normal distribution. The sampling distribution of the means when sampling from a normal population is normal independent of the size of n.

The form of the central limit theorem above is not the most general that can be given. Provided certain (weak) conditions on the third moments are obeyed, then the condition that the x_i values all have the same distribution can be relaxed, and it is possible to prove that the sampling distribution of *any* linear combination of independent random variables having arbitrary distributions with finite means and variances tends to normality for large samples. There are even circumstances under which the assumption of independence can be relaxed.

EXAMPLE 5.7

Five hundred resistors are found to have a mean value of 10.3 ohms and a standard deviation of 0.2 ohms. What is the probability that a sample of 100 resistors drawn at random from this population will have a combined value between 1027 and 1035 ohms?

For the sampling distribution of the means, $\mu_{\bar{x}} = \mu = 10.3$ and the standard deviation of this value is

$$\sigma_{\bar{x}} = \frac{\sigma}{\sqrt{n}} \sqrt{\frac{(N-n)}{(N-1)}} = \frac{0.2}{\sqrt{100}} \sqrt{\frac{(500-100)}{(500-1)}} = 0.018.$$

We seek the value of the probability such that $P[10.27 < x < 10.35]$. Using the central limit theorem, we can use the normal approximation. So, using standardized variables, this is equivalent to

$$P[-1.67 < z < 2.78] = N(2.78) + N(1.67) - 1 \approx 0.95.$$

The central limit theorem applies to both discrete and continuous distributions and is a remarkable theorem because nothing is said about the original density function, except that it has finite mean and variance. Although in practice these conditions are not usually restrictions, they are essential. Thus we have seen in Example 5.1 that the distribution of $\bar{x}_n$ for the Cauchy distribution is the same as for a single observation. The failure of the theorem in this case can be traced to the infinite variance of the Cauchy distribution and there are other examples, such as the details of the scattering of particles from nuclei, where the long 'tails' of distributions cause the theorem to fail. It is the central limit theorem that gives the normal distribution such a prominent position both in theory and in practice. In particular, it allows (approximate) quantitative probability statements to be made in experimental situations where the exact form of the underlying distribution is unknown. This was briefly mentioned in Section 1.4.

Just as we have been considering the sampling distribution of means we can also consider the sampling distribution of sums $T = \sum x_i$ of random variables of size n. If the random variable x is distributed with mean μ, and variance σ^2, then the sampling distribution of T has mean

$$\mu_T = n\mu, \tag{5.25}$$

and variance

$$\sigma_T^2 = \begin{cases} n\sigma^2 \left(\dfrac{N-n}{N-1}\right), \\ n\sigma^2 \end{cases} \tag{5.26}$$

where the first result is for sampling from a finite population of size N without replacement and the second result is otherwise.

We will conclude with some results on the properties of linear combinations of means, since up to now we have been concerned mainly with sampling distributions of a single sample mean.

Let

$$l = \sum_{i=1}^{n} a_i x_i, \tag{5.27}$$

where a_i are real constants, and the x_i values are random variables with means μ_i, variances σ_i^2, and covariances $\sigma_{ij}(i,j = 1, 2, \ldots n\,; i \neq j)$. (The index i now indicates different random variables, not a sample of a single random variable.) Then,

$$\mu_l = \sum_{i=1}^{n} a_i \mu_i \tag{5.28}$$

and

$$\sigma_l^2 = \sum_{i=1}^{n} a_i^2 \sigma_i^2 + 2 \sum_{i<j} a_i a_j \sigma_{ij}, \tag{5.29}$$

which reduces to

$$\sigma_l^2 = \sum_{i=1}^{n} a_i^2 \sigma_i^2 \tag{5.30}$$

if the x's are mutually independent. Note that the constants are squared in (5.30), so for example, the variance of $(x_1 + x_2)$ is the same as that of $(x_1 - x_2)$. (For the proof of these results, see Problem 5.3.)

A useful corollary to the above result is as follows. Let $\bar{x}_i (i = 1, 2, \ldots, n)$ be the means of a random sample of size n_i drawn from an infinite population with mean μ_i and variance σ_i. If $\bar{x}_1$ and $\bar{x}_2$ are independently distributed, then

$$\mu_{\bar{x}_1 + \bar{x}_2} = \mu_1 \pm \mu_2 \tag{5.31}$$

and

$$\sigma_{\bar{x}_1 + \bar{x}_2}^2 = \sum_{i=1}^{2} \left(\dfrac{\sigma_i^2}{n_i}\right). \tag{5.32}$$

These results follow immediately from (5.28) and (5.30), and the results (5.14) and (5.16a), by the substitutions $x_1 = \bar{x}_1$ and $x_2 = \bar{x}_2$, with $a_1 = a_2 = 1$ for the first case and $a_1 = -a_2 = 1$ for the second.

5.4. EXPERIMENTAL ERRORS

In the preceding sections we were concerned with theoretical statistics only. In this section we provide the link between theoretical statistics and experimental situations. This continues the discussion started in Chapter 1. In an experimental observation one can never measure the value of a quantity with absolute precision, that is, one can never reduce the statistical error on the measurement to zero, although we can reduce it by increasing n, i.e., taking more data. Recall in Section 1.4 we distinguished the *precision* of a measurement from its *accuracy*, that is, the deviation of the observation from the 'true' value, assuming that such a concept is meaningful. Thus there may exist, in addition to fluctuations in the measurement process that limit the precision, unknown systematic errors that limit the accuracy. In general, the only errors that we can deal with in detail here are the former type, and the conventional measure of this type of error is taken to be the standard error, defined above and which we have previously introduced in Section 4.2. This definition of the error is, of course, arbitrary, and formerly (but now only very rarely) the *probable error p*, defined by

$$\int_{\mu-p}^{\mu+p} f(x)dx = 1/2,$$

was used. Needless to say, multiplying errors by an arbitrary factor 'to be on the safe side' renders statistical analyses meaningless.

Consider, for example, an idealized nuclear counting experiment for a scattering process. The number of trials is very large, because the numbers of particles in the beam and target are large, but the probability of a scatter, p, is very small. In this situation the Poisson distribution is applicable, and as we have seen in equation (4.51), if $N_e = np$ is the total number of counts recorded then $\sigma = \sqrt{N_e}$. The result of the experiment would be given as

$$N = N_e \pm \Delta N, \tag{5.33}$$

where

$$\Delta N = \sqrt{N_e}. \tag{5.34}$$

If the population distribution is unknown, then we can consider the sampling distribution. For example, from a set of observations x_i, we know that an estimate of the mean is the sample mean

$$\bar{x} = \frac{1}{n}\sum_{i=1}^{n} x_i \tag{5.35}$$

and the laws of large numbers ensure that $\bar{x}$ is a good estimate for large n. The variance of $\bar{x}$ is

$$\sigma_{\bar{x}}^2 = \sigma^2/n, \tag{5.36}$$

so to calculate $\sigma_{\bar{x}}^2$, we need to estimate σ^2. We have seen that the sample variance is

$$s^2 = \frac{1}{n-1} \sum_{i=1}^{n} (x_i - \bar{x})^2, \tag{5.37}$$

and thus

$$\sigma_{\bar{x}}^2 = \frac{1}{n(n-1)} \sum_{i=1}^{n} (x_i - \bar{x})^2. \tag{5.38}$$

An experimental result would then be quoted as

$$x = \bar{x}_e \pm \Delta x, \tag{5.39a}$$

where

$$\Delta x = \sigma_{\bar{x}} = \left[\frac{1}{n(n-1)} \sum_{i=1}^{n} (x_i - \bar{x})^2 \right]^{1/2}. \tag{5.39b}$$

Now by the central limit theorem we know that the distribution of the sample means is approximately normal, and therefore (5.39) may be interpreted (compare Section 4.2) as

$$P[\bar{x}_e - \Delta x \leq x \leq \bar{x}_e + \Delta x] \simeq 68.3\%,$$
$$P[\bar{x}_e - 2\Delta x \leq x \leq \bar{x}_e + 2\Delta x] \simeq 95.4\%, \tag{5.40}$$
$$P[\bar{x}_e - 3\Delta x \leq x \leq \bar{x}_e + 3\Delta x] \simeq 99.7\%.$$

So even though the form of the underlying distribution of x is unknown, the central limit theorem enables an approximate quantitative statement to be made about the probability of the true value of x lying within a specified range.

Since we have moved away from mathematical statistics into the real world of experimental data, it is worth commenting on a situation that commonly arises. In calculating $\sigma_{\bar{x}}$ from (5.39b) one often finds that a few data points (referred to as 'outliers') are making very significant contributions to the summation. What, if anything, should one do about this? A general comment is that transforming the data can reduce the effect of outliers. For example, taking logarithms shrinks large values much more than smaller ones, but this is not always practical. In the light of (5.40), it might seem reasonable to ignore data that are, say, three standard deviations away from the mean, and tables exist giving criteria to select data for rejection. There is even a 'rule' for rejecting data (called Chauvenet's criterion), one version of which states that if we have n data points, the point x_i should be rejected if $P[x_i > \bar{x}] < 1/2n$. However, common sense tells you that the more data that you take, the more outliers will be found. So, if we expect a rare (but real) event with a probability $1/2n$ in a single trial, the probability of its occurrence at least once in n trials is

$$1 - \left(1 - \frac{1}{2n}\right)^n = 1 - \left\{ \left(1 - \frac{1}{2n}\right)^{2n} \right\}^{1/2} \approx 1 - e^{-1/2} = 0.39$$

when n becomes large, which is not negligible. If outliers are rejected, for whatever reason, and then $\bar{x}_e$ and $\sigma_{\bar{x}}$ recalculated, because $\sigma_{\bar{x}}$ will now be smaller, you may well find new points that satisfy the recalculated rejection criterion, and logically these should also be rejected. But if you were to repeat this process, you could converge to a value that seriously distorted the information in the origin data set. So blindly applying a rule, however reasonable it may appear, is dangerous and is to be discouraged.

The existence of outliers should alert you to possible problems and such points should be examined very carefully to see if there is any valid experimental reason why they should be rejected. But this should be done honestly, avoiding any temptation to 'massage the data' and must be defensible. In the absence of such reasons, there are only two alternatives: either include the outliers and accept that such statistical fluctuations do rarely occur or reject them and possibly miss the chance of finding some new phenomenon. You should certainly never use a rejection criterion on a data set more than once. Also, if you use n standard deviations as the criteria for rejection, you will have to decide on a value for n, and $n = 3$ would definitely be considered too small. The choice is yours. But whatever you do, it should be clearly stated when reporting the data.

5.4.1. Propagation of Errors

If we have a function y of the p variables $\theta_i (i = 1, 2, \ldots, p)$, i.e.

$$y \equiv y(\boldsymbol{\theta}) = y(\theta_1, \theta_2, \ldots, \theta_p),$$

then we are often interested in knowing the approximate error on y, given that we know the errors on θ_i. If the true values of θ_i are $\bar{\theta}_i$ (in practice, estimates of these quantities would usually have to be used) and the quantities $(\theta_i - \bar{\theta}_i)$ are small, then a Taylor expansion of $y(\boldsymbol{\theta})$ about the point $\boldsymbol{\theta} = \bar{\boldsymbol{\theta}}$ gives, to first order in $(\theta_i - \bar{\theta}_i)$,

$$y(\boldsymbol{\theta}) = y(\bar{\boldsymbol{\theta}}) + \sum_{i=1}^{p} (\theta_i - \bar{\theta}_i) \left. \frac{\partial y(\boldsymbol{\theta})}{\partial \theta_i} \right|_{\boldsymbol{\theta} = \bar{\boldsymbol{\theta}}}. \qquad (5.41)$$

Now

$$\operatorname{var} y(\boldsymbol{\theta}) = E\left[\left(y(\boldsymbol{\theta}) - E[y(\boldsymbol{\theta})]\right)^2\right] \simeq E\left[\{y(\boldsymbol{\theta}) - y(\bar{\boldsymbol{\theta}})\}^2\right] \qquad (5.42)$$

and using (5.41) in (5.42) gives

$$\operatorname{var} y(\boldsymbol{\theta}) \simeq \sum_{i=1}^{p} \sum_{j=1}^{p} \left. \frac{\partial y(\boldsymbol{\theta})}{\partial \theta_i} \right|_{\boldsymbol{\theta} = \bar{\boldsymbol{\theta}}} E\left[(\theta_i - \bar{\theta}_i)(\theta_j - \bar{\theta}_j)\right] \left. \frac{\partial y(\boldsymbol{\theta})}{\partial \theta_j} \right|_{\boldsymbol{\theta} = \bar{\boldsymbol{\theta}}}. \qquad (5.43)$$

But from Equation (3.28),

$$V_{ij} = E\left[(\theta_i - \bar{\theta}_i)(\theta_j - \bar{\theta}_j)\right],$$

is the variance matrix of the parameters θ_i. Thus, if we set

$$(\Delta y)^2 = \operatorname{var} y,$$

we have

$$(\Delta y)^2 = \sum_{i=1}^{p} \sum_{j=1}^{p} \left\{ \left. \frac{\partial y(\boldsymbol{\theta})}{\partial \theta_i} \right|_{\boldsymbol{\theta} = \bar{\boldsymbol{\theta}}} V_{ij} \left. \frac{\partial y(\boldsymbol{\theta})}{\partial \theta_j} \right|_{\boldsymbol{\theta} = \bar{\boldsymbol{\theta}}} \right\}. \qquad (5.44)$$

Equation (5.44) is often referred to as the *law of propagation of errors*. If the errors are uncorrelated (i.e. $\text{cov}(\theta_i, \theta_j) = 0$), then

$$V_{ij} = \begin{cases} (\Delta\theta_i)^2, & i = j \\ 0, & i \neq j \end{cases}$$

and (5.44) reduces to

$$(\Delta y)^2 = \sum_{i=1}^{p} \left[\frac{\partial y(\boldsymbol{\theta})}{\partial \theta_i} \bigg|_{\boldsymbol{\theta} = \bar{\boldsymbol{\theta}}} \Delta\theta_i \right]^2. \tag{5.45}$$

When using these expressions one should always ensure that the quantities $\Delta\theta_i \equiv \theta_i - \bar{\theta}_i$ are small enough to justify truncation of the Taylor series (5.41). In particular, care should be taken with functions that are highly nonlinear in the vicinity of the mean of a size comparable to the standard deviation of the parameters θ_i. Such situations are better dealt with by using the method of confidence intervals discussed in Chapter 9.

EXAMPLE 5.8

If s and t are two random variables with variances σ_s^2 and σ_t^2, respectively, and a covariance σ_{st}^2, what are the approximate errors on the function: (a) $x = as + bt$, (b) $x = ast$, and (c) $x = as/t$, where a and b are constants?

(a) Taking derivatives, we have $\partial x/\partial s = a$ and $\partial x/\partial t = b$. Also, the variance matrix is

$$V_{st} = \begin{pmatrix} \sigma_s^2 & \sigma_{st}^2 \\ \sigma_{ts}^2 & \sigma_t^2 \end{pmatrix}, \quad \text{with } \sigma_{ts}^2 = \sigma_{st}^2.$$

So, using (5.44) gives

$$\sigma_x^2 = a^2\sigma_s^2 + b^2\sigma_t^2 + 2ab\sigma_{st}^2,$$

and the approximate error on x is $\Delta x = \sigma_x$.

(b) Taking derivatives, $\partial x/\partial s = at$ and $\partial x/\partial t = as$, and using the same variance matrix as in (a) gives

$$\sigma_x^2 = (at\sigma_s)^2 + (as\sigma_t)^2 + 2a^2 st\sigma_{st}^2.$$

(c) Taking derivatives, $\partial x/\partial s = a/t$ and $\partial x/\partial t = -as/t^2$, and using the same variance matrix as in (a) gives

$$\sigma_x^2 = \left(\frac{as}{t}\right)^2 \left[\frac{\sigma_s^2}{s^2} + \frac{\sigma_t^2}{t^2} - 2\frac{\sigma_{st}^2}{st}\right].$$

The results (5.44) and (5.45) are for the case of a single function $y(\boldsymbol{\theta})$ that is a function of the p parameters $\theta_i (i = 1, 2, \ldots, p)$. They are easily generalized to the case where there are n

functions $y_k(k = 1, 2, ..., n) = y(\boldsymbol{\theta})$ that are functions of the same p parameters. Then (5.44) becomes the set of equations

$$\text{var}(y_k) = (\Delta y_k)^2 = \sum_{i=1}^{p} \sum_{j=1}^{p} \left\{ \frac{\partial y_k(\boldsymbol{\theta})}{\partial \theta_i} \bigg|_{\boldsymbol{\theta}=\bar{\boldsymbol{\theta}}} V_{ij} \frac{\partial y_k(\boldsymbol{\theta})}{\partial \theta_j} \bigg|_{\boldsymbol{\theta}=\bar{\boldsymbol{\theta}}} \right\}, \quad k = 1, 2, ..., n. \quad (5.46)$$

A new feature is that the various functions y_k will be correlated, because they are all formed from the same set of parameters $\theta_i (i = 1, 2, ..., p)$. This will be true whether or not the parameters are themselves correlated. Their covariances may be found from definition (3.28) and are

$$\text{cov}(y_k, y_l) = \sum_{i=1}^{n} \sum_{j=1}^{n} \left(\frac{\partial y_k}{\partial \theta_i} \right) \left(\frac{\partial y_l}{\partial \theta_j} \right) \text{cov}(\theta_i, \theta_j). \quad (5.47)$$

The two results (5.46) and (5.47) may be combined in the single matrix form

$$\mathbf{V}_y = \mathbf{G} \mathbf{V}_\theta \mathbf{G}^T, \quad (5.48)$$

where $\mathbf{V}_y$ is the $(n \times n)$ variance matrix of $\mathbf{y}$; $\mathbf{V}_\theta$ is the $(p \times p)$ variance matrix of $\boldsymbol{\theta}$; and $\mathbf{G}$ is an $(n \times p)$ matrix of derivatives with elements

$$G_{ki} = \frac{\partial y_k}{\partial \theta_i}.$$

EXAMPLE 5.9

Measurements are made of a particle's position in two dimensions using the independent Cartesian coordinates (x, y), with measurement errors σ_x and σ_y. What is the variance matrix for the corresponding cylindrical polar coordinates (r, ϕ), where

$$r = \sqrt{x^2 + y^2} \quad \text{and} \quad \tan \phi = y/x?$$

From the relationship between cylindrical polar and Cartesian coordinates, we have

$$G = \begin{pmatrix} \partial r/\partial x & \partial r/\partial y \\ \partial \phi/\partial x & \partial \phi/\partial y \end{pmatrix} = \begin{pmatrix} x/r & y/r \\ -y/r^2 & x/r^2 \end{pmatrix}.$$

Also

$$\mathbf{V}_{\text{Cartesian}} = \begin{pmatrix} \sigma_x^2 & 0 \\ 0 & \sigma_y^2 \end{pmatrix},$$

and so from (5.48)

$$\mathbf{V}_{\text{polar}} = \begin{pmatrix} x/r & y/r \\ -y/r^2 & x/r^2 \end{pmatrix} \begin{pmatrix} \sigma_x^2 & 0 \\ 0 & \sigma_y^2 \end{pmatrix} \begin{pmatrix} x/r & -y/r^2 \\ y/r & x/r^2 \end{pmatrix}.$$

Multiplying out gives

$$\mathbf{V}_{\text{polar}} = \begin{pmatrix} \left[(x/r)^2 \sigma_x^2 + (y/r)^2 \sigma_y^2 \right] & \left[(xy/r^3)(\sigma_y^2 - \sigma_x^2) \right] \\ \left[(xy/r^3)(\sigma_y^2 - \sigma_x^2) \right] & \frac{1}{r^2}\left[(x/r)^2 \sigma_x^2 + (y/r)^2 \sigma_y^2 \right] \end{pmatrix}.$$

Calculations involving a set of parameters that are uncorrelated are less complicated than those where correlations exist, because in the former case the variance matrix is diagonal. Given a set of variables θ_i that are correlated, it is always possible to find a new set of uncorrelated variables ω_i, for which the associated variance matrix is diagonal, in terms of the original set. This is achieved by a linear transformation of the form

$$\omega_i = \sum_{j=1}^{n} A_{ij}\theta_j, \tag{5.49}$$

which leads to a variance matrix U_{ij} for the set y_i, given by

$$\begin{aligned} U_{ij} &= \text{cov}(\omega_i, \omega_j) = \text{cov}\left(\sum_{k=1}^{n} A_{ik}\theta_k, \sum_{l=1}^{n} A_{jl}\theta_l\right) \\ &= \sum_{k,l=1}^{n} A_{ik}A_{jl}\,\text{cov}(\theta_k, \theta_l) = \sum_{k,l=1}^{n} A_{ik}V_{kl}A_{lj}^T, \end{aligned} \tag{5.50}$$

or in matrix notation $\mathbf{U} = \mathbf{A}\mathbf{V}\mathbf{A}^T$. Thus we need to find the matrix $\mathbf{A}$ that transforms the real symmetric matrix $\mathbf{V}$ to diagonal form. This is a standard technique in matrix algebra, but we will not pursue it further, because although this may simplify calculations, the transformed variables usually do not have a simple physical interpretation, and also if there are more than three variables, numerical techniques have to be used anyway.

So far we have implicitly assumed that the errors are statistical in origin, i.e., random, but since we are making the connection between mathematical statistics and experiments, this is a convenient place to return to the problem of systematic errors. In Chapter 1, the advice was to keep these separate from random errors and quote results in the form $x \pm \Delta_R \pm \Delta_S$, where Δ_R and Δ_S are the random and systematic errors, respectively. One reason for this is that it makes clear whether making more measurements is worthwhile. This is because taking more data will in general reduce the size of Δ_R, but will not change Δ_S, and there is no point in reducing Δ_R much below the value of Δ_S.

Nevertheless, we may still need to use both errors to calculate the overall error, or in general the variance matrix, for a function of x. This can be done using the general result (5.48) provided we know the variance matrix $\mathbf{V}_\theta$. For example, if we have two parameters θ_1 and θ_2 with random errors σ_1 and σ_2 and a common systematic error S, then we can consider each parameter to be the sum of two parts, θ_1^R with random error σ_1 and θ_1^S with a systematic error S, and similarly for θ_2. By construction, θ_1^R and θ_2^R are independent of each other, but θ_1^S and θ_2^S are totally correlated because they effect θ_1 and θ_2 in the same way. Then, using the definitions from Chapter 3,

$$\begin{aligned} \text{var}(\theta_1) &= E[\theta_1^2] - (E[\theta_1])^2 \\ &= E\left[(\theta_1^R + \theta_1^S)^2\right] - \left(E[\theta_1^R + \theta_1^S]\right)^2 = \sigma_1^2 + S^2, \end{aligned}$$

and similarly for θ_2; and

$$\begin{aligned} \text{cov}(\theta_1, \theta_2) &= E[\theta_1\theta_2] - E[\theta_1]E[\theta_2] \\ &= E[(\theta_1^R + \theta_1^S)(\theta_2^R + \theta_2^S)] - E[(\theta_1^R + \theta_1^S)]E[(\theta_2^R + \theta_2^S)] \\ &= \text{cov}(\theta_1^S, \theta_2^S) = S^2. \end{aligned}$$

So the variance matrix is

$$\mathbf{V}_\theta = \begin{pmatrix} \sigma_1^2 + S^2 & S^2 \\ S^2 & \sigma_2^2 + S^2 \end{pmatrix}. \tag{5.51}$$

This may be generalized in a straightforward way to the case where there are several sources of systematic errors that may be shared by subsets of the parameters.

PROBLEMS 5

5.1 A variable x is uniformly distributed in the interval $I \leq x \leq I+1$. If $\bar{x}_n$ is the mean of a random sample of size n drawn from the population, find an unbiased estimator for I in terms of $\bar{x}_n$. What is the mean squared error of the associated biased estimator?

5.2 Prove the relation (5.9).

5.3 A sample of size $n_1 = 6$ is drawn from a normal population with a mean $\mu_1 = 60$ a variance $\sigma_1^2 = 12$. A second sample, of size $n_2 = 4$, is selected, independent from the first sample, from a different normal population having a mean $\mu_2 = 50$ and variance $\sigma_2^2 = 8$. What is $P[(\bar{x}_1 - \bar{x}_2) < 7.5]$?

5.4 Prove the results given in (5.28) and (5.29).

5.5 A particular organism is repeatedly exposed to doses of radiation r_i that are normally distributed with mean 4 and variance 2 (in arbitrary units). It is found that the maximum cumulative dosage of radiation R that the organism can absorb without suffering permanent damage is normally distributed with mean 100 and variance 25 (in the same units). What is the maximum number of doses that the organism may absorb before the probability of damage exceeds 3%?

5.6 A power unit is manufactured by an identical process in several different factories A, B, C, etc. and the mean output of the unit across all factories is 50 watts with a standard deviation σ of 7 watts. A random sample of 100 units is taken from factory A and the sample mean is found to be 49 watts. Is the product from factory A up to the overall standard of manufacture?

5.7 A prospective purchaser of resistors decides to buy a sample of size n from the manufacturer to check that their average value does not vary by more than 3% from the average value $\bar{R}$ of all resistors from the same manufacturer, with a probability of 0.05. If the values of the resistors are normally distributed with a standard deviation that is 15% of the value $\bar{R}$, how many would have to be bought?

5.8 A beam of particles is incident on a target and F events are recorded where the particle scatters into the forward hemisphere and B events where it scatters into the backward hemisphere. What is the standard deviation on the 'forward–backward asymmetry' R, defined as $R \equiv (F-B)/(F+B)$?

5.9 $F(x,y,z)$ is a function of the three variables x,y,z with the form $F(x,y,z) = xy^2z^3$. If the variance matrix of x,y,z is

$$\mathbf{V} = \frac{1}{1000}\begin{pmatrix} 1 & -1 & 1 \\ -1 & 1 & 0 \\ 1 & 0 & 1 \end{pmatrix},$$

what is the percentage error on F when $x = 2$, $y = 1$ and $z = 1$?

CHAPTER 6

Sampling Distributions Associated with the Normal Distribution

OUTLINE

6.1	Chi-Squared Distribution	105	6.4 Relations Between χ^2, t, and F Distributions	119
6.2	Student's t Distribution	111		
6.3	F Distribution	116		

The special position held by the normal distribution, mainly by virtue of the central limit theorem, is reflected in the prominent positions of distributions resulting from sampling from the normal. In this chapter we consider the basic properties of three frequently used sampling distributions: the chi-squared, the Student's t,[1] and the F distributions. These are widely used in estimation problems, for finding both the best values of parameters and their optimal ranges, and in testing hypotheses, topics that will be discussed in detail in Chapters 7–11.

6.1. CHI-SQUARED DISTRIBUTION

If we wish to concentrate on a measure to describe the dispersion of a population, then we consider the sample variance. The chi-squared distribution is introduced for problems involving this quantity. It is defined as follows.

If $x_i (i = 1, 2, ..., n)$ is a sample of n random variables normally and independently distributed with means μ_i and variances σ_i^2, then the statistic

$$\chi^2 \equiv \sum_{i=1}^{n} \left(\frac{x_i - \mu_i}{\sigma_i} \right)^2 \tag{6.1}$$

---

[1]Confusingly, this not a distribution specifically designed for use by students. The name refers to its originator, W.S. Gosset, who published under the pseudonym 'Student'.

is distributed with density function

$$f(\chi^2, n) = \frac{1}{2^{n/2} \Gamma(n/2)} \chi^{2[(n/2)-1]} \exp(-\chi^2/2), \quad \chi^2 > 0. \tag{6.2}$$

This is known as the χ^2-*distribution (chi-squared)* with n *degrees of freedom*. It is another example of the general gamma distribution defined in equation (4.27), this time with $x = \chi^2$, $\alpha = n/2$ and $\lambda = 1/2$. The symbol Γ in (6.2) is the gamma function (used previously in Problem 3.1) defined by the integral

$$\Gamma(x) \equiv \int_0^\infty e^{-u} u^{x-1} du, \quad 0 < x < \infty. \tag{6.3}$$

It is frequently encountered in sampling distributions associated with the normal distribution.

The χ^2-distribution may be derived using characteristic functions, as follows. We first write χ^2 as

$$\chi^2 = \sum_{i=1}^n z_i^2,$$

where the z_i are distributed as the standard normal distribution $N(z_i; 0, 1)$. The quantities $u_i = z_i^2$ therefore have density functions

$$n(u_i) = \frac{1}{(2\pi u_i)^{1/2}} \exp(-u_i/2),$$

and the cf of u_i is

$$\phi_i(t) = \int_0^\infty \frac{1}{(2\pi u_i)^{1/2}} e^{(-u_i/2)} e^{itu_i} du_i = (1 - 2it)^{-1/2}, \quad (u_i \geq 0). \tag{6.4}$$

If $\phi(t)$ is the cf of χ^2, then since the random variables u_i are independently distributed, we know from the work of Section 3.2.3 that

$$\phi(t) = \sum_{i=1}^n \phi_i(t) = (1 - 2it)^{-n/2}. \tag{6.5}$$

Finally, the density function of χ^2 is obtained from the inversion theorem

$$f(\chi^2, n) = \frac{1}{2\pi} \int_0^\infty (1 - 2it)^{-n/2} e^{-i\chi^2 t} \, dt.$$

Using the definition of the gamma function, this yields (6.2), although the evaluation of the integral is rather lengthy.

If the variables x_i are not independent, but have a joint n-dimensional normal distribution with an associated variance matrix $\mathbf{V}$, as discussed in Section 4.3, then the variable to consider is

$$\mathbf{z} = (\mathbf{x} - \boldsymbol{\mu})^T \mathbf{V}^{-1} (\mathbf{x} - \boldsymbol{\mu}).$$

EXAMPLE 6.1

Use the result $\Gamma(1/2) = \sqrt{\pi}$ to verify (6.4).
Change variables in the integrand to $x = u\left(\dfrac{1}{2} - it\right)$. This gives

$$\phi_i(t) = \frac{1}{\sqrt{\pi}(1-2it)^{1/2}} \int_0^\infty x^{-1/2} e^{-x} dx,$$

and from (6.3)

$$\int_0^\infty x^{-1/2} e^{-x} dx = \Gamma(1/2) = \sqrt{\pi}.$$

Therefore

$$\phi_i(t) = (1-2it)^{-1/2}.$$

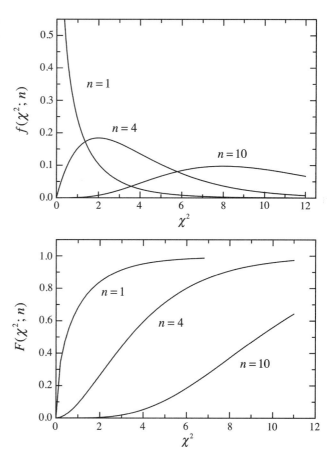

FIGURE 6.1 Graphs of the χ^2 density function $f(\chi^2, n)$ and its distribution function $F(\chi^2, n)$ for $n = 1, 4$ and 10.

The χ^2 distribution is one of the most important sampling distributions occurring in physical science. Its density and distribution functions are one-parameter families of curves. Examples of $f(\chi^2, n)$ and the distribution function $F(\chi^2, n)$ for $n = 1, 4$ and 10 are shown in Fig. 6.1. The distribution function $F(\chi^2, n)$ is tabulated in Appendix C, Table C.4 for a range of values of n. An alternative useful table may be constructed by calculating the proportion α of the area under the χ^2 curves to the *right* of χ^2_α, i.e., points such that

$$P[\chi^2 \geq \chi^2_\alpha] = \alpha = \int_{\chi^2_\alpha}^{\infty} f(\chi^2, n)\, d\chi^2. \tag{6.6}$$

Such points are called *percentage points*, or *critical values*, of the χ^2 distribution (recall the percentiles defined in Chapter 1) and may be deduced from Table C.4. They are shown graphically in Fig. 6.2. A point of interest about these curves is that for a fixed value of P, the ratio $\chi^2/n \to 1$ as $n \to \infty$.

EXAMPLE 6.2

(a) What is $P[\chi^2 \leq 30]$ when χ^2 is a random variable with 26 degrees of freedom? (b) If χ^2 is a random variable with 15 degrees of freedom, what is its value that corresponds to $\alpha = 0.05$?

(a) From Table C.4, we have to find an entry close to, but not more than 30 for $n = 26$. This is a little less than 0.75. The exact figure would have to be found by direct integration of the density function.

(b) From the definition (6.6), we need to find a value χ^2_c of χ^2 for 15 degrees of freedom such that $P[\chi^2 \geq \chi^2_c] = 0.05$, that is, a value χ^2_c such that $P[\chi^2 \leq \chi^2_c] = 0.95$. From Table C.4, this is $\chi^2_c = 25$.

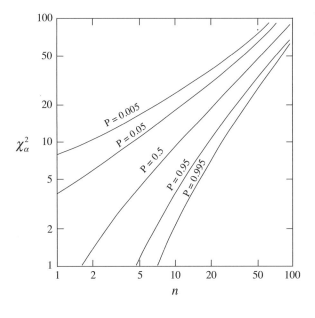

FIGURE 6.2 Percentage points of the chi-squared distribution, $P = P[\chi^2 \geq \chi^2_\alpha]$.

6.1. CHI-SQUARED DISTRIBUTION

The mgf of the χ^2 distribution is obtainable directly from (6.4) and is

$$M(t) = (1 - 2t)^{-n/2}. \tag{6.7}$$

It follows that the mean and variance are given by

$$\mu = n \text{ and } \sigma^2 = 2n. \tag{6.8}$$

The third and fourth moments about the mean may also be found from the mgf. They are

$$\mu_3 = 8n; \quad \mu_4 = 12n(n+4)$$

giving

$$\beta_1 = \frac{8}{n}, \quad \beta_2 = 3\left(1 + \frac{4}{n}\right),$$

which tend to the values for the normal distribution as $n \to \infty$, and the χ^2 distribution does indeed tend to normality for large samples.

This can be demonstrated by constructing the cf for the standardized variable

$$y \equiv \left(\frac{\chi^2 - \mu}{\sigma}\right) = \left(\frac{\chi^2 - n}{\sqrt{2n}}\right),$$

which from (6.4) is

$$\phi_y(t) = \exp\left[-\frac{int}{(2n)^{1/2}}\right]\left[1 - \frac{2it}{(2n)^{1/2}}\right]^{-n/2},$$

and taking logarithms gives

$$\ln \phi_y(t) \to -\frac{int}{(2n)^{1/2}} - \frac{n}{2}\ln\left[1 - \frac{2it}{(2n)^{1/2}}\right].$$

Finally, letting $n \to \infty$ and expanding the logarithm gives

$$\ln \phi_y(t) \to -\frac{int}{(2n)^{1/2}} - \frac{n}{2}\left[-\frac{2it}{(2n)^{1/2}} - \frac{1}{2}\left(\frac{2it}{(2n)^{1/2}}\right)^2 \cdots\right], \quad (n \to \infty),$$

implying

$$\phi_y(t) \to \exp(-t^2/2).$$

This is the cf of a standardized normal distribution and so, by the inversion theorem, the χ^2 distribution tends to normality as $n \to \infty$, although the rate of convergence is quite slow.

Because the χ^2 distribution is a one-parameter family of curves, it frequently happens that tabulated values do not exist for precisely the range one requires. In such cases a very useful statistic is $(2\chi^2)^{1/2}$, which can be shown to tend rapidly to normality with mean $\mu = (2n-1)^{1/2}$ and unit variance. The statistic

$$u = (2\chi^2)^{1/2} - (2n-1)^{1/2} \tag{6.9a}$$

TABLE 6.1 Values of $P[\chi^2 \geq \chi_\alpha^2]$ for $n = 5$, 10, and 20, and $\chi_\alpha^2 = 2, 5, 10, 20,$ and 30 using the exact χ^2 distribution function, and the normal approximation using the variable u of (6.9a)

n	5		10		20	
χ_α^2	exact	approx.	exact	approx.	exact	approx.
2	0.849	0.841	0.996	0.991		
5	0.416	0.436	0.891	0.885		
10	0.075	0.071	0.441	0.456	0.968	0.963
20	0.001	0.001	0.029	0.024	0.458	0.462
30			0.001	0.000	0.070	0.067

is therefore a standard normal variate for even quite moderate values of n, and so tables of the normal distribution may be used. Table 6.1 shows a comparison between the exact χ^2 distribution and the normal approximation based on the statistic $(2\chi^2)^{1/2}$ for a range of values of n and χ^2. Another statistic that converges to normality faster, but is more complicated to calculate, is $(\chi^2/n)^{1/3}$. This can be shown to tend very rapidly to normality with mean $1 - 2/(9n)$ and variance $2/(9n)$. Thus the statistic

$$u = \left\{ \left(\frac{\chi^2}{n}\right)^{1/3} + \frac{2}{9n} - 1 \right\} \left(\frac{9n}{2}\right)^{1/2} \tag{6.9b}$$

is a standard normal variate for even moderate values of n.

An important property of the χ^2 distribution is that the sum of m-independent random variables $\chi_1^2, \chi_2^2, \ldots, \chi_m^2$, each having chi-squared distributions with $n_1, n_2, \ldots, n_m$ degrees of freedom, respectively, is itself distributed as χ^2 with $n = n_1 + n_2 + \ldots + n_m$ degrees of freedom. This is called the *additive property of* χ^2 and may be proved by using the characteristic function (see Problem 6.3).

There are two other important results that we shall need later. The first concerns a sample $x_1, x_2, \ldots, x_n$ of size n drawn from a normal population with mean zero and unit variance. Then the statistic

$$u = \sum_{i=1}^{n} (x_i - \bar{x})^2, \tag{6.10}$$

is distributed as χ^2 with $(n-1)$ degrees of freedom. In general, if the parent population has variance σ^2, then

$$\chi^2 = \frac{1}{\sigma^2} \sum_{i=1}^{n} (x_i - \bar{x})^2, \tag{6.11}$$

is distributed as χ^2 with $(n-1)$ degrees of freedom. Moreover, since the sample variance is

$$s^2 = \frac{\sigma^2 \chi^2}{n-1}, \tag{6.12}$$

it follows that $(n-1)s^2/\sigma^2$ is distributed as χ^2 with $(n-1)$ degrees of freedom, *independent* of the sample mean $\bar{x}$. Thus the sample mean and sample variance are independent random variables when sampling from normal populations. This somewhat surprising result is very important in practice and we shall use it later to construct the sampling distribution known as the Student's t distribution.

If we assume that a sample is drawn at random from a single normal population with mean μ and variance σ^2, then from (6.1)

$$\chi^2 = \frac{1}{\sigma^2}\sum_{i=1}^{n}(x_i - \mu)^2. \qquad (6.13)$$

However, since the mean of the population is rarely known, in these cases it is more useful to use the result that the quantity u in (6.10) is distributed as χ^2 with $(n-1)$ degrees of freedom. In that case χ^2 defined in (6.11) is distributed with $(n-1)$ degrees of freedom if $\bar{x}$ is used instead of μ. This illustrates an important general result: *the number of degrees of freedom must be reduced by one for each parameter estimated from the data.*

EXAMPLE 6.3

Points are plotted randomly in a two-dimensional plane using Cartesian coordinates (x,y) and the distance from a fixed point (x_0,y_0) measured. If the differences

$$\Delta_x = x_0 - x \quad \text{and} \quad \Delta_y = y_0 - y$$

are independent random variables, normally distributed with zero means and standard deviations 2.1, what is the probability that the distance between the points (x,y) and (x_0,y_0) exceeds 3.5?

The distance d between the points (x,y) and (x_0,y_0) is given by

$$d^2 = \Delta_x^2 + \Delta_y^2,$$

and because the quantities $z_{x,y} = \Delta_{x,y}/\sigma = \Delta_{x,y}/2.1$ are standard normal variates,

$$P\left[d^2 > (3.5)^2 = 12.25\right] = P\left[z_x^2 + z_y^2 > (12.25/(2.1)^2 = 2.78\right]$$

$$= P[\chi^2 > 2.78] = 1 - P[\chi^2 < 2.78] = 0.25.$$

6.2. STUDENT'S t DISTRIBUTION

The central limit theorem tells us that the distribution of the sample mean $\bar{x}$ is approximately normal with mean μ (the population mean) and variance σ^2/n (where σ^2 is the population variance and n is the sample size). Thus, in standard measure, the statistic

$$u = \left(\frac{\bar{x} - \mu}{\sigma_n}\right),$$

where $\sigma_n = \sigma/\sqrt{n}$, is approximately normally distributed with mean zero and unit variance for large n. However, in experimental situations neither the mean nor the population variance may be known, in which case they must be replaced by estimates from the sample. While σ^2 can be safely replaced by the sample variance s^2 for large $n \geq 30$, for small n the statistic u will not be approximately normally distributed and serious loss of meaning in the interpretation will occur. So we have to consider the distribution of the variable

$$t = \left(\frac{\bar{x} - \mu}{s/\sqrt{n}}\right),$$

where $s = \hat{\sigma}$ is an estimator for σ. If we write t as

$$t = \left(\frac{\bar{x} - \mu}{\sigma/\sqrt{n}}\right)\left(\frac{\hat{\sigma}}{\sigma}\right)^{-1},$$

we see from the central limit theorem that the numerator is distributed like a standard normal variable and the denominator is distributed like a χ^2 variable with either $(n-1)$ or n degrees of freedom, depending whether or not μ is estimated from the data (see equation (5.16a)). The distribution of t is called the *Student's t distribution*. It enables one to use the sample variance, as well as the sample mean, to make statements about the population mean. The discussion will concentrate around three important results, but firstly we will derive the density function of t.

Let u have a normal distribution with mean zero and unit variance. Further, let w have a χ^2 distribution with n degrees of freedom, and let u and $\sqrt{w}$ be independently distributed. Because u and $\sqrt{w}$ are independently distributed, their joint density is the product of their individual densities. Thus from the form of the chi-squared distribution (6.2), and the standardized normal distribution (4.10), the joint density function of u and w is

$$f(u,w;n) = \frac{1}{(2\pi)^{1/2}}e^{-u^2/2}\frac{1}{\Gamma(n/2)2^{n/2}}w^{(n-2)/2}e^{-w/2}. \tag{6.14}$$

If we substitute

$$u = t\left(\frac{w}{n}\right)^{1/2},$$

then (6.14) becomes

$$f(t,w;n) = \frac{e^{-t^2 w/2n}e^{-w/2}w^{(n-2)/2}}{(2\pi)^{1/2}\Gamma(n/2)2^{n/2}},$$

and $f(t;n)$ is the marginal distribution of t, i.e.

$$f(t;n) = \int_0^\infty f(t,w;n)dw.$$

This integral may be evaluated directly using the definition of the gamma function (6.3) with result that the random variable

$$t = \frac{u}{(w/n)^{1/2}},$$

has a density function

$$f(t;n) = \frac{\Gamma[(n+1)/2]}{(\pi n)^{1/2}\Gamma(n/2)}\left[1 + \frac{t^2}{n}\right]^{-(n+1)/2}, \quad -\infty < t < \infty. \tag{6.15}$$

The statistic t is said to have a *Student's t distribution with n degrees of freedom*. It tends to a standard normal distribution for $n \to \infty$, as we will prove below, but for small n the tails are wider than those of the latter, and for $n = 1$ the distribution is of the Cauchy form.

Like the χ^2 distribution, the Student's t distribution is a one-parameter family of curves. The distribution function is tabulated in Table C.5, and in using it one can use the fact that

$$P[t < -t_\alpha(n)] = P[t > t_\alpha(n)] = \alpha,$$

since the distribution is symmetrical about $t = 0$. Percentage points for the distribution are shown graphically in Fig. 6.3.

EXAMPLE 6.4

(a) What is $P[t \le 0.7]$ when t is a random variable with 16 degrees of freedom? (b) If t is a random variable with 5 degrees of freedom, what is its value that corresponds to $\alpha = 0.05$?

(a) From Table C.5, we have to find an entry close to, but not more than, 0.7 for $n = 16$. This is very close to 0.75.

(b) From the definition analogous to (6.6), we need to find a value t_c of t for 5 degrees of freedom such that $P[t \ge t_c] = 0.05$, that is, a value t_c such that $P[t \le t_c] = 0.95$. From Table C.5, this is approximately $t_c = 2$.

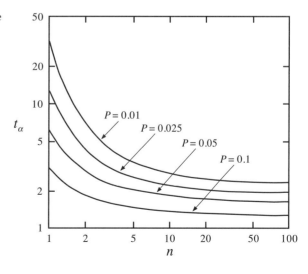

FIGURE 6.3 Percentage points of the Student's t distribution, $P = P[t > t_\alpha]$.

The mean and variance as usual can be found from the mgf. From (6.13) and the definitions (3.11) and (3.13) we can show that moments of order r only exist for $r < n$ and are zero by symmetry for odd moments. For even moments, direct integration gives

$$\mu_{2r} = n^r \frac{\Gamma(r+1/2)\Gamma(n/2-r)}{\Gamma(1/2)\Gamma(n/2)}, \quad 2r < n. \tag{6.16}$$

The mean and variance follow from (6.16): they are

$$\mu = 0; \quad \sigma^2 = \frac{n}{n-2}, \quad (n > 2) \tag{6.17}$$

We now return to the three basic results mentioned earlier. The first of these specifies the distribution of the difference of the sample mean and the population mean with respect to the sample variance. Let $x_i (i = 1, 2, \ldots, n)$ be a random sample of size n drawn from a normal population with mean μ and variance σ^2. Then the statistic

$$u = \left(\frac{\bar{x} - \mu}{\sigma/\sqrt{n}}\right)$$

is distributed as $N(u; 0, 1)$. Furthermore, from (6.12) we know that the statistic $w = (n-1)s^2/\sigma^2$, where as usual s^2 is the sample variance, is distributed as χ^2 with $(n-1)$ degrees of freedom. Therefore, from the form of the Student's t distribution, the statistic

$$t = \frac{u}{\left[w/(n-1)\right]^{1/2}} = \frac{\sqrt{n}}{s}(\bar{x} - \mu) \tag{6.18}$$

is distributed as t with $(n-1)$ degrees of freedom.

EXAMPLE 6.5

A group of 9 students entering the physics department of university A has a mean score of 78% in a national science examination, with a standard deviation of 5%. The national average for all students taking the same examination is 75%. What can be said about whether university A is getting significantly better than average students?

Using $\bar{x} = 78$, $s = 5$, and $\mu = 75$, we have $t = \sqrt{n}(\bar{x} - \mu)/s = 1.8$, and this value is for $n - 1 = 8$ degrees of freedom. From Table C.5, $F(t, n) = 0.95$ for $t = 1.86$ and $n = 8$. Thus the probability of getting a sample mean at least as large as this is approximately 5%.

The second result concerns the asymptotic behavior of the t distribution. As the number of degrees of freedom of the distribution approaches infinity, the distribution tends to the normal distribution in standard form. This follows by using Stirling's approximation,

$$\Gamma(n+1) \to (2\pi)^{1/2} n^{n+1/2} e^{-n}, \quad n \to \infty$$

in the gamma functions in (6.16). Then the moments (6.16) become

$$\mu_{2r} \to \frac{(2r)!}{2^r r!}. \tag{6.19}$$

However (6.19) is the form for the moments of the normal distribution expressed in standard measure (see equation (4.7)). Therefore the Student's t distribution tends to a normal distribution with mean zero and unit variance.

The final result concerns the t distribution when two normal populations are involved. Let random samples $x_{11}, x_{12}, \ldots, x_{1n_1}$ and $x_{21}, x_{22}, \ldots, x_{2n_2}$ of sizes n_1 and n_2, respectively, be independently drawn from two normal populations 1 and 2 with means μ_1 and μ_2, and the same variance σ^2; and define the statistic t by

$$t \equiv \frac{(\bar{x}_1 - \bar{x}_2) - (\mu_1 - \mu_2)}{[S_P^2(1/n_1 + 1/n_2)]^{1/2}}, \tag{6.20}$$

where

$$\bar{x}_i = \frac{1}{n_i} \sum_{j=1}^{n_i} x_{ij}, \quad i = 1, 2,$$

and S_P^2, the pooled sample variance, is given by

$$S_P^2 = \frac{\sum_{i=1}^{2} \sum_{j=1}^{n_i} (x_{ij} - \bar{x}_i)^2}{n_1 + n_2 - 2} = \frac{(n_1 - 1)s_1^2 + (n_2 - 1)s_2^2}{n_1 + n_2 - 2}. \tag{6.21}$$

Then, using (6.12) and the additive property of χ^2, the quantity

$$w = S_P^2(n_1 + n_2 - 2)/\sigma^2, \tag{6.22}$$

is distributed as χ^2 with $(n_1 + n_2 - 2)$ degrees of freedom. Furthermore, we know, from equations (5.31) and (5.32), that $\bar{x} = \bar{x}_1 - \bar{x}_2$ is normally distributed with mean $\mu = \mu_1 - \mu_2$ and variance

$$\sigma_d^2 = \frac{\sigma^2}{n_1} + \frac{\sigma^2}{n_2}.$$

Thus the quantity

$$u = \frac{(\bar{x}_1 - \bar{x}_2) - (\mu_1 - \mu_2)}{[\sigma^2(1/n_1 + 1/n_2)]^{1/2}} = \frac{\bar{x} - \mu}{\sigma_d}, \tag{6.23}$$

is normally distributed with mean zero and unit variance. But we showed in Section 6.1 that the sample mean and sample variance are independent variables when sampling randomly from a normal population, so $\bar{x}$ and u are independent random variables. Thus the quantity

$$t = \frac{u}{[w/(n_1 + n_2 - 2)]^{1/2}}, \tag{6.24}$$

has a t distribution with $(n_1 + n_2 - 2)$ degrees of freedom. Substituting (6.22) and (6.23) into (6.24) gives (6.20) and completes the proof.

EXAMPLE 6.6

The table shows the scores in a certain examination of two groups of students, A and B. The students in group A attended revision classes to prepare for the exam. Has this significantly improved the mean score of the group compared to that of group B?

n	1	2	3	4	5	6	7
A	71	75	79	71	70	73	72
B	70	68	72	73	67		

From the data we can calculate, $\bar{x}_A = 73$, $\bar{x}_B = 70$, $(n_A - 1)s_A^2 = 58$, and $(n_B - 1)s_B^2 = 26$, where $n_A = 7$ and $n_B = 5$, so from (6.21), $S_p^2 = 8.4$ We can now test whether $\mu_A = \mu_B$ by calculating t from (6.20). Thus, assuming $\mu_A = \mu_B$,

$$t = (\bar{x}_A - \bar{x}_B) \left[S_p^2 \left(\frac{1}{n_A} + \frac{1}{n_B} \right) \right]^{-1/2} = 1.77.$$

From Table C.5, the probability of getting a value of t at least as great as this for 10 degrees of freedom is less than 5%.

6.3. F DISTRIBUTION

The F distribution is designed for use in situations where we wish to compare two variances, or more than two means, situations for which the χ^2 and the Student's t distributions are not appropriate.

We begin by constructing the form of the F density function. Let two independent random variables $u = \chi_1^2$ and $v = \chi_2^2$ be distributed as χ^2 with m and n degrees of freedom, respectively. Then the joint density of u and v is, from (6.2),

$$g(u,v) = \frac{u^{(m-2)/2} v^{(n-2)/2}}{\Gamma(m/2)\Gamma(n/2) 2^{(m+n)/2}} \exp\left[-\frac{1}{2}(u+v)\right].$$

The statistic F is defined by

$$F = F(m,n) \equiv \frac{\chi_1^2/m}{\chi_2^2/n} = \frac{u/m}{v/n} \tag{6.25}$$

and so substituting

$$u = \left(\frac{m}{n}\right) vF,$$

into $g(u,v)$ gives the joint density function of F and v as

$$f(F,v) = \frac{v^{(n-2)/2}}{\Gamma(m/2)\Gamma(n/2) 2^{(m+n)/2}} \left(\frac{nv}{m}\right) \left(\frac{mvF}{n}\right)^{(n-2)/2} \exp\left[-\frac{v}{2}\left(1 + \frac{m}{n}F\right)\right].$$

6.3. F DISTRIBUTION

The density function of F is then obtained by integrating out the dependence on v. Thus

$$f(F;m,n) = \frac{F^{(n-2)/2}}{\Gamma(m/2)\Gamma(n/2)2^{(m+n)/2}}\left(\frac{m}{n}\right)^{m/2} I(F;m,n).$$

where, using the definition of the gamma function,

$$I(F;m,n) = \int_0^\infty v^{(m+n-2)/2} \exp\left[-\frac{v}{2}\left(1+\frac{m}{n}F\right)\right] dv$$

$$= \frac{\Gamma[(m+n)/2]2^{(m+n)/2}}{(1+mF/n)^{(m+n)/2}}.$$

So, finally, the density function for the statistic F is

$$f(F;m,n) = \frac{\Gamma[(m+n)/2]}{\Gamma(m/2)\Gamma(n/2)}\left(\frac{m}{n}\right)^{m/2} \frac{F^{(m-n)/2}}{(1+mF/n)^{(m+n)/2}}, \quad F \geq 0 \qquad (6.26)$$

with m and n being degrees of freedom.

The mgf may be deduced in the usual way from its definition. The moments of order r exist only for $2r < n$ and are given by

$$\mu'_r = \left(\frac{n}{m}\right)^r \frac{\Gamma(r+m/2)\Gamma(n/2-r)}{\Gamma(m/2)\Gamma(n/2)}. \qquad (6.27)$$

The mean and variance follow directly from (6.27) and are

$$\mu = \frac{n}{n-2}, \quad n > 2,$$

and

$$\sigma^2 = \frac{2n^2(m+n-2)}{m(n-2)^2(n-4)}, \quad n > 4.$$

Equation (6.27) may also be used to calculate β_1 and β_2 and the result shows that the F distribution is always skewed. The pdf of the F distribution is more complicated than those of the χ^2 and t distributions in being a two-parameter family of curves.

The distribution function of F is tabulated in Table C.6. Percentage points are defined in the same way as for the χ^2 distribution. Thus,

$$P[F \geq F_\alpha] = \alpha = \int_{F_\alpha}^\infty f(F;m,n)\,dF.$$

Right-tailed percentage points may be obtained from Table C.6, and should left-tailed percentage points be needed they may be obtained from the relation

$$F_{1-\alpha}(m,n) = [F_\alpha(n,m)]^{-1}. \qquad (6.28)$$

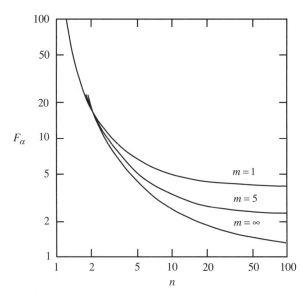

FIGURE 6.4 Percentage points of the F distribution, $P = P[F > F_\alpha] = 0.05$.

As an example, the percentage points for $P = 0.05$ are shown graphically in Fig. 6.4.

EXAMPLE 6.7

(a) *Find the critical value $F_{0.05}$ for $m = 6$ and $n = 14$. (b) Find the critical value $F_{0.975}$ for $m = 9$ and $n = 30$.*

(a) This is the value for which $P[F > F_{0.05}] = 0.05$, or equivalently $P[F \leq F_{0.05}] = 0.95$. From Table C.6, with $m = 6$ and $n = 14$, this is $F_{0.05} = 2.85$.

(b) We first find the critical value $F_{0.025}$ for $m = 30$ and $n = 9$. This is the value for which $P[F > F_{0.025}] = 0.025$, or equivalently $P[F \leq F_{0.025}] = 0.975$. From Table C.6, with $m = 30$ and $n = 9$, this is $F_{0.025} = 3.56$. Now we can use (6.28) to give

$$F_{0.975}(30,9) = [F_{0.025}(9,30)]^{-1} = 0.281.$$

One use of the F distribution is to compare two variances, for example to see whether two variances are equal so that the conditions for applying the Student's t test are satisfied. Let s_1^2 and s_2^2 be the variances of two independent random samples of sizes m and n, respectively, whose populations are assumed to be normal with variances σ_1^2 and σ_2^2. Then from the definition of the sample variance, we may write

$$s^2 = \frac{1}{n-1} \sum_{i=1}^{n} (x_i - \bar{x})^2 = \frac{\chi^2}{n-1},$$

where χ^2 is a chi-squared random variable with $(n-1)$ degrees of freedom. Thus, if we assume that $\sigma_1^2 = \sigma_2^2$, the ratio $F = s_1^2/s_2^2$ is distributed as an F random variable with $(m-1)$ and $(n-1)$ degrees of freedom.

EXAMPLE 6.8

Do the data of Example 6.6 justify the use of the t distribution to compare the mean scores of the two groups of students?

From the results of Example 6.6, we have $s_A^2 = 9.67$ and $s_B^2 = 6.50$, so that $F(6,4) = 1.49$. From Table C.6, the value of $F(6.4)$ is 4.01 for a critical value of 10%. As the value found from the data is well below this, the use of the Student's t distribution is compatible with the data.

6.4. RELATIONS BETWEEN χ^2, t, AND F DISTRIBUTIONS

The F distribution is related to the χ^2 distribution as follows. It is straightforward to show that as $n \to \infty$,

$$P\left[\left|\chi^2/n - 1\right|\right] \to 0.$$

Thus

$$F(m, \infty) = \chi_1^2/m, \tag{6.29}$$

(see for example Fig. 6.2), and the distribution of χ_1^2/m with m degrees of freedom is a special case of the F distribution with m and ∞ degrees of freedom. So for any α,

$$F_\alpha(m, \infty) = \frac{\chi_\alpha^2(m)}{m}, \tag{6.30}$$

which may be directly verified by the use of a set of tables. If we consider the limit as $m \to \infty$, we have

$$F(\infty, n) = \frac{n}{\chi^2(n)}, \tag{6.31}$$

and so

$$F_\alpha(\infty, n) = \frac{n}{\chi_{1-\alpha}^2(n)}. \tag{6.32}$$

Thus the left-tailed percentage points of the χ^2/n are special cases of the right-tailed percentage points of $F(\infty, n)$.

The F distribution is also related to the Student's t distribution. This can be seen by noting that when $m = 1$, then $\chi^2/m = u^2$, where u is a standard normal variate. We may thus write

$$F(1, n) = \frac{u^2}{(\chi_2^2/n)}. \tag{6.33}$$

But the variate

$$t = \frac{u}{(\chi_2^2/n)^{1/2}}, \qquad (6.34)$$

is distributed as the Student's t distribution with n degrees of freedom, so (6.33) may be written as

$$F(1,n) = t^2(n). \qquad (6.35)$$

Using (6.35),

$$P[F(1,n) < F_\alpha(1,n)] = 1 - \alpha$$

is equivalent to

$$P[-(F_\alpha(1,n))^{1/2} < t(n) < (F_\alpha(1,n))^{1/2}] = 1 - \alpha,$$

and using the symmetry of the t distribution about $t = 0$ we have

$$P[t(n) < -(F_\alpha(1,n))^{1/2}] = P[t(n) > (F_\alpha(1,n))^{1/2}] = \alpha/2. \qquad (6.36)$$

But

$$P[t(n) > t_{\alpha/2}(n)] = \alpha/2$$

and so

$$t_{\alpha/2}(n) = F_\alpha[(1,n)]^{1/2},$$

or

$$F_\alpha(1,n) = t^2_{\alpha/2}(n). \qquad (6.37)$$

Similarly, we can show that for $n = 1$

$$F(m,1) = [t^2(m)]^{-1}, \qquad (6.38)$$

and

$$F_\alpha(m,1) = [t^2_{(1+\alpha)/2}(m)]^{-1}. \qquad (6.39)$$

Finally, if $m = 1$ and $n \to \infty$

$$F_\alpha(1, \infty) = u^2_{\alpha/2}, \qquad (6.40)$$

and if $n = 1$ and $m \to \infty$

$$F_\alpha(\infty, 1) = [u^2_{(1+\alpha/2)/2}]^{-1}, \qquad (6.41)$$

where u_α is a point of the standard normal variate such that

$$P[u > u_\alpha] = \alpha.$$

TABLE 6.2 Percentage points F_α of the $F(m,n)$ distribution and their relation to the χ^2 and Student's t distributions.

		m	
n	1	m	∞
1	$t^2_{\alpha/2}(1) = \dfrac{1}{t^2_{(1+\alpha)/2}(1)}$	$\dfrac{1}{t^2_{(1+\alpha)/2}(m)}$	$\dfrac{1}{u^2_{(1+\alpha)/2}}$
n	$t^2_{\alpha/2}(n)$	$F_\alpha(m,n)$	$\dfrac{n}{\chi^2_{1-\alpha}(n)}$
∞	$u^2_{\alpha/2}$	$\dfrac{\chi^2_\alpha(m)}{m}$	1

The various relationships above are summarized in Table 6.2.

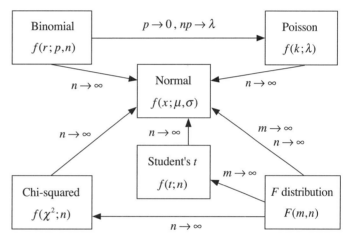

FIGURE 6.5 The relationships between the population distributions (Binomial, Poisson, and normal) and the sampling distributions (chi-squared, Student's t and F) as their parameters tend to certain limits.

The relationships between these three distributions in certain limiting situations are shown in Fig. 6.5, together with their relationships to the three most important population distributions discussed in Chapter 4.

PROBLEMS 6

6.1 Find the value of the 90th percentile of the χ^2 distribution for $n = 100$ degrees of freedom.

6.2 Prove the result stated in the text that for a sample $x_1, x_2, \ldots, x_n$ of size n drawn from a normal population with mean zero and unit variance, the statistic

$$u = \sum_{i=1}^{n}(x_i - \bar{x})^2,$$

is distributed as χ^2 with $(n-1)$ degrees of freedom.

6.3 Prove the additive property of χ^2.

6.4 Twenty measurements taken from a normal population that was assumed to have mean 14.5 yielded values of the sample mean and sample variance of 15 and 2.31, respectively. Use the Students's t distribution to test the assumption.

6.5 A target is to be located in an n-dimensional hyperspace by measuring its coordinates from a fixed point. If the coordinate errors are normal variates with mean zero and variance $4/3$, what is the largest value of n for which the method can be used if the probability that the distance from the fixed point to the target D exceeds 4 is 10%.

6.6 Prove the relation (6.28).

6.7 A random sample of size 20 is selected from a normal distribution and the t statistic calculated. Find the value of k that satisfies $P[k < t < -1.328] = 0.075$.

6.8 Find the number h such that $P[-h < t < h] = 0.90$ where t is a random variable with a Student's t distribution with 15 degrees of freedom.

6.9 Resistors of a given value are manufactured by two machines A and B, and for consistency the outputs from both machines should have equal variances. A random sample of $n_A = 16$ resistors from machine A has $s_A^2 = 15$ and an independent random sample of size $n_B = 21$ from machine B has $s_B^2 = 5$. Are the machines making the resistors consistently?

CHAPTER 7

Parameter Estimation I: Maximum Likelihood and Minimum Variance

OUTLINE

7.1 Estimation of a Single Parameter 123
7.2 Variance of an Estimator 128
   7.2.1 Approximate methods 130
7.3 Simultaneous Estimation of Several Parameters 133
7.4 Minimum Variance 136
   7.4.1 Parameter Estimation 136
   7.4.2 Minimum Variance Bound 137

In previous chapters we have encountered the problem of estimating the values of the parameters of a population from a sample. For example, we have used the sample mean and the sample variance as estimators of the corresponding population parameters. These choices satisfy the desirable general properties of point estimators discussed in Chapter 5 and are supported by the laws of large numbers. In this chapter and in the one that follows we turn to a discussion of specific practical methods of point estimation, starting with the so-called *maximum likelihood method*. We briefly met the likelihood function in Chapters 2 and 5, but in this chapter we will consider its use in estimation problems. Of all the possible methods of parameter estimation, that of maximum likelihood is, in a sense to be discussed below, the most general, and is widely used in practice.

7.1. ESTIMATION OF A SINGLE PARAMETER

The likelihood function has been defined in (5.11). If the dependence of L on x_i is suppressed, then for a sample of size n,

$$L(\theta) = \prod_{i=1}^{n} f(x_i; \theta), \qquad (7.1)$$

where $f(x;\theta)$ is the density function of the parameter population.[1] The *maximum likelihood* *(ML) estimator* of a population parameter θ is defined as that statistic $\hat{\theta}$ which maximizes $L(\theta)$ for variations of θ, that is, the solution (if it exists) of the equations[2]

$$\frac{\partial L(\theta)}{\partial \theta} = 0, \quad \frac{\partial^2 L(\theta)}{\partial \theta^2} < 0. \tag{7.2}$$

Since $L(\theta) > 0$, the first equation is equivalent to

$$\frac{1}{L}\frac{\partial L(\theta)}{\partial \theta} = \frac{\partial \ln L(\theta)}{\partial \theta} = 0, \tag{7.3}$$

which is the form more often used in practice. It is clear from (7.2) that the solution obtained by estimating the parameter θ is the same as estimating a function of θ, e.g., $F(\theta)$ since

$$\frac{\partial \ln L}{\partial \theta} = \frac{\partial \ln L(F)}{\partial F}\frac{\partial F}{\partial \theta}, \tag{7.4}$$

and the two sides vanish together. This is a useful invariance property of maximum likelihood estimators, but it does not extend to their variances. This is readily seen by considering the probability

$$P[\theta_1 \leq \theta \leq \theta_2] = \left[\int_{\theta_1}^{\theta_2} L(\theta)\mathrm{d}\theta\right]\left[\int_{-\infty}^{\infty} L(\theta)\mathrm{d}\theta\right]^{-1}, \tag{7.5}$$

i.e., the probability that an interval (θ_1, θ_2) will contain the true value θ. For example, if this probability is chosen to be 0.68, then for a normal distribution it would correspond to a standard error of one standard deviation in the usual sense. If we now use a function $F(\theta)$ to estimate the parameter θ, then

$$P[\theta_1 \leq \theta \leq \theta_2] = \left[\int_{F_1}^{F_2} L\frac{\partial \theta}{\partial F}\mathrm{d}F\right]\left[\int_{-\infty}^{\infty} L(\theta)\mathrm{d}\theta\right]^{-1},$$

which in general will not be equal to the value obtained from (7.5).

A practical consideration when using the maximum likelihood method is that the data do not have to be binned. However, this strength can become a weakness in the case of very large samples, because a complicated function may have to be evaluated at many points. In this case it is usual to apply the method to binned data. If N observations, distributed with a density $f(x;\theta)$, are divided into m bins, with $n_j (i = 1, 2, ..., m)$ entries in bin j, then

$$e_j(\theta) = N \int_{x_j^{min}}^{x_j^{max}} f(x;\theta)\mathrm{d}x$$

---

[1] We have assumed that $f(x;\theta)$ is a function of the single random variable x, but all the results of this section, and Section 7.2, may be generalized in a straightforward way to the case of estimating a single parameter from a multivariate distribution.

[2] A brief review of maxima and minima is given in Appendix A, Section A.2.

is the expectation value of the number of entries in the jth bin having lower and upper limits x_j^{min} and x_j^{max}, respectively. If we take the probability to be in bin j as (e_j/N), the joint probability is given by the multinomial distribution defined in Section 4.7,

$$f_{joint} = \frac{N!}{n_1! n_1! \ldots n_m!} \left(\frac{e_1}{N}\right)^{n_1} \left(\frac{e_2}{N}\right)^{n_2} \cdots \left(\frac{e_m}{N}\right)^{n_m}$$

and

$$\ln L(\theta) = \sum_{j=1}^{m} n_j \ln e_j(\theta) + C, \tag{7.6}$$

where C does not depend on the parameter θ. The ML estimator $\hat{\theta}$ is now obtained by maximizing (7.6) with respect to θ, usually by numerical means. It is evident from this expression that the method has no difficulty in accommodating bins that have no data.

The importance of ML estimators stems from their properties. It can be shown that they are generally consistent and have minimum variance. If a sufficient estimator for a parameter exists then it is a function of the ML estimator. The latter follows directly from the factorization condition (5.12), because maximizing L is equivalent to choosing $\hat{\theta}$ to maximize $L_1(\hat{\theta};\theta)$ in that equation. Another important property is that for large samples, ML estimators have a distribution that tends to normality. There are situations where these results do not hold and the ML estimator is a poor estimator, but for the common distributions met in practice they are valid.

To prove the normality property, we set $\ln L(\theta) = h(\theta)$, so that the ML estimator is defined by the solution of $dh(\theta)/d\theta \equiv h'(\theta) = 0$. Then, providing $h'(\theta)$ can be differentiated further, we can expand it about the point $\hat{\theta}$ to give

$$h'(\theta) = h'(\hat{\theta}) + (\theta - \hat{\theta})h''(\hat{\theta}) + \cdots \tag{7.7}$$

where, setting $f_i = f(x_i; \theta)$ and using (7.1),

$$h'(\hat{\theta}) = \sum_{i=1}^{n} \left\{\frac{f'_i}{f_i}\right\}\bigg|_{\theta=\hat{\theta}} \quad \text{and} \quad h''(\hat{\theta}) = \sum_{i=1}^{n} \left\{\left(\frac{f'_i}{f_i}\right)'\right\}\bigg|_{\theta=\hat{\theta}},$$

and the primes denote differentiation with respect to θ. For large samples we know that $E[h'(\hat{\theta})] = 0$, i.e.

$$E[h'(\hat{\theta})] = \int_{-\infty}^{\infty} \frac{f'(x)}{f(x)} f(x) dx = 0.$$

Differentiating this result again, and writing out in full gives

$$\int_{-\infty}^{\infty} \left\{\frac{f'^2(x)}{f(x)} + f(x)\left(\frac{f'(x)}{f(x)}\right)'\right\} dx = \int_{-\infty}^{\infty} \left\{\left(\frac{f'(x)}{f(x)}\right)^2 + \left(\frac{f'(x)}{f(x)}\right)'\right\} f(x) dx$$

$$= E\left[\left(\frac{f'(x)}{f(x)}\right)^2\right] + E\left[\left(\frac{f'(x)}{f(x)}\right)'\right] = 0,$$

that is,

$$E\left[\{h'(\hat{\theta})\}^2\right] = -E[h''(\hat{\theta})] = 1/c^2, \quad (7.8)$$

where c^2 depends on the density f and the estimator $\hat{\theta}$. Substituting (7.8) into (7.7) and integrating gives

$$h(\theta) - h(\hat{\theta}) = -\frac{1}{2}\left(\frac{\theta - \hat{\theta}}{c}\right)^2,$$

where we have used $h'(\hat{\theta}) = 0$. Finally, taking exponentials,

$$L(\theta) = k \exp\left\{-\frac{1}{2}\left(\frac{\theta - \hat{\theta}}{c}\right)^2\right\},$$

where k is a constant. So we have proved that ML estimators are asymptotically distributed as a normal distribution.

One has to be a little careful about using (7.4), because if the result of maximizing L, or its logarithm, results in an unbiased estimator, it does not always follow that the estimator obtained by maximizing a function of L is also unbiased. So one has to balance the convenience of the invariance property of ML estimators against the fact that the resulting estimator may not be unbiased. For example, for the exponential distribution

$$f(t; \tau) = \frac{1}{\tau}e^{-t/\tau},$$

that among other things describes the decay of an unstable quantum state, the mean of the measurements t_i is an unbiased ML estimator for the lifetime τ, i.e.

$$\hat{\tau} = \frac{1}{n}\sum_{i=1}^{n} t_i \quad (7.9)$$

for all n (see in Problem 7.1). However, the estimator for any function of τ may be found by evaluating the function using $\hat{\tau}$. So if we were to take the function to be $R = 1/\tau$ (the rate of decay), then from (7.9),

$$\hat{R} = \frac{1}{\hat{\tau}} = n\left(\sum_{i=1}^{n} t_i\right)^{-1},$$

but it is straightforward to show (for example by the method used in Problem 7.1) that

$$E[\hat{R}] = \frac{n}{n-1} R,$$

and so $\hat{R}$ is not an unbiased estimator for R, except asymptotically when n is large. Fortunately, this latter condition is usually satisfied in practical applications of the maximum likelihood method.

7.1. ESTIMATION OF A SINGLE PARAMETER

The use of the maximum likelihood method for estimating one parameter is illustrated by the following example. The related problem of finding the ML estimator for the variable σ^2 in a normal population is left to Problem 7.2.

EXAMPLE 7.1

Find the ML estimator $\hat{\mu}$ for the parameter μ in the normal population

$$f(x; \mu, \sigma^2) = \frac{1}{(2\pi\sigma^2)^{1/2}} \exp\left[-\frac{1}{2}\left(\frac{x-\mu}{\sigma}\right)^2\right],$$

for samples of size n, where σ is known and $-\infty \leq x \leq \infty$. Is the estimator unbiased?

From (7.1) we have

$$\ln L(\mu, \sigma^2) = -n \ln\left[(2\pi\sigma^2)^{1/2}\right] - \frac{1}{2\sigma^2}\sum_{i=1}^{n}(x_i - \mu)^2.$$

The ML estimator of μ is found by maximizing $\ln L$ with respect to μ, that is, the solution of

$$\frac{\partial \ln L(\mu)}{\partial \mu} = \frac{1}{\sigma^2}\sum_{i=1}^{n}(x_i - \mu) = 0.$$

Thus

$$\hat{\mu} = \frac{1}{n}\sum_{i=1}^{n} x_i = \bar{x}.$$

Therefore the sample mean is the ML estimator of the parameter μ. One could shown in a straightforward way that $\hat{\mu}$ is an unbiased estimator for μ by calculating its expectation value using the joint probability distribution for the x_i. (This is done in full for a related exercise in Problem 7.2.) However it also follows from the general result that the sample mean is an unbiased estimator of the mean for any probability density function.

EXAMPLE 7.2

Find the ML estimator for the parameter θ in a population with a density function

$$f(x; \theta) = (1 + \theta)x^\theta, \quad (0 \leq x \leq 1).$$

The likelihood function is

$$L(\theta) = \prod_{i=1}^{n}(1 + \theta)x_i^\theta,$$

with

$$\ln L(\theta) = n \ln(1 + \theta) + \theta \Sigma, \quad \text{where} \quad \Sigma \equiv \sum_{i=1}^{n} \ln x_i.$$

Taking the derivative gives

$$\partial \ln L(\theta)/\partial \theta = n/(1+\theta) + \Sigma = 0,$$

and so

$$\hat{\theta} = -(n+\Sigma)/\Sigma.$$

In the above discussion, we have made the usual assumption that n is a fixed known number. However, there are often circumstances where the number of events observed in an experiment is itself a random variable, typically with a Poisson distribution with mean λ. In these cases, the overall likelihood function is the product of the probability of finding a given value of n (given by equation (4.47)) and the usual likelihood function for the n values of x. So the combined likelihood function is

$$L(n,\theta) = \frac{\lambda^n}{n!} e^{-\lambda} \prod_{i=1}^{n} f(x_i; \theta).$$

This is called the *extended likelihood function*. It differs from the usual likelihood function only in that it is taken to be a function of both n and the sample values x_i. Much of the standard formalism of the maximum likelihood method carries over to $L(n,\theta)$ and we will not pursue it further here, except to say that the extended likelihood method usually results in smaller variances for estimators $\hat{\theta}$ because the method exploits the statistical information contained in n as well as that in the sample.

7.2. VARIANCE OF AN ESTIMATOR

The likelihood function $L(\theta)$ may be *formally* regarded as a probability density function for the parameter θ viewed as a random variable. Thus we can define the variance of the estimator as

$$\operatorname{var} \hat{\theta} = \int_{-\infty}^{\infty} (\theta - \hat{\theta})^2 L(\theta) d\theta \bigg/ \int_{-\infty}^{\infty} L(\theta) d\theta, \qquad (7.10)$$

and, by analogy with the work of Section 5.4, an estimate from experimental data would be quoted as

$$\theta = \hat{\theta}_e \pm \Delta\hat{\theta}_e \qquad (7.11a)$$

where $\hat{\theta}_e$ is the ML estimator obtained from the data and

$$\Delta\hat{\theta}_e = (\operatorname{var} \hat{\theta}_e)^{1/2}. \qquad (7.11b)$$

The interpretation of (7.11a) is that if the experiment were to be repeated many times, with the same number of measurements in each experiment, one would expect the standard deviation of the distribution of the estimates of θ to be $\Delta\hat{\theta}_e$.

From the normality property of ML estimators, it follows that for large samples, the form of $L(\theta)$ is

$$L(\theta) = \frac{1}{(2\pi v)^{1/2}} \exp\left[-\frac{1}{2}\frac{(\theta - \hat{\theta})^2}{v}\right], \qquad (7.12)$$

where $v = \text{var}\,\hat{\theta}$. Then

$$\ln L(\theta) = -\ln\left[(2\pi v)^{1/2}\right] - \frac{1}{2}\frac{(\theta - \hat{\theta})^2}{v}, \qquad (7.13)$$

and

$$\frac{\partial^2 \ln L(\theta)}{\partial \theta^2} = -\frac{1}{v}$$

Thus

$$\text{var}\,\hat{\theta} = \left[-\frac{\partial^2 \ln L(\theta)}{\partial \theta^2}\right]^{-1}\bigg|_{\theta=\hat{\theta}}. \qquad (7.14)$$

This is the most commonly used form for the variance of an ML estimator when making numerical calculations.

EXAMPLE 7.3

Find the ML estimator $\hat{\mu}$ and its variance for the parameter μ in the same normal population as in Example 7.1, but now for a set of experimental observations of the same quantity x_i with associated experimental errors Δx_i.

The density function is

$$f(x, \Delta x; \mu) = \frac{1}{\sqrt{2\pi}\Delta x}\exp\left[-\frac{1}{2}\left(\frac{x-\mu}{\Delta x}\right)^2\right],$$

from which

$$\ln L(\mu) = -\ln\left[(2\pi)^{1/2}\sum_{i=1}^{n}\Delta x_i\right] - \frac{1}{2}\sum_{i=1}^{n}\left(\frac{x_i-\mu}{\Delta x_i}\right)^2,$$

and

$$\frac{\partial \ln L(\mu)}{\partial \mu} = \sum_{i=1}^{n}\left[\frac{x_i-\mu}{(\Delta x_i)^2}\right].$$

Setting this last expression to zero gives

$$\hat{\mu} = \sum_{i=1}^{n}(x_i/\Delta x_i^2) \bigg/ \sum_{i=1}^{n}(1/\Delta x_i^2).$$

This result is called the *weighted mean* of a set of observations. The variance of $\hat{\mu}$ may be found using the second derivative

$$\frac{\partial^2 \ln L(\mu)}{\partial \mu^2} = -\sum_{i=1}^{n}\left[\frac{1}{(\Delta x_i)^2}\right]$$

in (7.14). It is

$$\operatorname{var} \hat{\mu} = (\Delta\hat{\mu})^2 = \left[\sum_{i=1}^{n}\left(\frac{1}{\Delta x_1}\right)^2\right]^{-1}.$$

The formula for the weighted mean $\hat{\mu}$, although formally correct, should be used with care. This is because the experimental errors Δx_i are only estimates of the population standard deviation and we must be sure that they are mutually consistent; that is, we must be sure that the measurements all come from the same normal distribution. We will return to this question in Section 11.1, where we discuss ways of testing whether a set of data does indeed come from the same population distribution.

7.2.1. Approximate methods

If an experiment has 'good statistics' then the likelihood function will indeed be a close approximation to a normal distribution and the method above for estimating the variance will be valid. However, many effects may be present which could produce a function that is clearly not normal and in this case the use of (7.14) usually produces an underestimate for $\Delta\hat{\theta}$. In these circumstances a more realistic estimate is to average $\partial^2 \ln L(\theta)/\partial\theta^2$ over the likelihood function, so that

$$\frac{1}{(\Delta\hat{\theta})^2} = -\left(\overline{\frac{\partial^2 \ln L(\theta)}{\partial\theta^2}}\right) = \left[\int_{-\infty}^{\infty}\left(-\frac{\partial^2 \ln L(\theta)}{\partial\theta^2}\right)L(\theta)d\theta\right]\left[\int_{-\infty}^{\infty} L(\theta)d\theta\right]^{-1}, \quad (7.15)$$

where the overbar denotes an average.

A related method that partially deals with the problem of non-invariance is to use the function

$$S(\theta) = \left[-\left(\overline{\frac{\partial^2 \ln L(\theta)}{\partial\theta^2}}\right)\right]^{-1/2} \frac{\partial \ln L(\theta)}{\partial\theta},$$

called the *Bartlett S function*, which can be shown to have a mean $\mu = 0$ and variance $\sigma^2 = 1$. In the case of a normal distribution, $S(\theta)$ is a straight line passing through zero when $\theta = \hat{\theta}$ and the values at $\pm n$ standard deviations are found from the points where $S = \pm n$. For non-normal functions, the solutions of the equations $S(\theta_\pm) = \mp 1$ determine the 'one-standard deviation' quantities $\theta_\pm$ so that the result would be quoted as

$$\theta = \hat{\theta}^{+\theta_+}_{-\theta_-}.$$

Alternatively, a direct graphical method can be used to estimate the variance. A plot of $L(\theta)$ is made and the two values found where it falls to $e^{1/2}$ of its maximum value, i.e., the two values that would correspond to one standard deviation in the case of a normal distribution. Reverting to using $\ln L(\theta)$, we can expand this in a Taylor series about the ML estimate $\hat{\theta}$ to give

$$\ln L(\theta) = \ln L(\hat{\theta}) + \left[\frac{\partial \ln L}{\partial \theta}\right]_{\theta=\hat{\theta}}(\theta - \hat{\theta}) + \frac{1}{2!}\left[\frac{\partial^2 \ln L}{\partial \theta^2}\right]_{\theta=\hat{\theta}}(\theta - \hat{\theta})^2 + \cdots \quad (7.16)$$

From the definition of $\hat{\theta}$, we know that $\ln L(\hat{\theta}) = \ln L_{\max}$, and the second term is zero because $\partial \ln L(\theta)/\partial \theta = 0$ for $\theta = \hat{\theta}$. So, if we ignore terms of higher order than those shown in (7.16), we have

$$\ln L(\theta) - \ln L_{\max} = \frac{(\theta - \hat{\theta})^2}{2 \operatorname{var} \hat{\theta}}, \quad (7.17a)$$

where we have used (7.14), and var $\hat{\theta}$ is evaluated at $\theta = \hat{\theta}$. Equation (7.17a) implies that

$$\ln L(\hat{\theta} \pm \hat{\sigma}) = \ln L_{\max} - \frac{1}{2}. \quad (7.17b)$$

Thus a change in the value of θ of one standard deviation from its ML estimate $\hat{\theta}$ corresponds to a decrease in the value of ½ from its maximum value. Likewise a decrease of 2 defines the points where θ changes by two standard deviations from its ML estimate, and so on. In the case where the likelihood function has an approximate normal distribution, $\ln L(\theta)$ will be approximately parabolic.

This is illustrated in Fig. 7.1 for a case where $\hat{\theta} = 10.0$ and $\ln L_{\max} = -50$. In this case the two points where $\hat{\theta}$ changes by one standard deviation can be found from the figure and lead to error estimates $\hat{\sigma}_- = 0.52$ and $\hat{\sigma}_+ = 0.58$. These are close enough that it is reasonable to average them and quote the final result as $\hat{\sigma} = 10.0 \pm 0.55$. Alternatively, the asymmetric

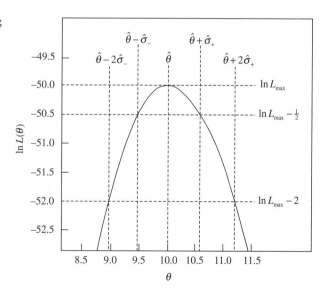

FIGURE 7.1 Graphical method for finding errors on an ML estimate.

errors could be quoted, i.e., $\hat{\sigma} = 10.0^{+0.58}_{-0.52}$, as would have to be done if the curve were not approximately parabolic.

Another useful formula for $\Delta\hat{\theta}$ may be derived for situations where one wants to answer the question: how many data are required to establish a particular result to a specified accuracy? The problem is to find a value for $\partial^2 \ln L(\theta)/\partial\theta^2$ averaged over many repeated experiments consisting of n events each. Since

$$\ln L(x; \theta) = \sum_{i=1}^{n} \ln f(x_i; \theta),$$

we have

$$\overline{\left(\frac{\partial^2 \ln L(\theta)}{\partial\theta^2}\right)} = n \int \frac{\partial^2 \ln f(x;\theta)}{\partial\theta^2} f(x;\theta) dx = nE\left[\frac{\partial^2 \ln f(x;\theta)}{\partial\theta^2}\right]. \tag{7.18a}$$

This form may be used in (7.15) directly, or it may be expressed in terms of first derivatives by writing

$$\frac{\partial^2 \ln f}{\partial\theta^2} = \frac{1}{f}\frac{\partial^2 f}{\partial\theta^2} - \frac{1}{f^2}\left(\frac{\partial f}{\partial\theta}\right)^2 = \frac{1}{f}\frac{\partial^2 f}{\partial\theta^2} - \left(\frac{\partial \ln f}{\partial\theta}\right)^2,$$

and then taking expectation values to give

$$E\left[\frac{\partial^2 \ln f}{\partial\theta^2}\right] = -E\left[\frac{\partial \ln f}{\partial\theta}\right]^2 + E\left[\frac{1}{f}\frac{\partial^2 f}{\partial\theta^2}\right] = -\int \frac{1}{f}\left(\frac{\partial f}{\partial\theta}\right)^2 dx, \tag{7.18b}$$

since the second term is

$$E\left[\frac{1}{f}\frac{\partial^2 f}{\partial\theta^2}\right] = \frac{\partial^2}{\partial\theta^2}\int f(\theta)dx = 0.$$

From (7.18a) and (7.18b),

$$\int \frac{\partial^2 \ln f(x;\theta)}{\partial\theta^2} f(x;\theta) dx = -\int \left(\frac{\partial f(x;\theta)}{\partial\theta}\right)^2 \frac{1}{f(x;\theta)} dx, \tag{7.19}$$

and so

$$\Delta\hat{\theta} = \frac{1}{\sqrt{n}}\left[\int \left(\frac{\partial f(x;\theta)}{\partial\theta}\right)^2 \frac{1}{f(x;\theta)} dx\right]^{-1/2}. \tag{7.20}$$

This result also confirms that to increase the precision of the experiment n-fold requires n^2 as many events.

EXAMPLE 7.4

Consider the density function

$$f(x;\theta) = \frac{1}{2}(1 + \theta x), \quad -1 \le x \le 1.$$

How many events would be required to determine θ to a precision of 1% for a value of $\hat{\theta} = 0.5$?
We have

$$\frac{\partial f(x;\theta)}{\partial \theta} = \frac{x}{2},$$

and

$$\int_{-1}^{1} \left(\frac{\partial f(x;\theta)}{\partial \theta}\right)^2 \frac{1}{f(x;\theta)} dx = \frac{1}{2\theta^3}\left[\frac{1}{2}(1+\theta x)^2 - 2(1+\theta x) + \ln(1+\theta x)\right]_{-1}^{1}$$

$$= \frac{1}{2\theta^3}\left[\ln\left(\frac{1+\theta}{1-\theta}\right) - 2\theta\right].$$

Thus from (7.20),

$$\left(\frac{\Delta\hat{\theta}}{\hat{\theta}}\right) = \left(\frac{2\hat{\theta}}{n}\right)^{1/2}\left[\ln\left(\frac{1+\hat{\theta}}{1-\hat{\theta}}\right) - 2\hat{\theta}\right]^{-1/2}.$$

Setting $(\Delta\hat{\theta}/\hat{\theta}) = 0.01$ for $\hat{\theta} = 0.5$ gives $n \simeq 1.01 \times 10^5$.

7.3. SIMULTANEOUS ESTIMATION OF SEVERAL PARAMETERS

If we wish to estimate simultaneously several parameters then the preceding results generalize in a straightforward manner. The maximum likelihood equation becomes the set of simultaneous equations

$$\frac{\partial \ln L(\theta_1, \theta_2, \ldots, \theta_i, \ldots, \theta_n)}{\partial \theta_i} = 0, \quad i = 1, 2, \ldots, n. \tag{7.21}$$

Also the analogous properties of the ML estimators for a single parameter hold. As an example, consider the generalization of the normality property. This states that the ML estimators $\hat{\theta}_i (i = 1, 2, \ldots, \theta_p)$ for the parameters of a density function $f(x; \theta_1, \theta_2, \ldots, \theta_p)$ from samples of size n are, for large samples, approximately distributed as the multivariate normal distribution with means $\theta_1, \theta_2, \ldots, \theta_p$ and a variance matrix $\mathbf{V}$ where

$$M_{ij} = (V_{ij})^{-1} = -nE\left[\frac{\partial^2 \ln f(x; \theta_1, \theta_2, \ldots, \theta_p)}{\partial \theta_i \partial \theta_j}\right]. \tag{7.22}$$

The use of (7.21) and (7.22) is illustrated in the following example.

EXAMPLE 7.5

Find the simultaneous ML estimators for the parameters μ and σ of the normal population

$$f(x; \mu, \sigma) = \frac{1}{(2\pi\sigma^2)^{1/2}} \exp\left[-\frac{1}{2}\left(\frac{x-\mu}{\sigma}\right)^2\right],$$

and find the form of the joint distribution of the estimators for large samples.

From (7.21),

$$\frac{\partial \ln L(\mu, \sigma)}{\partial \mu} = \frac{1}{\sigma^2} \sum_{j=1}^{n} (x_j - \mu) = 0,$$

and

$$\frac{\partial \ln L(\mu, \sigma)}{\partial \sigma} = \frac{1}{2\sigma^4} \sum_{j=1}^{n} (x_j - \mu)^2 - \frac{n}{2\sigma^2} = 0,$$

giving

$$\hat{\mu} = \bar{x}; \qquad \hat{\sigma}^2 = \frac{1}{n} \sum_{j=1}^{n} (x_j - \bar{x})^2.$$

Note that $\hat{\sigma}^2$ is a biased estimator of σ^2. This is often the case with ML estimators, but fortunately there usually exists a constant c, in this case $n/(n-1)$, such that multiplying the ML estimator by c produces an unbiased estimator.

Because the two estimators $\hat{\mu}$ and $\hat{\sigma}$ are approximately normally distributed with mean μ and a matrix $\mathbf{M}$ given by (7.22), we have, with $\mu = \theta_1$ and $\sigma = \theta_2$,

$$M_{11} = -nE\left[-\frac{1}{\sigma^2}\right] = \frac{n}{\sigma^2}, \qquad M_{22} = -nE\left[-\frac{3(x-\mu)^2}{\sigma^4} + \frac{1}{\sigma^2}\right] = \frac{2n}{\sigma^2}$$

and

$$M_{12} = M_{21} = -nE\left[-\frac{2(x-\mu)}{\sigma^3}\right] = 0.$$

Thus, the variance matrix is

$$V_{ij} = (M^{-1})_{ij} = \begin{pmatrix} \sigma^2/n & 0 \\ 0 & \sigma^2/2n \end{pmatrix},$$

and the variance and covariances are given by

$$\frac{1}{n}\sigma_{ij} = V_{ij}.$$

Finally, from (4.17) the form of the distribution of the estimators is

$$S(\hat{\mu}, \hat{\sigma}) = \frac{\sqrt{2n}}{2\pi\sigma^2} \exp\left\{-\frac{n}{2}\left[2\left(\frac{\hat{\sigma} - \sigma}{\sigma}\right)^2 + \left(\frac{\hat{\mu} - \mu}{\sigma}\right)^2\right]\right\}.$$

There is one point that should be remarked about the simultaneous estimation of several parameters, which is illustrated by reference to Example 7.4. If we know μ, the estimation of σ^2 alone gives (see Problem 7.2)

$$\hat{\sigma}^2 = \frac{1}{n} \sum_{i=1}^{n} (x_i - \mu)^2, \tag{7.23a}$$

whereas in Example 7.5, we found from the *simultaneous* estimation of μ and σ^2 that

$$\hat{\sigma}^2 = \frac{1}{n}\sum_{j=1}^{n}(x_j - \bar{x})^2. \tag{7.23b}$$

However, from the results found in Example 7.1 we see that we can estimate μ, independent of any possible knowledge of σ^2, to be $\bar{x}$. Thus, if we now find the estimator of σ^2 that maximizes the likelihood for all samples giving the estimated value of $\mu = \bar{x}$, it might be thought that the result (7.23b) would ensue, whereas in fact in this latter case

$$\hat{\sigma}^2 = \frac{1}{n-1}\sum_{i=1}^{n}(x_i - \bar{x})^2. \tag{7.23c}$$

The difference between (7.23b) and (7.23c) is that in the former case we have considered the variations of $\ln L(\mu, \sigma)$ over all samples of size n, whereas in the latter the constraint that $\sum x$ is a constant has been imposed, and thus the number of degrees of freedom has been lowered by one. For large n the difference is of little importance, but it is a useful reminder that every parameter estimated from the sample (i.e., every constraint applied) lowers the number of degrees of freedom by one.

The maximum likelihood method has the disadvantage that in order to estimate a parameter the form of the distribution must be known. Furthermore, it often happens that $L(\theta)$ is a highly nonlinear function of the parameters, and so maximizing the likelihood function may be a difficult problem.[3] Finally, if the data under study are normally distributed, then maximizing $L(\theta)$ is equivalent to minimizing

$$\chi^2 = \sum_{i=1}^{n}\left(\frac{x_i - \mu_i}{\sigma_i}\right)^2,$$

which may be more useful in practice, as we shall illustrate when we consider the method of minimum chi-squared in Chapter 8.

We will conclude with a few brief remarks on the interpretation of maximum likelihood estimators. Bayes' theorem tells us that maximizing the likelihood does not necessarily maximize the posterior probability of an event. This is only true if the prior probabilities are equal or somehow 'smooth'. Thus, ML estimators (and of course other estimators) should always be interpreted in the light of prior knowledge. In Chapter 8 we shall see how such knowledge can formally be included in the estimation procedure. However, because in general it is difficult to reduce prior knowledge to the required form, the actual method of estimation is not always of practical use. An alternative method is to form a likelihood function that is the product of the likelihood functions for all previous related experiments and use this function to make a new estimate of the parameter.

---

[3] A brief discussion of optimizing non-linear functions is given in Appendix B.

7.4. MINIMUM VARIANCE

The requirement that an estimator has minimum variance can also, in principle, be used as a criterion for parameter estimation and this is illustrated below.

7.4.1. Parameter Estimation

Consider the problem of estimating the population parameter μ, where samples are drawn from n populations, each with the same mean μ but with different variances. The estimate will be obtained by combining the sample means $\bar{x}_i$ and the corresponding sample variances. Since $\bar{x}_i$ is an unbiased estimate of μ, we have seen in Example 5.3 that the quantity

$$\bar{x} = \sum_{i=1}^{n} a_i \bar{x}_i, \qquad (7.24a)$$

with

$$\sum_{i=1}^{n} a_i = 1, \qquad (7.24b)$$

is also an unbiased estimate, regardless of the values of the coefficients a_i, so the problem is one of selecting a suitable set of a_i. This will be done by choosing the set a_i such that $\bar{x}$ has minimum variance. Thus we seek to minimize

$$\mathrm{var}(\bar{x}) = \mathrm{var}\left(\sum_{i=1}^{n} a_i \bar{x}_i\right)$$
$$= \sum_{i=1}^{n} a_i^2 \, \mathrm{var}(\bar{x}_i) = \sum_{i=1}^{n} a_i^2 \sigma_i^2,$$

subject to the constraint (7.24b). To do this we use the method of Lagrange multipliers.[4]
If we introduce a multiplier λ, then the variational function is

$$L = \sum_{i=1}^{n} a_i^2 \sigma_i^2 + \lambda \left(\sum_{i=1}^{n} a_i - 1\right).$$

and

$$\frac{\partial L}{\partial a_i} = 0 = 2 a_i \sigma_i^2 + \lambda.$$

Thus $a_i = -\lambda / 2\sigma_i^2$, and since the sum of the a_i is unity,

$$\lambda = -2 \left[\sum_{j=1}^{n} (1/\sigma_j^2)\right]^{-1}.$$

---
[4] Readers unfamiliar with this technique are referred to Appendix A.

Hence

$$a_i = (1/\sigma_i^2)\left[\sum_{j=1}^{n}(1/\sigma_j^2)\right]^{-1},$$

giving

$$\bar{x} = \left[\sum_{i=1}^{n}(\bar{x}_i/\sigma_i^2)\right]\left[\sum_{i=1}^{n}(1/\sigma_i^2)\right]^{-1} \quad \text{and} \quad \text{var}(\bar{x}) = \left[\sum_{j=1}^{n}(1/\sigma_j^2)\right]^{-1}.$$

In this example, the minimum variance estimator is the weighted mean, identical to the estimator obtained using the maximum likelihood method (cf. Example 7.2), where the population distribution was assumed to be normal. For other population densities, the results of the two methods will differ.

7.4.2. Minimum Variance Bound

In many cases it is not possible to find the variance of an estimator analytically (i.e., exactly) and even to do so numerically, for example by using the Monte Carlo method, involves a great deal of computation. In this situation, a very useful result that puts a lower limit on the variance may be used. This has various names, such as the *Cramér–Rao*, or *Fréchet, inequality*, or simply the *minimum variance bound*, and is true in general and not just for ML estimators.

Consider the ML estimator $\hat{\theta}$ of a parameter θ that is a function of the sample $x = x_1, x_2, \ldots, x_n$, with a joint pdf given by (7.1). The expectation value of $\hat{\theta}$ is

$$E[\hat{\theta}] = \int \hat{\theta} L(\mathbf{x}; \theta) d\mathbf{x}. \tag{7.25}$$

Differentiating with respect to θ, gives

$$\begin{aligned}\frac{dE[\hat{\theta}]}{d\theta} &= \int \hat{\theta} \frac{dL(\mathbf{x}; \theta)}{d\theta} d\mathbf{x} \\ &= \int \hat{\theta} \frac{d \ln L(\mathbf{x}; \theta)}{d\theta} L(\mathbf{x}; \theta) d\mathbf{x} = E\left[\hat{\theta} \frac{d \ln L(\theta)}{d\theta}\right].\end{aligned} \tag{7.26}$$

However in general,

$$E[\hat{\theta}] = \theta + b(\theta), \tag{7.27}$$

where b is the bias. Differentiating (7.27) and suppressing the dependence of L on $\mathbf{x}$ give

$$1 + \frac{db(\theta)}{d\theta} = E\left[\hat{\theta} \frac{d \ln L(\theta)}{d\theta}\right]. \tag{7.28}$$

The right-hand side of (7.28) may be evaluated by firstly differentiating the normalization condition

$$\int L(\mathbf{x}; \theta) d\mathbf{x} = 1$$

to give

$$\int \frac{dL(\theta)}{d\theta} dx = \int \frac{d \ln L(\theta)}{d\theta} L(\theta) dx = E\left[\frac{d \ln L(x; \theta)}{d\theta}\right] = 0. \quad (7.29)$$

Multiplying (7.28) by $E[\hat{\theta}]$ and subtracting the result from (7.27) give

$$E\left[\hat{\theta} \frac{d \ln L(\theta)}{d\theta}\right] - E[\hat{\theta}] E\left[\frac{d \ln L(\theta)}{d\theta}\right] = E\left[\left(\hat{\theta} - E[\hat{\theta}]\right) \frac{d \ln L(\theta)}{d\theta}\right] = 1 + \frac{db(\theta)}{d\theta}. \quad (7.30)$$

The second step is to use a form of the so-called *Schwarz inequality*, which for two random variables x and y, such that x^2 and y^2 have finite expectation values, takes the form[5]

$$\{E[xy]\}^2 \leq E[x^2] E[y^2]. \quad (7.31)$$

Applying this to (7.30) gives

$$\left(1 + \frac{db(\theta)}{d\theta}\right)^2 \leq E\left[(\hat{\theta} - E[\hat{\theta}])^2\right] E\left[\left(\frac{d \ln L(\theta)}{d\theta}\right)^2\right]. \quad (7.32)$$

The first factor on the right-hand side is $\text{var}(\hat{\theta}) = \sigma^2(\hat{\theta})$. To evaluate the second factor, we have

$$E\left[\left(\frac{d \ln L(\theta)}{d\theta}\right)^2\right] = nE\left[\left(\frac{df(x; \theta)/d\theta}{f(x; \theta)}\right)^2\right] \equiv I(\theta) \geq 0, \quad (7.33)$$

where the quantity $I(\theta)$ is called the *information of the sample with respect to* θ, or simply the *information*. It may also be written in the form (cf (7.19))

$$I(\theta) = E\left[-\frac{d^2 \ln L(\theta)}{d\theta^2}\right].$$

So, finally, for a single parameter θ having an estimator $\hat{\theta}$ with a bias b, the bound is

$$\text{var}(\hat{\theta}) = \sigma^2(\hat{\theta}) \geq \left(1 + \frac{db}{d\theta}\right)^2 \left\{E\left[-\frac{d^2 \ln L}{d\theta^2}\right]\right\}^{-1}. \quad (7.34)$$

It is worth noting that in the derivation no assumption has been made about the estimator. If the equality holds in (7.34), the estimator is efficient.

---

[5]The result (7.31) may be proved as follows. For any value λ, $E[(\lambda x + y)^2] = \lambda^2 E[x^2] + 2\lambda E[xy] + E[y^2] \geq 0$ and the solutions for λ in the case of the equality are

$$\lambda_\pm = -\frac{E[xy]}{E[x^2]} \pm \left\{\left(\frac{E[xy]}{E[x^2]}\right)^2 - \left(\frac{E[y^2]}{E[x^2]}\right)\right\}^{1/2}.$$

So the inequality holds only if $\{E[xy]\}^2 \leq E[x^2] E[y^2]$.

7.4. MINIMUM VARIANCE

There remains the question of what conditions are necessary for the minimum variance bound to be attained, that is, for the equality in (7.31) to hold. This is valid if $(\lambda x + y) = 0$, because only then is $E[(\lambda x + y)^2] = 0$ for all values of λ, x, and y. Applying this to (7.31) with $x = \hat{\theta} - E[\hat{\theta}]$ and $y = d \ln L(\theta)/d\theta$ gives

$$\frac{d \ln L(\theta)}{d\theta} = A(\theta)(\hat{\theta} - E[\hat{\theta}]), \qquad (7.35)$$

where $A(\theta)$ does not depend on the sample $x_1, x_2, \ldots, x_n$. Finally, integrating (7.35) gives the condition

$$\ln L(\theta) = B(\theta)\hat{\theta} + C(\theta) + D, \qquad (7.36)$$

where B and C are functions of θ, and D is independent of θ. Thus an estimator $\hat{\theta}$ will have a variance that satisfies the minimum bound if the associated likelihood function has the structure (7.36). The actual value of the minimum variance bound may be found by using (7.34) in (7.32). For an unbiased estimator, this gives

$$\sigma^2(\hat{\theta}) = \frac{1}{[A(\theta)]^2 E\left[(\hat{\theta} - E[\hat{\theta}])^2\right]} = \frac{1}{[A(\theta)]^2 \sigma^2(\hat{\theta})},$$

and so

$$\sigma^2(\hat{\theta}) = |A(\theta)|^{-1}. \qquad (7.37)$$

EXAMPLE 7.6

Find the ML estimator for the parameter p of the binomial distribution and show that it is an unbiased minimum variance estimator.

The binomial probability is given by equation (4.34) as

$$f(r; p, n) = \binom{n}{r} p^r (1-p)^{n-r},$$

and so

$$L(p) = \prod_{i=1}^{n} \binom{n}{r_i} p^{r_i} (1-p)^{n-r_i}.$$

Then, taking logarithms,

$$\ln L(p) = \sum_{i=1}^{n} \left\{ r_i \ln p + (n - r_i) \ln(1-p) + \ln\binom{n}{r_i} \right\}.$$

and

$$\frac{d \ln L(p)}{dp} = \frac{h}{p} - \frac{n-h}{1-p},$$

where

$$h = \sum_{i=1}^{n} r_i.$$

Setting $d \ln L(p)/dp = 0$ shows that $\hat{p} = h/n = \bar{r}$ is an unbiased estimator for the parameter p. Moreover, using $\hat{p} = \bar{r}$ we can write $\ln L(p)$ as

$$\ln L(p) = n\hat{p}[\ln p - \ln(1-p)] + n^2 \ln(1-p) + \sum_{i=1}^{n} \binom{n}{r_i},$$

which is of the form (7.36), so $\bar{r}$ is a minimum variance estimator for p. The value of the variance is found by writing

$$\frac{d \ln L(p)}{dp} = \frac{h}{p} - \frac{n-h}{1-p} = \frac{n(\hat{p}-p)}{p(1-p)},$$

which is of the form (7.35), with $A(p) = n/[(p(1-p)]$, and so

$$\sigma^2(\hat{p}) = A^{-1}(p) = p(1-p)/n.$$

PROBLEMS 7

7.1 Find the ML estimator for the parameter τ (the lifetime) in the exponential density $f(t; \tau) = e^{-t/\tau}/\tau$ and show that it is an unbiased estimator. Also find its variance.

7.2 Find the ML estimator $\hat{\sigma}^2$ for the parameter σ^2 in the normal population

$$f(x; \mu, \sigma^2) = \frac{1}{(2\pi\sigma^2)^{1/2}} \exp\left[-\frac{1}{2}\left(\frac{x-\mu}{\sigma}\right)^2\right],$$

for samples of size n. Is the estimator unbiased?

7.3 Find the ML estimator for the parameter λ of the Poisson distribution (see equation (4.47)) and show that it is an unbiased minimum variance estimator.

7.4 Find equations for the ML estimators of the constants α and β in the Weibull distribution of Section 4.4.

7.5 Find the ML estimator for the parameter k for a sample of size n from a population having a density function

$$f(x) = \begin{cases} a(k+2)^3 x^k & 0 \leq x \leq 1 \\ 0 & \text{otherwise} \end{cases}$$

where a is a constant.

7.6 A data set is subject to two independent scans. In the first scan, n_1 events of a given type x are identified and in the second, n_2 events of the same type are found. If there are n_{12} events in common in the two scans, what is the efficiency E_1 of the first scan and what is its standard deviation? Estimate the total number of events of type x.

7.7 A set of n independent measurements $E_i (i = 1, 2, \ldots, n)$ is made of the energy of a quantum system in the vicinity of an excited state of energy E_0 and width Γ described by the Breit–Wigner density of Section 4.5. If $|E_i - E_0| \ll \Gamma$, show that the mean energy $\bar{E}$ is the ML estimator of E_0.

7.8 Find the unbiased minimum variance bound (MVB) for the parameter θ in the distribution

$$f(x; \theta) = \frac{1}{\pi} \frac{1}{\left[1 + (x - \theta)^2\right]}.$$

Note the integral:

$$\int_0^\infty \frac{(1 - x^2)}{(1 + x^2)^3} dx = \frac{\pi}{8}.$$

CHAPTER 8

Parameter Estimation II: Least-Squares and Other Methods

OUTLINE

8.1 Unconstrained Linear Least Squares 143
   8.1.1 General Solution for the Parameters 145
   8.1.2 Errors on the Parameter Estimates 149
   8.1.3 Quality of the Fit 151
   8.1.4 Orthogonal Polynomials 152
   8.1.5 Fitting a Straight Line 154
   8.1.6 Combining Experiments 158

8.2 Linear Least Squares with Constraints 159

8.3 Nonlinear Least Squares 162

8.4 Other Methods 163
   8.4.1 Minimum Chi-Square 163
   8.4.2 Method of Moments 165
   8.4.3 Bayes' Estimators 167

The method of least squares is an application of minimum variance estimators, which were introduced in Section 7.4, to the multivariate problem and is widely used in situations where a functional form is known (or assumed) to exist between the observed quantities and the parameters to be estimated. This may be dictated by the requirements of a theoretical model of the data, or may be chosen arbitrarily to provide a convenient interpolation formula for use in other situations. We will firstly consider the technique for the situation where it is most used; where the data depend *linearly* on the parameters to be estimated. In this form the least-squares method is frequently used in curve-fitting problems.

8.1. UNCONSTRAINED LINEAR LEAST SQUARES

Initially the method will be formulated as a general procedure for finding estimators $\hat{\theta}_i (i = 1, 2, ..., p)$ of parameters $\theta_i (i = 1, 2, ..., p)$ which minimize the function

$$S = \sum_{i=1}^{n}(y_i - \hat{\eta}_i)^2 = \sum_{i=1}^{n} r_i^2, \tag{8.1}$$

where

$$\hat{\eta}_i = f(x_{1i}, x_{2i}, \ldots, x_{ki}; \hat{\theta}_1, \hat{\theta}_2, \ldots, \hat{\theta}_p), \tag{8.2}$$

with $x_{1i}, x_{2i}, \ldots, x_{ki}$ being the ith set of observations on $(k+1)$ variables, of which only y_i is random. Relation (8.2) is called the *equation of the regression curve of best fit*, or simply the *best-fit curve*. The word 'regression' comes from an early investigation that showed that tall fathers tended to have tall sons, although not on average as tall as themselves — referred to as 'regression to the norm' — and some authors prefer that regression is used to describe situations like this one where only qualitative statements can be made about the relationship between two variables.

We shall consider firstly the general case where the observations are correlated and have different 'weights' that are proportional to their experimental errors.[1] Later we will look at simpler cases, which follow easily from the general situation. Suppose we make observations of a quantity y that is a function $f(x; \theta_1, \theta_2, \ldots, \theta_p)$ of one variable x and p parameters $\theta_i (i = 1, 2, \ldots, p)$. Note that x is *not* a random variable and f is *not* a density function. The observations y_i are made at points x_i and are subject to experimental errors e_i. If the n observations y_i depend *linearly* on the p parameters then the observational equations may be written as

$$y_i = \sum_{k=1}^{p} \theta_k \phi_k(x_i) + e_i, \quad i = 1, 2, \ldots, n \tag{8.3}$$

where $\phi_k(x)$ are any linearly independent functions of x. The word 'linear' here refers to the coefficients θ_k, that is, they contain no powers, square roots, trigonometric functions, etc. Many situations that at first sight look nonlinear can be transformed so that the linear least-squares method may be used. For example, by taking logarithms of the equation $y = ae^{\lambda x}$, we get $\ln y = \ln a + \lambda x$, which is a linear relationship between $\ln y$ and x (see Example 8.1). On the other hand, the fitting functions $\phi_k(x)$ can be nonlinear provided they only depend on the variables x_i. In matrix notation[2] (8.3) may be written as

$$\mathbf{Y} = \mathbf{\Phi}\mathbf{\Theta} + \mathbf{E}, \tag{8.4}$$

where $\mathbf{Y}$ and $\mathbf{E}$ are $(n \times 1)$ column vectors, $\mathbf{\Theta}$ is a $(p \times 1)$ column vector, and $\mathbf{\Phi}$ is the $(n \times p)$ matrix (known as the *design matrix*):

$$\mathbf{\Phi} = \begin{pmatrix} \phi_1(x_1) & \phi_2(x_1) & \cdots & \phi_p(x_1) \\ \phi_1(x_2) & \phi_2(x_2) & \cdots & \phi_p(x_2) \\ \vdots & \vdots & & \vdots \\ \phi_1(x_n) & \phi_2(x_n) & \cdots & \phi_p(x_n) \end{pmatrix}.$$

---

[1] The least-squares method can also be formulated when both x and y have errors, but is more complicated. As it is not the usual situation met in practice, it will not be discussed here.

[2] A brief review of matrix algebra is given in Appendix A, Section A.1.

8.1.1. General Solution for the Parameters

The problem is to obtain estimates $\hat{\theta}_k$ for the parameters. For $n = p$ a unique solution exists and is obtainable directly from (8.4) by a simple matrix inversion, but for the more practical case where $n > p$ the system of equations is over-determined. In this situation, no general unique solution exists, and so what we seek is a 'best average solution' in a sense that will be discussed later. Thus we seek to approximate the experimental points y_i by a series of degree p, i.e.

$$f_i \equiv f(x_i; \theta_1, \theta_2, \ldots, \theta_p) = \sum_{k=1}^{p} \theta_k \phi_k(x_i). \tag{8.5}$$

Since the experimental errors are assumed to be random we would expect them to have a joint distribution with zero mean, i.e.,

$$E[\mathbf{Y}] \equiv \mathbf{Y}^0 = \mathbf{\Phi} \mathbf{\Theta}, \tag{8.6}$$

and an associated variance matrix

$$V_{ij} = \begin{pmatrix} \sigma_1^2 & \sigma_{12} & \cdots & \sigma_{1n} \\ \sigma_{21} & \sigma_2^2 & \cdots & \sigma_{2n} \\ \vdots & \vdots & & \vdots \\ \sigma_{n1} & \sigma_{n2} & \cdots & \sigma_n^2 \end{pmatrix}, \tag{8.7}$$

where

$$\sigma_i^2 = E[e_i^2] = \text{var}(y_i),$$

and

$$\sigma_{ij} = \sigma_{ji} = E[e_i e_j] = \text{cov}(y_i, y_j).$$

Note that we have only assumed that the population distribution of the errors has a *finite second moment*. In particular, it is *not* necessary to assume that the distribution is normal. However, *if* the errors are normally distributed, as is often the case, then the least-squares method gives the same results as the maximum likelihood method.

The quantities r_i of (8.1), called the *residuals*, are now replaced by

$$r_i \equiv y_i - \hat{f}_i = y_i - \sum_{k=1}^{p} \theta_k \phi_k(x_i), \tag{8.8}$$

and we will minimize the weighted sum

$$S = \sum_{i=1}^{n} \sum_{j=1}^{n} r_i r_j (V^{-1})_{ij} = \mathbf{R}^T \mathbf{V}^{-1} \mathbf{R}, \tag{8.9}$$

where $\mathbf{R}$ is an $(n \times 1)$ column vector of residuals.

To minimize S with respect to $\mathbf{\Theta}$, we set $\partial S/\partial \mathbf{\Theta} = 0$, giving the solution

$$\hat{\mathbf{\Theta}} = (\mathbf{\Phi}^T \mathbf{V}^{-1} \mathbf{\Phi})^{-1} \mathbf{\Phi}^T \mathbf{V}^{-1} \mathbf{Y}, \tag{8.10}$$

or, in nonmatrix notation,

$$\hat{\theta}_k = \sum_{l=1}^{p}(E^{-1})_{kl}\sum_{i=1}^{n}\sum_{j=1}^{n}\phi_l(x_i)(V^{-1})_{ij}y_j, \qquad (8.11)$$

where

$$E_{kl} = \sum_{i=1}^{n}\sum_{j=1}^{n}\phi_k(x_i)(V^{-1})_{ij}\phi_l(x_j). \qquad (8.12)$$

These are the so-called *normal equations* for the parameters. Note that to find the estimators for the parameters only requires knowledge of the *relative* errors on the observations, because any scale factor in V would cancel in (8.10). However, this is not true for the variances of the parameters, as we shall see in Section 8.1.2 later.

EXAMPLE 8.1

The table below shows the values of data y_i ($i = 1, 2, \ldots, 7$) with uncorrelated errors σ_i taken at the points x_i. Use the general formulation of the least-squares method to find estimators for the parameters a and b in a fit to the data of the form $y = a\exp(bx)$ and calculate the predictions for $\hat{y}_i$. Plot the data and the best-fit line.

i	1	2	3	4	5	6	7
x_i	1	2	3	4	5	6	7
y_i	4	5	8	16	30	38	70
σ_i	2	2	3	3	4	5	5

By taking logarithms of the fitting function, the problem can be converted to the linear form $y' = a' + b'x'$, where $y' = \ln y$, $a' = \ln a$, $b' = b$ and $x' = x$. The errors on y' follow from (5.45) for the propagation of errors, that is

$$\sigma' = \frac{d\ln y}{dy}\sigma = \frac{\sigma}{y}.$$

A new table can then be constructed as follows:

i	1	2	3	4	5	6	7
x'_i	1	2	3	4	5	6	7
y'_i	1.386	1.609	2.079	2.773	3.401	3.638	4.248
σ'_i	0.500	0.400	0.375	0.188	0.133	0.105	0.071

Using the notation above, the various matrices needed for the primed quantities are

$$\Phi'^T = \begin{pmatrix} 1 & 1 & 1 & 1 & 1 & 1 & 1 \\ 1 & 2 & 3 & 4 & 5 & 6 & 7 \end{pmatrix}$$

$$\mathbf{Y}'^T = (\,1.386 \quad 1.609 \quad 2.079 \quad 2.773 \quad 3.401 \quad 3.638 \quad 4.248\,),$$

and

$$\mathbf{V}' = 10^{-3} \begin{pmatrix} 250.00 & 0 & 0 & 0 & 0 & 0 & 0 \\ 0 & 160.00 & 0 & 0 & 0 & 0 & 0 \\ 0 & 0 & 140.63 & 0 & 0 & 0 & 0 \\ 0 & 0 & 0 & 35.34 & 0 & 0 & 0 \\ 0 & 0 & 0 & 0 & 17.69 & 0 & 0 \\ 0 & 0 & 0 & 0 & 0 & 11.03 & 0 \\ 0 & 0 & 0 & 0 & 0 & 0 & 5.05 \end{pmatrix}.$$

These can be used to calculate the matrices $(\mathbf{\Phi}'^T\mathbf{V}'^{-1}\mathbf{\Phi}')^{-1}$ and $(\mathbf{\Phi}'^T\mathbf{V}'^{-1}\mathbf{Y}')$ and hence $\hat{\mathbf{\Theta}}$ from (8.11), where $\hat{\theta}_1 = \hat{a}' = \ln \hat{a}$ and $\hat{\theta}_2 = \hat{b}' = \hat{b}$. The result is $\hat{a} = 2.101$ and $\hat{b} = 0.498$. From these we can calculate the fitted values from $\hat{y}_i = \hat{a}\exp(\hat{b}x_i)$ and they are given below:

i	1	2	3	4	5	6	7
$\hat{y}_i$	3.46	5.69	9.37	15.42	23.38	41.78	68.76

A plot of the data and the fitted function is shown in Fig. 8.1.

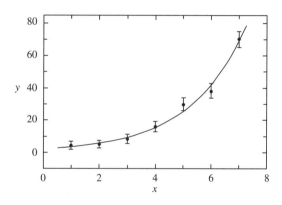

FIGURE 8.1 Best fit to the data using $y = 2.101\exp(0.498x)$.

In (8.9) the sums are over all the data points, but the least-squares method can also be applied to binned data. In this case we will assume that the fitting function *is* a probability density, and for simplicity is a function of a single parameter θ to be estimated. Using the notation in the analogous discussion in Section 7.1 about fitting binned data using the maximum likelihood method, we assume there are N observations of a random variable x independently distributed with a density function $f(x;\theta)$, and divided between m bins. If the observed number of entries in the jth bin is o_j, then the predicted (expected) number of entries for that bin is

$$e_j(\theta) = E[o_j] = N \int_{x_j^{\text{xmin}}}^{x_j^{\text{xmax}}} f(x;\theta)dx = Np_j(\theta), \tag{8.13}$$

where x_j^{xmin} and x_j^{xmax} define the bin limits and $p_j(\theta)$ is the probability of having an entry in the jth bin. Then, by analogy with (8.9), the least-squares estimators are found by numerically minimizing the quantity

$$S = \sum_{j=1}^{m} \frac{[o_j - e_j(\theta)]^2}{\sigma_j^2}, \qquad (8.14)$$

where σ_j^2 are the variances on the observed number of entries in the jth bin. If the mean number of entries in each bin is small compared to N, the entries in each bin are defined by the Poisson distribution, for which the variance is equal to the mean, $\sigma_j^2 = e_i$, so in this case

$$S = \sum_{j=1}^{m} \frac{[o_j - e_j(\theta)]^2}{e_j(\theta)} = \sum_{j=1}^{m} \frac{[o_j - Np_j(\theta)]^2}{Np_j(\theta)}. \qquad (8.15)$$

Sometimes, for reasons of computational simplicity, the variance of the number of entries in a bin is replaced by the number of entries actually observed o_j, rather than the predicted number e_j, so that S becomes

$$S = \sum_{j=1}^{m} \frac{[o_j - e_j(\theta)]^2}{o_j} = \sum_{j=1}^{m} \frac{[o_j - Np_j(\theta)]^2}{o_j}, \qquad (8.16)$$

but this is only valid if the number of entries in each bin is large; if for example any of the o_j were zero, clearly S is undefined.

The estimates $\hat{\theta}_k$ from (8.11) have been obtained by minimizing the sum of the residuals, and although this has an intrinsic geometrical appeal, it still might be considered rather arbitrary. However, the importance of least-squares estimates stems from their minimum variance properties, which are summarized by the statement that the least-squares estimates $\hat{\theta}_k$ of the parameters θ_k minimize the variance of any linear combination of the parameters. To prove this, consider the general sum

$$L = \mathbf{C}^T \mathbf{\Theta}, \qquad (8.17)$$

where $\mathbf{C}$ is a $(p \times 1)$ vector of known constant coefficients. Let $\mathbf{G}$ be any $(n \times 1)$ vector such that

$$\mathbf{C}^T = \mathbf{G}^T \mathbf{\Phi}. \qquad (8.18)$$

The problem of minimizing the variance of L is now equivalent to minimizing the variance of $\mathbf{G}^T \mathbf{Y}$ subject to the constraint (8.18). To do this, we use the method of Lagrange multipliers, as used in Section 7.4.1 when discussing the maximum likelihood method. Since $\mathbf{G}$ is a constant vector,

$$\text{var}(\mathbf{G}^T \mathbf{Y}) = \mathbf{G}^T (\text{var } \mathbf{Y}) \, \mathbf{G} = \mathbf{G}^T \mathbf{V} \mathbf{G},$$

which is easily proved from the definition of the variance matrix, and we can construct a variational function

$$F = \mathbf{G}^T \mathbf{V} \mathbf{G} - \mathbf{\Lambda}(\mathbf{\Phi}^T \mathbf{G} - \mathbf{C}), \qquad (8.19)$$

where $\mathbf{\Lambda}$ is a $(p \times 1)$ vector of Lagrange multipliers. Setting $dF = 0$ gives

$$\mathbf{G}^T = \mathbf{\Lambda}^T \mathbf{\Phi}^T \mathbf{V}^{-1}, \tag{8.20}$$

and so

$$\mathbf{\Lambda}^T = \mathbf{G}^T \mathbf{\Phi} \, (\mathbf{\Phi}^T \mathbf{V}^{-1} \mathbf{\Phi})^{-1}. \tag{8.21}$$

Eliminating $\mathbf{\Lambda}^T$ between (8.20) and (8.21) gives

$$\mathbf{G}^T = \mathbf{G}^T \mathbf{\Phi} \, (\mathbf{\Phi}^T \mathbf{V}^{-1} \mathbf{\Phi})^{-1} \mathbf{\Phi}^T \mathbf{V}^{-1}. \tag{8.22}$$

If we now multiply (8.22) on the right by $\mathbf{Y}$ and use (8.10), we have

$$\mathbf{G}^T \mathbf{Y} = (\mathbf{G}^T \mathbf{\Phi}) \, \hat{\mathbf{\Theta}} = \mathbf{C}^T \hat{\mathbf{\Theta}}. \tag{8.23}$$

Thus we have shown that the value of $\hat{\mathbf{\Theta}}$ which minimizes the variance of any linear combination of the parameters is the least-squares estimate, a result originally due to Gauss.

8.1.2. Errors on the Parameter Estimates

Having obtained the least-squares estimates $\hat{\theta}_k$, we can now consider their variances and covariances. As mentioned above, this cannot be done with only knowledge of the relative errors on the observations, but instead requires the absolute values of these quantities. It is therefore convenient at this stage to allow for the possibility that the variance matrix may only be determined up to a scale factor w by writing

$$\mathbf{V} = w \mathbf{W}^{-1}, \tag{8.24}$$

where $\mathbf{W}$ is the so-called *weight matrix* of the observations. In this case (8.9) becomes

$$S = \frac{1}{w} (\mathbf{Y} - \mathbf{\Phi}\mathbf{\Theta})^T \mathbf{W} (\mathbf{Y} - \mathbf{\Phi}\mathbf{\Theta}). \tag{8.25}$$

and the solution of the normal equations is

$$\hat{\mathbf{\Theta}} = (\mathbf{\Phi}^T \mathbf{W} \mathbf{\Phi})^{-1} \mathbf{\Phi}^T \mathbf{W} \mathbf{Y}. \tag{8.26}$$

We have previously used the result that for any linear combination of y_i, say $\mathbf{P}^T \mathbf{Y}$, with $\mathbf{P}$ a constant vector

$$\text{var}(\mathbf{P}^T \mathbf{Y}) = \mathbf{P}^T \text{var}(\mathbf{Y}) \, \mathbf{P}. \tag{8.27}$$

Applying (8.27) to $\hat{\mathbf{\Theta}}$ as given by (8.26), we have

$$\text{var}(\hat{\mathbf{\Theta}}) = (\mathbf{\Phi}^T \mathbf{W} \mathbf{\Phi})^{-1} \mathbf{\Phi}^T \mathbf{W} \, \text{var}(\mathbf{Y}) \mathbf{W} \mathbf{\Phi} (\mathbf{\Phi}^T \mathbf{W} \mathbf{\Phi})^{-1},$$

and using

$$\text{var}(\mathbf{Y}) = \mathbf{V} = w \, \mathbf{W}^{-1}.$$

gives

$$\mathbf{E} = \text{var}(\hat{\mathbf{\Theta}}) = w (\mathbf{\Phi}^T \mathbf{W} \mathbf{\Phi})^{-1}. \tag{8.28}$$

This is the variance matrix of the parameters and is given by a quantity that appears in the solution (8.26) for the parameters themselves. The matrix **E** is also called the *error matrix*, and the errors on the parameters are

$$\Delta \hat{\theta}_i = \hat{\sigma}_i = (E_{ii})^{1/2}.$$

It is sometimes useful to know which linear combinations of parameter estimates have zero covariances. Since **E** is a real, symmetric matrix, it can be diagonalized by a unitary matrix **U**. This same matrix then transforms the parameter estimates into the required linear combination.

Finally, if w is unknown, we need to find an estimate for it. This may be done by returning to (8.25) and finding the expected value of the weighted sum of residuals S:

$$wE[S] = E[\mathbf{R}^T\mathbf{W}\mathbf{R}]. \tag{8.29}$$

When $\boldsymbol{\Theta} = \hat{\boldsymbol{\Theta}}$, the right-hand side of (8.29) becomes

$$E[\mathbf{R}^T\mathbf{W}(\mathbf{Y} - \boldsymbol{\Phi}\hat{\boldsymbol{\Theta}})] = E[\mathbf{R}^T\mathbf{W}\mathbf{Y}],$$

since

$$\mathbf{R}^T\mathbf{W}\boldsymbol{\Phi}\boldsymbol{\Theta} = 0,$$

is equivalent to the statement of the normal equations. Furthermore,

$$\mathbf{R}^T\mathbf{W}\mathbf{Y} = (\mathbf{Y}^T - \hat{\boldsymbol{\Theta}}^T\boldsymbol{\Phi}^T)\mathbf{W}\mathbf{Y} = (\mathbf{Y}^T\mathbf{W}\mathbf{Y}) - (\hat{\boldsymbol{\Theta}}^T\mathbf{N}\hat{\boldsymbol{\Theta}}), \tag{8.30}$$

where

$$\mathbf{N} = \boldsymbol{\Phi}^T\mathbf{W}\boldsymbol{\Phi}.$$

By using the normal equations once again, (8.30) may be reduced to

$$(\mathbf{Y} - \mathbf{Y}^0)^T\mathbf{W}(\mathbf{Y} - \mathbf{Y}^0) - (\hat{\boldsymbol{\Theta}} - \boldsymbol{\Theta})^T\mathbf{N}(\hat{\boldsymbol{\Theta}} - \boldsymbol{\Theta}),$$

where $\mathbf{Y}^0$ is defined in (8.6), and thus we have arrived at the result that

$$\begin{aligned}E[S] &= E[\mathbf{R}^T\mathbf{V}^{-1}\mathbf{R}] \\ &= E[(\mathbf{Y} - \mathbf{Y}^0)^T\mathbf{V}^{-1}(\mathbf{Y} - \mathbf{Y}^0) - (\hat{\boldsymbol{\Theta}} - \boldsymbol{\Theta})^T\mathbf{M}^{-1}(\hat{\boldsymbol{\Theta}} - \boldsymbol{\Theta})],\end{aligned} \tag{8.31}$$

where

$$\mathbf{M} = w\,\mathbf{N}^{-1}.$$

The result (8.31) is the variance matrix of the parameters.

Consider the first term in (8.31). The quantity $(\mathbf{Y} - \mathbf{Y}^0)$ is a vector of random variables distributed with mean zero and variance matrix **V**. Thus

$$\begin{aligned}E[(\mathbf{Y} - \mathbf{Y}^0)^T\mathbf{V}^{-1}(\mathbf{Y} - \mathbf{Y}^0)] &= E[\mathrm{Tr}\,\{(\mathbf{Y} - \mathbf{Y}^0)^T\mathbf{V}^{-1}(\mathbf{Y} - \mathbf{Y}^0)\}] \\ &= \mathrm{Tr}\{E[(\mathbf{Y} - \mathbf{Y}^0)^T(\mathbf{Y} - \mathbf{Y}^0)\mathbf{V}^{-1}]\} \\ &= \mathrm{Tr}\,(\mathbf{V}\mathbf{V}^{-1}) = n,\end{aligned}$$

where Tr denotes the trace of a matrix. Similarly, since $\mathbf{M}$ is the variance matrix of $\hat{\Theta}$,

$$E[(\hat{\Theta} - \Theta)^T \mathbf{M}^{-1}(\hat{\Theta} - \Theta)] = p.$$

Thus, from (8.31) we have

$$E[\mathbf{R}^T \mathbf{V}^{-1} \mathbf{R}] = n - p,$$

and so an unbiased estimate for w is

$$\hat{w} = \frac{\mathbf{R}^T \mathbf{W} \mathbf{R}}{n - p},$$

and consequently an unbiased estimate for the variance matrix of $\hat{\Theta}$ is

$$\mathbf{E} = \frac{\mathbf{R}^T \mathbf{W} \mathbf{R}}{n - p} (\mathbf{\Phi}^T \mathbf{W} \mathbf{\Phi})^{-1} = \frac{\mathbf{R}^T \mathbf{V}^{-1} \mathbf{R}}{n - p} (\mathbf{\Phi}^T \mathbf{V}^{-1} \mathbf{\Phi})^{-1}. \tag{8.32}$$

Equation (8.32) looks rather complicated, but $\mathbf{R}^T \mathbf{W} \mathbf{R}$ can be calculated in a straightforward way from

$$\mathbf{R}^T \mathbf{W} \mathbf{R} = (\mathbf{Y} - \mathbf{\Phi}\hat{\Theta})^T \mathbf{W}(\mathbf{Y} - \mathbf{\Phi}\hat{\Theta}),$$

using the measured and fitted values. In the common case where the values y_i are random variables normally distributed about f_i, then $\mathbf{R}^T \mathbf{V}^{-1} \mathbf{R}$ is the chi-squared value for the fit and $(n - p)$ is the number of degrees of freedom n_{df}. In this case (8.32) becomes

$$\mathbf{E} = \frac{\chi^2}{n_{df}} (\mathbf{\Phi}^T \mathbf{V}^{-1} \mathbf{\Phi})^{-1}. \tag{8.33}$$

EXAMPLE 8.2

Calculate the errors on the best-fit parameters in Example 8.1.

These follow immediately using the matrices calculated in Example 8.1. For the primed quantities defined in Example 8.1, the error matrix is

$$\mathbf{E}' = (\mathbf{\Phi}'^T \mathbf{V}'^{-1} \mathbf{\Phi}')^{-1} = 10^{-2} \begin{pmatrix} 6.053 & -0.959 \\ -0.959 & 0.159 \end{pmatrix},$$

from which

$$\sigma(a) = \hat{a}\sigma(a') = 2.101 \times \sqrt{0.06053} = 0.517$$

and

$$\sigma(b) = \sigma(b') = \sqrt{0.00159} = 0.040$$

8.1.3. Quality of the Fit

To examine how well the predictions of the least-squares method fit the data we have to assume a distribution for the y_i, and this will be taken to be normal about f_i, with the errors on the observations used to define the weights of the data, i.e., $w = 1$, which is the usual situation in practice. In this case, we have seen above that the weighted sum of residuals S, of (8.9), is distributed as χ^2 with $n - p$ degrees of freedom. Thus for a fit of given order p, one can calculate the probability P_p that the expected value S_e is smaller than the observed values S_o. The order of the fit is then increased until this probability reaches any desired level. To increase p below the point where $\chi^2 \sim (n - p)$ would result in apparently better fits to the data. However, to do so would ignore the fact that y_i are random variables and as such contain only a limited amount of information. The fit of Example 8.1 has a χ^2 value of 2.72 for 5 degrees of freedom, which is acceptable because $P[\chi_5^2 < 2.72]$ is approximately 0.25.

What should one do if a satisfactory value of $\chi^2 \sim (n - p)$ cannot be achieved using a reasonable order p (for example if p is dictated by the model), that is, if $\chi^2 \gg n_{df}$? Firstly, one should examine the data to see whether there are isolated data points that contribute substantially higher-than-average values to χ^2. If this is the case, then these points should be carefully examined to see if there are any genuine reasons why they should be rejected, but as emphasized in Section 5.4 this must be done honestly, avoiding any temptation to 'massage the data', and must be defensible. In the absence of such reasons, one may have to conclude that the errors on the data have been underestimated and/or contain systematic errors. In this situation, one possibility is to scale the experimental errors by choosing a value of w so that $\chi^2 \sim n_{df}$. This will not change the values of the estimated parameters of the best fit, but will increase their variances to better reflect the spread of the data. Conversely, if $\chi^2 \ll n_{df}$, the errors should be examined to see whether they have been overestimated.

Another test that can be used to supplement the χ^2 test is based on the F distribution of Section 6.3. This procedure can test the significance of adding additional terms in expansion (8.5), that is, to answer the question: is θ_k different from zero? If S_p and S_{p-1} denote the values of S for fits of order p and $p - 1$, respectively, then from the additive property pf χ^2, the quantity $(S_{p-1} - S_p)$ obeys a χ^2 distribution with one degree of freedom, and which is distributed *independently* of S_p itself. Thus the statistic

$$F = \frac{S_{p-1} - S_p}{S_p/(n - p)}$$

obeys an F distribution with 1 and $(n - p)$ degrees of freedom. From tables of the F distribution we can now find the probability P that the observed value F_o is greater than the expected value F_e. Thus if P_p corresponds to $F_o(n - p)$ then we may assume $\theta_p = 0$ with a probability P_p of being correct. It is still possible that even though $\theta_p = 0$, higher terms are nonzero, but in this case the χ^2 test would indicate that a satisfactory fit had not yet been achieved. These points will be discussed in more detail in Chapter 10, when we discuss hypothesis testing.

8.1.4. Orthogonal Polynomials

The solutions for the parameters $\hat{\Theta}$ and their error matrix $\mathbf{E}$ both require the inversion of the matrix $(\mathbf{\Phi}^T \mathbf{V}^{-1} \mathbf{\Phi})$. In the discussion so far we have not specified the functions $\phi_k(x)$ except that they form a linearly independent set. If simple powers of x are used for $\phi_k(x)$, then the matrix is ill-conditioned for even quite moderate values of k, and the degree of ill-conditioning increases as k becomes larger. Ill-conditioning simply means that the large differences in the size of the elements of the matrix to be inverted can lead to serious rounding errors in the inverted matrix, and these can lead to errors in $\hat{\Theta}$ as calculated from (8.10). If a power series, or similar form, is dictated by the requirements of a particular model, the parameters of which are required to be estimated, then one can only hope to circumvent the problem by a judicious choice of method to invert the matrix. Such techniques are to be found in books on numerical methods. However, if all that is required is *any* form that gives an adequate representation of the data then it would clearly be advantageous to choose functions such that the matrix $(\Phi^T V^{-1} \Phi)$ is diagonal. Such functions are called *orthogonal polynomials* and their construction is briefly described here.

We will assume that the observations are uncorrelated (this is the usual situation met in practice) and denote the diagonal elements of the weight matrix $W = w V^{-1}$ for the data as $W(x_j)$ ($j = 1, 2, \ldots n$). Then if we fit using polynomials $\psi_k(x)$ ($k = 1, 2, \ldots, p$), the matrix of the normal equations will be diagonal if

$$\sum_{j=1}^{n} W(x_j)\, \psi_r(x_j)\, \psi_s(x_j) = 0, \tag{8.34a}$$

for $r \neq s$. In this case, the least-squares estimate $\hat{\Theta}$ from (8.10) is

$$\hat{\theta}_k = \frac{\sum_{j=1}^{n} W(x_j) y_j \psi_k(x_j)}{\sum_{j=1}^{n} W(x_j) \psi_k^2(x_j)}, \quad k = 1, 2, \ldots, p. \tag{8.35}$$

A valuable feature of using orthogonal polynomials is seen if we calculate the weighted sum of squared residuals at the minimum. From (8.9) this is, using p polynomials,

$$S_p = \frac{1}{w} \sum_{j=1}^{n} W(x_j) \left[y_j^2 - \sum_{k=1}^{p} \hat{\theta}_k^2\, \psi_k^2(x_j) \right].$$

If we now perform a new fit using $p+1$ polynomials, S_p is reduced by

$$\frac{1}{w} \hat{\theta}_{p+1}^2 \sum_{j=1}^{n} W(x_j) \psi_{p+1}^2(x_j),$$

and the first p coefficients $\hat{\theta}_k$ ($k = 1, 2, \ldots, p$) are unchanged.

To construct the polynomials we will assume for convenience that the values of x are normalized to lie in the interval $(-1, 1)$, and since it is desirable that none of the $\psi_k(x)$ has a large absolute value, we will arrange that the leading coefficient of $\psi_k(x)$ is 2^{k-2}. In this case it can be shown that the polynomials satisfy the following recurrence relations, the derivation of which may be found in many textbooks on numerical analysis:

$$\psi_1(x) = 1/2,$$
$$\psi_2(x) = (2x + \beta_1)\psi_1(x),$$

and for $r \geq 2$,

$$\psi_{r+1}(x) = (2x + \beta_r)\psi_r(x) + \gamma_{r-1}\psi_{r-1}(x).$$

To calculate the coefficients β_r and γ_r, we apply the orthogonality condition to ψ_s and ψ_{r+1}, that is,

$$\sum_{j=1}^{n} W(x_j)\,\psi_s(x_j)\psi_{r+1}(x_j) = 0, \qquad s \neq r+1. \tag{8.34b}$$

Then using the recurrence relations in (8.34b) and setting first $s = j$ and then $s = j-1$ leads immediately to the results

$$\beta_r = -\frac{\sum_{j=1}^{n} W(x_j) x_j \psi_r^2(x_j)}{\sum_{j=1}^{n} W(x_j)\psi_r^2(x_j)}, \qquad r = 1, 2, \ldots, \tag{8.36a}$$

and

$$\gamma_{r-1} = -\frac{\sum_{j=1}^{n} W(x_j)\psi_r^2(x_j)}{\sum_{j=1}^{n} W(x_j)\psi_{r-1}^2(x_j)}, \qquad r = 2, 3, \ldots. \tag{8.36b}$$

8.1.5. Fitting a Straight Line

Because the least-squares method has been formulated above for any linear functions and allows for the data to have correlated errors, the resulting formulas look a little forbidding, so it is instructive to derive explicit formulas for the simple case where the errors are uncorrelated, the situation often met in practice, and are fitted by a linear form containing just two parameters. It is worth re-emphasizing that 'linear' refers to the parameters and that the fitting functions do not have to be linear, so even the two-parameter case can be far from trivial and is widely used (see Example 8.1). To make things even simpler, we shall assume that the fitting function is the straight line $y = a + bx$. In this case, $p = 2$, with $\theta_1 = a$, $\theta_2 = b$, $\phi_1(x) = 1$, and $\phi_2(x) = x$, and the variances of the data values will be used to construct the weights, i.e., we will set the scale factor $w = 1$. It is then straightforward, if rather tedious, to show from the general equations that for data with uncorrelated errors,

$$\hat{a} = \frac{\overline{x^2}\,\overline{y} - \overline{x}\,\overline{xy}}{\overline{x^2} - \overline{x}^2} \quad \text{and} \quad \hat{b} = \frac{\overline{xy} - \overline{x}\,\overline{y}}{\overline{x^2} - \overline{x}^2}, \tag{8.37a}$$

where the overbars as usual denote averages, but in this case taking account of the errors on the measurements. For example,

$$\overline{y} \equiv \frac{\sum_{i=1}^{n} y_i/\sigma_i^2}{\sum_{i=1}^{n} 1/\sigma_i^2} \to \frac{1}{n}\sum_{i=1}^{n} y_i \quad \text{if the errors are all equal.} \tag{8.37b}$$

The denominator is the total weight and acts as a normalization factor. Thus, denoting the denominator in (8.37b) as N, (8.37a) for $\hat{b}$ written out in full is

$$\hat{b} = \frac{N\sum_{i=1}^{n} x_i y_i/\sigma_i^2 - \sum_{i=1}^{n} x_i/\sigma_i^2 \sum_{i=1}^{n} y_i/\sigma_i^2}{N\sum_{i=1}^{n} x_i^2/\sigma_i^2 - \left(\sum_{i=1}^{n} x_i/\sigma_i^2\right)^2}. \tag{8.38a}$$

8.1. UNCONSTRAINED LINEAR LEAST SQUARES

A similar expression can be derived for $\hat{a}$, but in practice, it is easier to calculate $\hat{a}$ from

$$\hat{a} = \bar{y} - \hat{b}\bar{x} \tag{8.38b}$$

once $\hat{b}$ has been found. The final result $y = \hat{a} + \hat{b}x$ can be used to interpolate to points where there are no measured data. In principle it can also be used to extrapolate to points outside the region where measurements exist, but care should be taken if this done, because no data have been used in these regions to constrain the parameters, and the results can rapidly become unreliable as one moves away from the fitted region.

To find the variances and covariance for the fitted parameters for the simple case of a straight-line fit we could again return to the general result (8.26). However it is simpler to use the results for $\hat{a}$ and $\hat{b}$ given in (8.37a). For example, $\hat{b}$ may be written as

$$\hat{b} = \frac{\overline{xy} - \bar{x}\bar{y}}{\overline{x^2} - \bar{x}^2} = \sum_{i=1}^{n} \frac{1}{n} \frac{(x_i - \bar{x})}{(\overline{x^2} - \bar{x}^2)} y_i. \tag{8.39}$$

Setting $\sigma_i = \sigma$ for simplicity, and using the results in Section 5.4.1 for combining errors, gives

$$\operatorname{var}(\hat{b}) = \sum_{i=1}^{n} \left[\frac{1}{n} \frac{(x_i - \bar{x})}{(\overline{x^2} - \bar{x}^2)}\right]^2 \sigma^2 = \frac{\sigma^2}{n(\overline{x^2} - \bar{x}^2)}. \tag{8.40}$$

Finally, if the errors on the data are independent but unequal, we make substitutions analogous to those in (8.37b), including setting

$$\sigma^2 \to \bar{\sigma}^2 = \frac{\sum_{i=1}^{n} \sigma_i^2/\sigma_i^2}{\sum_{i=1}^{n} 1/\sigma_i^2} = \frac{n}{\sum_{i=1}^{n} 1/\sigma_i^2}. \tag{8.41}$$

Then, writing out the result for the variance in full gives

$$\operatorname{var}(\hat{b}) = N\left[N\sum_{i=1}^{n} x_i^2/\sigma_i^2 - \left(\sum_{i=1}^{n} x_i/\sigma_i^2\right)^2\right]^{-1}. \tag{8.42a}$$

In a similar way we can show that

$$\operatorname{var}(\hat{a}) = \sum_{j=1}^{n} x_j^2/\sigma_j^2 \left[N\sum_{i=1}^{n} x_i^2/\sigma_i^2 - \left(\sum_{i=1}^{n} x_i/\sigma_i^2\right)^2\right]^{-1}, \tag{8.42b}$$

and

$$\operatorname{cov}(\hat{b}, \hat{a}) = -\sum_{j=1}^{n} x_j/\sigma_j^2 \left[N\sum_{i=1}^{n} x_i^2/\sigma_i^2 - \left(\sum_{i=1}^{n} x_i/\sigma_i^2\right)^2\right]^{-1}, \tag{8.42c}$$

with a common factor appearing on the right-hand side of all three expressions.

To find the error on the fitted value of f we can use (8.5), leading to

$$(\Delta f)^2 \equiv \operatorname{var} f(x) = \sum_{k=1}^{p} \sum_{l=1}^{p} \phi_k(x) E_{kl} \phi_l(x), \tag{8.43}$$

which could also have been obtained from (5.44). For the straight-line fit $y = a + bx$, this reduces to

$$\text{var}(y) = \text{var}(\hat{a}) + x^2 \text{var}(\hat{b}) + 2x \text{cov}(\hat{b}, \hat{a}). \tag{8.44}$$

It is essential that the covariance term is included in (8.44). Without it, the value of var (y) could be seriously in error.

EXAMPLE 8.3

The table below shows the values of data y_i ($i = 1, 2, \ldots, 7$) with uncorrelated errors σ_i taken at the points x_i. Use the specific formulas for a straight-line fit $y = a + bx$ to find estimators for the parameters a and b and their error matrix. Calculate the predictions for $\hat{y}_i$ and plot the data and the best-fit line. What is the predicted error at the point $x = 1.5$?

i	1	2	3	4	5	6	7
x_i	−3	−2	−1	0	1	2	3
y_i	0	1	2	6	6	10	12
σ_i	1	1	1	1	2	2	2

Using the notations above,

$$N = \sum_{i=1}^{7} 1/\sigma_i^2 = 4.75, \quad \sum_{i=1}^{7} x_i/\sigma_i^2 = -4.5, \quad \sum_{i=1}^{7} y_i/\sigma_i^2 = 16.0,$$

$$\sum_{i=1}^{7} x_i^2/\sigma_i^2 = 17.5, \quad \left(\sum_{i=1}^{7} x_i/\sigma_i^2\right)^2 = 20.25, \quad \sum_{i=1}^{7} x_i y_i/\sigma_i^2 = 11.5.$$

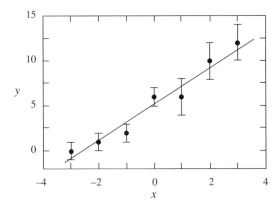

FIGURE 8.2 Least-squares fit to the data using $y = 5.276 + 2.014x$.

Substituting these numbers into (8.38a) gives $\hat{b} = 2.014$. Then from (8.38b) $\hat{a}$ is given by $\hat{a} = \bar{y} - \hat{b}\bar{x} = 5.276$. To find the error matrix we substitute into equations (8.42) to find the variances and the covariance. This gives

$\text{var}(\hat{b}) = 0.2783$, $\text{var}(\hat{a}) = 0.0755$ and $\text{covar}(\hat{b}, \hat{a}) = 0.0716$,

and hence the error matrix is

$$\mathbf{E} = \begin{pmatrix} 0.2783 & 0.0716 \\ 0.0716 & 0.0755 \end{pmatrix}.$$

From the values $\hat{a}$ and $\hat{b}$ we can calculate the values of $\hat{y}_i = \hat{a} + \hat{b}x_i$ as:

i	1	2	3	4	5	6	7
$\hat{y}_i$	−0.77	1.25	3.26	5.28	7.29	9.30	11.32

A plot of the data and the fitted function is shown in Fig. 8.2. To calculate the predicted error at the point $x = 1.5$, we use (8.44). This gives the variance as 0.9165 and hence the error on the fitted point is 0.96.

Because in the foregoing discussion we have not in general assumed a specific distribution, the basic formulation given above (but not necessarily that involving the χ^2 values of the fit) can be generalized to the case where the observations have both random and systematic errors. To illustrate this in principle, we will consider the simple example of a straight-line fit to data that have independent random errors σ_i and a systematic error ω that is common to all data points. We have shown in Section 5.4.1, equation (5.51), that in this case the variance matrix of the observations y_i has the form

$$V_{ij} = \begin{cases} \sigma_i^2 + \omega^2 & i = j \\ \omega^2 & i \neq j \end{cases}. \tag{8.45}$$

We now repeat the steps that led to (8.42a). Thus, setting $\sigma_i = \sigma$, (8.40) becomes

$$\text{var}(\hat{b}) = \frac{1}{n^2(\overline{x^2} - \overline{x}^2)^2} \sum_{i=1}^{n} \sum_{j=1}^{n} (x_i - \overline{x})(x_j - \overline{x}) \, \text{cov}(y_i, y_j),$$

which using (8.45) is

$$\text{var}(\hat{b}) = \frac{1}{n^2(\overline{x^2} - \overline{x}^2)^2} \left[\sum_{i=1}^{n} (x_i - \overline{x})^2 \sigma^2 + \sum_{i=1}^{n} \sum_{j=1}^{n} (x_i - \overline{x})(x_j - \overline{x}) \omega^2 \right]. \tag{8.46}$$

From the definition of the mean, the second term in (8.46) is zero. So after relaxing the condition $\sigma_i = \sigma$, (8.46) reduces to (8.40), or written in full to (8.42a), and hence $\text{var}(\hat{b})$ is unchanged. This is in accord with common sense, because if the systematic error is the same for all data points, then they will all move in parallel and the slope of the fitted straight line will not change. In a similar way, we can show that

$$\text{var}(\hat{a}) = \frac{1}{n^2(\overline{x^2} - \overline{x}^2)^2} \sum_{i=1}^{n} \sum_{j=1}^{n} (\overline{x^2} - \overline{x} \, x_i)(\overline{x^2} - \overline{x} \, x_j) \, \text{cov}(y_i, y_j).$$

Again, using (8.45), we see that the term in σ^2 will reproduce the former result (8.42b), but the term in ω^2 in this case does not cancel. Thus, over all, the effect of the systematic error in this simple case is to modify the error on the slope of the best-fit line, with the random and systematic errors adding in quadrature to the variance of $\hat{a}$.

8.1.6. Combining Experiments

The least-squares results may be used in a simple way to combine the results of several experiments measuring the same quantities. This was considered in Example 7.2 for the simple case of repeated measurements $y_i (i = 1, 2, \ldots, n)$ of a single quantity y each having independent errors $\sigma_i (i = 1, 2, \ldots, n)$. The result was the so-called *weighted mean*,

$$\hat{y} = \frac{\sum_{i=1}^{n} y_i/\sigma_i^2}{\sum_{i=1}^{n} 1/\sigma_i^2}, \quad \text{with} \quad \text{var}(\hat{y}) = \frac{1}{\sum_{i=1}^{n} 1/\sigma_i^2}. \tag{8.47}$$

It also follows directly from the general solution (8.11) for the simple case where the fitted function is a constant and $\mathbf{V}$ is a diagonal matrix. Thus (8.47) are the least-squares estimators. Knowing this, we can easily generalize the result to the case where the measurements are not independent, which would occur, for example if they were based in part on the same data set. Then the expression for S of (8.9) becomes

$$S(\lambda) = \sum_{i,j=1}^{n} (y_i - \lambda)(\mathbf{V}^{-1})_{ij}(y_j - \lambda),$$

and we seek an estimator $\hat{\lambda}$ for the true value λ, given a set of measurements y_i of λ. As usual, this is found by setting the derivative of S with respect to λ equal to zero and gives

$$\hat{\lambda} = \sum_{i=1}^{n} w_i y_i, \tag{8.48a}$$

where the weights are now given by

$$w_i = \sum_{j=1}^{n} (\mathbf{V}^{-1})_{ij} \left[\sum_{k,l=1}^{n} (\mathbf{V}^{-1})_{kl} \right]^{-1}, \tag{8.48b}$$

with the variance of $\hat{\lambda}$ given by

$$\text{var}(\hat{\lambda}) = \sum_{i,j=1}^{n} w_i V_{ij} w_j. \tag{8.48c}$$

Formulas (8.48) reduce to (8.47) if the errors are uncorrelated.

EXAMPLE 8.4

Three measurements of a quantity λ yield the results 3, 3.5, and 4 with a variance matrix

$$\mathbf{V} = \begin{pmatrix} 2 & 0 & 1 \\ 0 & 3 & 1 \\ 1 & 1 & 4 \end{pmatrix}.$$

Find the least-squares estimate for λ and its variance.

From the variance matrix, we have

$$\mathbf{V}^{-1} = \frac{1}{19} \begin{pmatrix} 11 & 1 & -3 \\ 1 & 7 & -2 \\ -3 & -2 & 6 \end{pmatrix}$$

and so

$$\left[\sum_{k,l=1}^{3} (\mathbf{V}^{-1})_{kl} \right]^{-1} = \frac{19}{16}.$$

Thus from (8.48b) the weights are

$$w_1 = 9/16, \quad w_2 = 6/16, \quad w_3 = 1/16$$

and from (8.84a), $\hat{\lambda} = 52/16 = 3.25$. The variance is found from (8.48c) and is 19/16. Thus $\hat{\lambda} = 3.3 \pm 1.1$.

Combining data from different experiments has to be done with care if it is to be meaningful, because the various experiments may not be compatible, something we mentioned briefly in Section 7.2. Thus a test, such as that based on the Student's t distribution, or on χ^2, should be used to establish compatibility. For example, the results in Example 8.4 yield a value $\chi^2 = 0.2$. Even if the data are compatible, averaging highly correlated data is difficult because a small error in the covariance matrix can result in a large error in the estimated value $\hat{\lambda}$ and an incorrect estimate of its variance. Also, the relative weights of the observations may not be what they might seem at first sight. For example, when counting the decay particles from a long-lived radioactive atom, assumed to be a Poisson process, one might be tempted to assume that the errors were the square root of the number of counts. However this is only true for the expected number of counts, which is a constant for a given time interval. So in this case the weights of different counts are the same, although unknown.

The above discussion can be generalized to situations where we wish to combine data from experiments that measure combinations of quantities λ_1, λ_2, etc. An example is given in Problem 8.3.

8.2. LINEAR LEAST SQUARES WITH CONSTRAINTS

It sometimes happens in practice that one has *some* information that can be used to refine the fit. As an example, we will generalize the discussion of Section 8.1 by considering the situation where the additional information takes the form of a set of *linear constraint equations on the parameters* of the form

$$C_{lp}\theta_p = Z_l,$$

or, in matrix notation

$$\mathbf{C\Theta} = \mathbf{Z}, \tag{8.49}$$

where the rank of $\mathbf{C}$ is l. Thus we have to now minimize the sum of residuals S given by (8.9), subject to the constraint (8.49). This problem can be solved if we introduce an ($l \times 1$) vector of Lagrange multipliers $\mathbf{\Lambda}$. Then the variation function that we have to consider is

$$L = (\mathbf{R}^T \mathbf{V}^{-1} \mathbf{R}) - 2\mathbf{\Lambda}^T(\mathbf{C\Theta} - \mathbf{Z}),$$

and the minimum of S subject to (8.49) is found by setting the total differential $dL = 0$, which gives

$$dL = 0 = 2[-\mathbf{Y}^T \mathbf{V}^{-1}\mathbf{\Phi} + \hat{\mathbf{\Theta}}_c^T(\mathbf{\Phi}^T \mathbf{V}^{-1}\mathbf{\Phi}) - \mathbf{\Lambda}^T \mathbf{C}]\, d\mathbf{\Theta},$$

i.e.

$$\mathbf{\Lambda}^T \mathbf{C} = \hat{\mathbf{\Theta}}_c^T(\mathbf{\Phi}^T \mathbf{V}^{-1}\mathbf{\Phi}) - \mathbf{Y}^T \mathbf{V}^{-1}\mathbf{\Phi}, \tag{8.50}$$

where $\hat{\mathbf{\Theta}}_c$ is the vector of estimates under the constraints.

Earlier we have seen that

$$(\mathbf{Y}^T \mathbf{V}^{-1}\mathbf{\Phi}) = \hat{\mathbf{\Theta}}^T(\mathbf{\Phi}^T \mathbf{V}^{-1}\mathbf{\Phi}), \tag{8.51}$$

where $\hat{\mathbf{\Theta}}$ is the estimate without the constraints, and using this relation in equation (8.50) gives

$$\mathbf{\Lambda}^T \mathbf{C} = (\hat{\mathbf{\Theta}}_c - \hat{\mathbf{\Theta}})^T(\mathbf{\Phi}^T \mathbf{V}^{-1}\mathbf{\Phi}), \tag{8.52}$$

If, as before, we set

$$\mathbf{M} = w(\mathbf{\Phi}^T \mathbf{V}^{-1}\mathbf{\Phi}) = (\mathbf{\Phi}^T \mathbf{W}\mathbf{\Phi}), \tag{8.53}$$

then

$$w\mathbf{\Lambda}^T \mathbf{C}\mathbf{M}^{-1}\mathbf{C}^T = (\hat{\mathbf{\Theta}}_c - \hat{\mathbf{\Theta}})^T \mathbf{C}^T = \mathbf{Z}^T - \hat{\mathbf{\Theta}}^T \mathbf{C}^T,$$

from which we obtain the result for $\mathbf{\Lambda}^T$:

$$w\mathbf{\Lambda}^T = (\mathbf{Z}^T - \hat{\mathbf{\Theta}}^T \mathbf{C}^T)(\mathbf{C}\mathbf{M}^{-1}\mathbf{C}^T)^{-1}. \tag{8.54}$$

Substituting (8.54) into (8.52) and solving for $\hat{\mathbf{\Theta}}_c$ gives

$$\hat{\mathbf{\Theta}}_c^T = \hat{\mathbf{\Theta}}^T + (\mathbf{Z}^T - \hat{\mathbf{\Theta}}^T \mathbf{C}^T)(\mathbf{C}\mathbf{M}^{-1}\mathbf{C}^T)^{-1}\mathbf{C}\mathbf{M}^{-1}, \tag{8.55}$$

This is the solution for the least-squares estimate of $\mathbf{\Theta}$ under the constraints, and like the unconstrained problem it only depends on the relative variances of the observations, because any scale factor in $\mathbf{V}$, and hence in $\mathbf{M}$, cancels in (8.55).

8.2. LINEAR LEAST SQUARES WITH CONSTRAINTS

To find the variance matrix for the estimates $\hat{\boldsymbol{\Theta}}_c$ does require knowledge of the full variance matrix of the observations, so if we use a scale factor w as defined in (8.24), then from (8.55),

$$\text{var}(\hat{\boldsymbol{\Theta}}_c) = w \left[\mathbf{M}^{-1} - \mathbf{M}^{-1}\mathbf{C}^T(\mathbf{C}\mathbf{M}^{-1}\mathbf{C}^T)^{-1}\mathbf{C}\mathbf{M}^{-1} \right], \tag{8.56}$$

and we are again left with the problem of finding an estimate for w. This may be done in a similar way to the unconstrained problem. Thus we consider the expected value of the weighted sum of the residues under the constraints. This is

$$E[S] = E[(\mathbf{R}^T \mathbf{V}^{-1} \mathbf{R}) + (\hat{\boldsymbol{\Theta}}_c - \hat{\boldsymbol{\Theta}})^T (\boldsymbol{\Phi}^T \mathbf{V}^{-1} \boldsymbol{\Phi})(\hat{\boldsymbol{\Theta}}_c - \hat{\boldsymbol{\Theta}})], \tag{8.57}$$

where $\mathbf{R}$ is the matrix of residuals without constraints as defined in (8.9). Using the same technique previously used in Section 8.1.2, we can show that the second term has an expected value of l, the rank of the constraint matrix $\mathbf{C}$, and we have already shown that the expected value of the first term is $(n-p)$. So an unbiased estimate of w is

$$\hat{w} = \frac{(\mathbf{R}^T \mathbf{W} \mathbf{R}) + (\hat{\boldsymbol{\Theta}}_c - \hat{\boldsymbol{\Theta}})^T (\boldsymbol{\Phi}^T \mathbf{W} \boldsymbol{\Phi})(\hat{\boldsymbol{\Theta}}_c - \hat{\boldsymbol{\Theta}})}{n - p + l}. \tag{8.58}$$

The second term may be written in a form that is independent of $\hat{\boldsymbol{\Theta}}_c$ by using (8.55) for $(\hat{\boldsymbol{\Theta}}_c - \hat{\boldsymbol{\Theta}})$. This gives

$$\hat{w} = \frac{(\mathbf{R}^T \mathbf{W} \mathbf{R}) + (\mathbf{Z} - \mathbf{C}\hat{\boldsymbol{\Theta}})^T (\mathbf{C}\mathbf{M}^{-1}\mathbf{C}^T)^{-1} (\mathbf{Z} - \mathbf{C}\hat{\boldsymbol{\Theta}})}{n - p + l}. \tag{8.59}$$

Finally, the error matrix for the parameters $\hat{\boldsymbol{\Theta}}_c$ is given by (8.56) with $\hat{w}$ given by (8.59). An example using these results is given in Problem 8.4.

Analogous formulas to those above may be derived for situations where the constraints are directly on the measurements themselves. As before, we will only consider the simple case of a set of linear constraint equations of the form

$$\mathbf{B}\hat{\boldsymbol{\eta}} = \mathbf{Z}$$

analogous to (8.49). Then repeating the steps that led to (8.55) gives the constrained solution

$$\hat{\mathbf{Y}}_c^T = \hat{\mathbf{Y}}^T + (\mathbf{Z}^T - \hat{\mathbf{Y}}^T \mathbf{B}^T)(\mathbf{B}\mathbf{V}\mathbf{B}^T)^{-1}\mathbf{B}^T\mathbf{V}, \tag{8.60}$$

with an associated variance matrix

$$\text{var}(\hat{\mathbf{Y}}_c^T) = \text{var}(\hat{\mathbf{Y}}^T) - \mathbf{V}\mathbf{B}^T(\mathbf{B}\mathbf{V}\mathbf{B}^T)^{-1}\mathbf{B}\mathbf{V}, \tag{8.61}$$

where for simplicity we have set $w = 1$. The use of (8.60) and (8.61) is illustrated in the following example.

EXAMPLE 8.5

Independent measurements of the three angles $y_i (i = 1, 2, 3)$ of a triangle yield (in degrees) the values $89 \pm 1, 33 \pm 2$, and 64 ± 2. Find the least-squares estimate for the angles and their variance matrix, subject to the constraint that the sum of the angles is exactly 180 degrees.

The various matrices we will need are

$$\mathbf{V} = \begin{pmatrix} 1 & 0 & 0 \\ 0 & 4 & 0 \\ 0 & 0 & 4 \end{pmatrix},$$

$$\mathbf{B} = (1 \ \ 1 \ \ 1), \quad \mathbf{Y} = (89 \ \ 33 \ \ 64) \quad \text{and} \quad Z = 180.$$

Then

$$(\mathbf{BVB}^T)^{-1} = 1/9 \quad \text{and} \quad \mathbf{B}^T \mathbf{V} = (1 \ \ 4 \ \ 4)$$

and so from (8.60),

$$\hat{\mathbf{Y}}_c^T = (89 \ \ 33 \ \ 64) - \frac{2}{3}(1 \ \ 4 \ \ 4),$$

and hence

$$\hat{y}_1 = 88\tfrac{1}{3}, \quad \hat{y}_2 = 30\tfrac{1}{3} \quad \text{and} \quad \hat{y}_3 = 61\tfrac{1}{3}.$$

As expected, the 'excess' of 6 in the measured sum of the angles has been divided unequally, with least being subtracted from y_1 because it is more precisely determined than the other angles. The variance matrix follows from (8.61) and is

$$\text{var}(\hat{\mathbf{Y}}_c^T) = \frac{1}{9}\begin{pmatrix} 8 & -4 & -4 \\ -4 & 20 & 0 \\ -4 & 0 & 20 \end{pmatrix},$$

so that $\hat{y}_1 = 88.3 \pm 0.9$, $\hat{y}_2 = 30.3 \pm 1.5$, and $\hat{y}_3 = 61.3 \pm 1.5$. Imposing the constraint has improved the precision of the angles, as expected.

The above discussion may be extended in several ways, for example to situations where there are constraints on both the data and the parameters to be estimated from them, or where the constraints are nonlinear. The general formalism is considerably more complicated, and in the nonlinear case the solution can usually only be obtained by iteration.

8.3. NONLINEAR LEAST SQUARES

If the fitting functions $\mathbf{F}(\boldsymbol{\Theta})$ are not linear in the parameters, then the weighted sum of residuals to be minimized is

$$S = [\mathbf{Y} - \mathbf{F}(\boldsymbol{\Theta})]^T \mathbf{W}[\mathbf{Y} - \mathbf{F}(\boldsymbol{\Theta})], \tag{8.62}$$

and differentiating S with respect to $\boldsymbol{\Theta}$ and setting the result to zero leads to a set of nonlinear simultaneous equations and consequently present a difficult problem to be solved. In practice S is minimized directly by an iterative procedure, starting from some initial estimates for $\boldsymbol{\Theta}$, which may be suggested by the theoretical model or in extreme situations may be little more than educated guesses. We will illustrate how such a scheme might *in principle* be applied.

The method is based on trying to convert the nonlinear problem to a series of linear ones. Let the initial estimate of Θ be Θ_0, Then if Θ_0 is close enough to the 'true' value Θ, we may expand the quantity $[Y - F(\Theta)]$ in a Taylor series about Θ_0 and keep only the first term. The technique relies on the truncation of the series being valid. Thus,

$$\Delta_0 = Y - F(\Theta_0) \simeq \frac{\partial F(\Theta_0)}{\partial \Theta_0} \delta_0, \qquad (8.63)$$

where δ_0 is a vector of small increments of Θ. The problem of calculating δ_0 is now reduced to one of linear least squares, since both Δ_0 and the design matrix

$$\Phi_0 = \frac{\partial F(\Theta_0)}{\partial \Theta}$$

are obtainable. Given a solution for δ_0 from the normal equations, a new approximation

$$F(\Theta_1) = F(\Theta_0 + \delta_0)$$

may be calculated. This in turn will lead to a new design matrix

$$\Phi_1 = \frac{\partial F(\Theta_1)}{\partial \Theta}$$

and a new vector Δ_1 and hence, via the normal equations, to a new incremental vector δ_1. This linearization procedure may now be iterated until the changes in Θ from one iteration to the next one are very small. At the close of the iterations the variance matrix for the parameters is again taken to be the inverse of the matrix of the normal equations.

As we have emphasized, the above procedure is only to illustrate a possible method of finding the minimum of S. In practice several difficulties could occur, for example the initial estimates Θ_0 could be such as to invalidate the truncation of the Taylor series at its first term. In general such a method is not sure of converging to any value, let alone to values representing a true minimum of S.

The problem of minimizing S is an example of a more general class of problems that come under the heading of 'optimization of a function of several variables' and in Appendix B there is a brief review of the methods that have proved to be successful in practice.

8.4. OTHER METHODS

Estimation using maximum likelihood, as described in Chapter 7, is a very general technique and is widely used in practical work, as is the method of least-squares described above. But several other methods are also in common use, and may be more suitable for certain applications. Three of them are briefly described below.

8.4.1. Minimum Chi-Square

Consider the case in which all the values of a population fall into k mutually exclusive categories $c_i (i = 1, 2, \ldots, k)$ and let p_i denote the proportion of values falling into category c_i, where

$$\sum_{i=1}^{k} p_i = 1. \tag{8.64}$$

Furthermore, in a random sample of n observations, let o_i and $e_i = np_i$ denote the *observed* and *expected* frequencies in category c_i, where

$$\sum_{i=1}^{k} o_i = \sum_{i=1}^{k} e_i = n. \tag{8.65}$$

Now in Section 4.7 we considered the multinomial distribution with density function:

$$f(r_1, r_2, \ldots, r_{k-1}) = n! \sum_{i=1}^{k} p_i^{r_i} \left(\sum_{i=1}^{k} r_i! \right)^{-1}, \tag{8.66}$$

where r_i denotes the frequency of observations in the ith category in which the true proportion of observations is $p_i (i = 1, 2, \ldots, k)$. We recall that the multinomial density function gives exact probabilities for any set of observed frequencies

$$r_1 = o_1, \quad r_2 = o_2, \quad \ldots, \quad r_k = o_k. \tag{8.67}$$

Each r_i is distributed binomially and we have seen in Section 4.8 that the binomial distribution tends rapidly to a Poisson distribution with both mean and variance equal to np_i. The Poisson distribution in turn tends to a normal distribution as np_i increases. Conventionally the Poisson distribution is considered approximately normal if the mean $\mu \geq 9$. Thus if $np_i \geq 9$, r_i is approximately normally distributed with mean and variance np_i. By converting to standard measure, it follows that the statistic

$$u_i = \frac{r_i - np_i}{(np_i)^{1/2}} \tag{8.68}$$

is approximately normally distributed with mean zero and unit variance. Furthermore,

$$\chi^2 = \sum_{i=1}^{k} u_i^2 = \sum_{i=1}^{k} \frac{(r_i - np_i)^2}{np_i} = \sum_{i=1}^{k} \frac{(o_i - e_i)^2}{e_i} \tag{8.69}$$

is distributed as χ^2 with $(k-1)$ degrees of freedom. Equation (8.69) can be used to test whether data are consistent with a specific distribution. We will return to this use of chi-squared in Chapter 11 when we discuss hypothesis testing.

A more common situation that arises in practice is where the generating density function is not completely specified, but instead contains a number of unknown parameters. If the observed frequencies are used to provide estimates of the p_i, then the quantity analogous to χ^2 of (8.69) is

$$\chi'^2 = \sum_{i=1}^{k} \frac{(o_i - n\hat{p}_i)^2}{n\hat{p}_i}. \tag{8.70}$$

There now arise two questions: (1) what is the best way of estimating p_i and (2) what is the distribution of χ'^2? There are clearly many different methods available to estimate the p_i, but one which is widely used is to choose values which minimize χ'^2. This may in general be a difficult problem and is another example of the general class of optimization problems mentioned above, and which are briefly discussed in Appendix B. It can be shown that for a wide class of methods of estimating the p_i, including that of minimum chi-square, χ'^2 is asymptotically distributed as χ^2 with $(k-1-c)$ degrees of freedom where c is the number of independent parameters of the distribution used to estimate the p_i.

In general, if x_i is a sample of size n from a multinomial population with mean $\mu(\theta)$ and variance matrix $V(\theta)$, where θ is to be estimated, then the value $\hat{\theta}$ $(x_1, x_2, \ldots, x_n)$ which minimizes

$$\chi^2 = \frac{1}{n}[\bar{x} - \mu(\theta)]^T [V(\theta)]^{-1} [\bar{x} - \mu(\theta)],$$

i.e., the minimum χ^2 estimate of θ, is known to be consistent, asymptotically efficient, and asymptotically normal distributed if x is distributed like the binomial, Poisson, or normal distribution (and many others).

EXAMPLE 8.6

A method for generating uniformly distributed random integers in the range 0–9 has been devised and tested by generating 1000 digits with results shown below.

Digit	0	1	2	3	4	5	6	7	8	9
Frequency	106	89	85	110	123	93	82	110	91	111

Do these results support the idea that the method of generation is suitable?

If the digits were uniformly distributed, then the expected frequencies would all be 100. So, using (8.69), we find $\chi^2 = 16.86$ and this is for 9 degrees of freedom. From Table C.4, $P[\chi^2 \geq 16.9]$ for 9 degrees of freedom is 0.05. So although it cannot be ruled out, as this is a fairly low probability, it raises some doubt that the method really is producing uniformly distributed integers. (Such statements will be made more precise when hypothesis testing is discussed in Chapter 11.)

The minimum chi-squared method of estimation can be used in a range of other situations, including those where the parameters are subject to constraints. An example is given in Problem 8.5.

8.4.2. Method of Moments

In Section 3.2.3 we saw that two distributions with a common moment generating function were equal. This provides a method for estimating the parameters of a distribution by estimating the moments of the distribution.

Let $f(x; \theta_1, \theta_2, \ldots, \theta_p)$ be a univariate density function with p parameters $\theta_i (i = 1, 2, \ldots, p)$, and let the first p algebraic moments be

$$\mu'_j(\theta_1, \theta_2, \ldots, \theta_p) = \int_{-\infty}^{\infty} x^j f(x; \theta_1, \theta_2, \ldots, \theta_p) \, dx, \qquad j = 1, 2, \ldots, p. \qquad (8.71)$$

Let x_n be a random sample of size n drawn from the density f. The first p sample algebraic moments are given by

$$m'_j = \frac{1}{n}\sum_{i=1}^{n} x_i^j. \qquad (8.72)$$

The estimators $\hat{\theta}_i$ of the parameters θ_i are obtained from the solutions of the p equations

$$m'_j = \mu'_j, \quad j = 1, 2, \ldots, p. \qquad (8.73)$$

EXAMPLE 8.7

Use the method of moments to find the estimators for the mean and variance of a normal distribution. We have previously seen (equation (4.6)) that for a normal distribution,

$$\mu'_1 = \mu; \quad \mu'_2 = \sigma^2 + \mu^2.$$

The sample moments are

$$m'_1 = \frac{1}{n}\sum_{i=1}^{n} x_i; \quad m'_2 = \frac{1}{n}\sum_{i=1}^{n} x_i^2.$$

Applying (8.73) gives

$$\hat{\mu} = \frac{1}{n}\sum_{i=1}^{n} x_i = \bar{x},$$

and

$$\hat{\sigma}^2 + \hat{\mu}^2 = \frac{1}{n}\sum_{i=1}^{n} x_i^2,$$

i.e.

$$\hat{\sigma}^2 = \frac{1}{n}\left[\sum_{i=1}^{n} x_i^2 - n\bar{x}^2\right] = \frac{1}{n}\sum_{i=1}^{n}(x_i - \bar{x})^2.$$

Thus, the estimators obtained by the method of moments are, for this example, the same as those obtained by the maximum likelihood method.

In some applications where the population density function is not completely known it may be advantageous to use particular linear combinations of moments. Consider, for example, a density function $f(x; \theta_1, \theta_2, \ldots, \theta_p)$, which is unknown but may be expanded in the form

$$f(x; \theta_1, \theta_2, \ldots, \theta_p) = \sum_{j=1}^{p} \theta_j P_j(x), \qquad (8.74)$$

where $P_j(x)$ is a set of orthogonal polynomials normalized such that

$$\int P_i(x)P_j(x)dx = \begin{cases} \phi_j, & i = j \\ 0, & i \neq j. \end{cases} \quad (8.75)$$

The population moments deduced from (8.74) are

$$\mu'_i = \int \sum_{j=1}^{p} \theta_j P_j(x)\, x^i \, dx. \quad (8.76)$$

However, we may also consider the linear combination of moments given by

$$\Omega_i = \int \sum_{j=1}^{p} \theta_j P_j(x) P_i(x) \, dx, \quad (8.77)$$

which by (8.75) is

$$\Omega_i = \theta_i \phi_i. \quad (8.78)$$

The equivalent sample moments are

$$m_i = \frac{1}{n} \sum_{j=1}^{n} P_i(x_j), \quad (8.79)$$

and so, by equating the two, we have

$$\hat{\theta}_i = \frac{1}{n\phi_i} \sum_{j=1}^{n} P_i(x_j). \quad (8.80)$$

This method is useful, for example, for finding the angular distribution coefficients a_j in the expansion of a differential cross-section in particle scattering problems. In this case, the differential cross-section $d\sigma/d\cos\theta$ is

$$\frac{d\sigma}{d\cos\theta} = \sum_j a_j P_j(\cos\theta), \quad (8.81)$$

where P_j are Legendre polynomials and the coefficients are

$$\hat{a}_j = \left(\frac{2j+1}{2n}\right) \sum_{i=1}^{n} P_j(x_i).$$

The modifications necessary to the above simple account in order to apply it to binned data are similar to those that have been discussed for the maximum likelihood and least-squares methods, and so we will not discus these further. Under quite general conditions, it can be shown that estimators obtained by the method of moments are consistent, but not in general most efficient.

8.4.3. Bayes' Estimators

In Section 2.3.2 we discussed the Bayesian interpretation of probability. There are several advantages of the Bayesian viewpoint. Foremost of these is that it can incorporate prior information about the parameter to be estimated. However, we saw from Bayes' theorem that to maximize the posterior probability requires knowledge of prior probabilities, and in general these are not known completely. Nevertheless, cases do occur where partial information is available, and in these circumstances it would clearly be advantageous to include it in the estimation procedure if possible. The objection to the Bayesian approach is that one has to choose a prior pdf and as this is necessarily subjective, different choices can lead to different outcomes. The Bayesian answer to this objection is that it is a fact of life that different people will have different views about data and so it is entirely reasonable that different interpretations should exist. There is no definite answer to this question, but it can make it difficult to compare different inferences drawn from comparable data sets.

We will consider the case where the prior information about the parameter is such that the parameter itself can be *formally* regarded as a random variable with a prior density $f_{\text{prior}}(\theta)$, as in the maximum likelihood method. There has been much theoretical work done on the question of how to choose a prior density, but all suggestions have problems. Empirically, the form for $f_{\text{prior}}(\theta)$ could be obtained, for example, by plotting all previous estimates of θ. This will very often be found to be an approximately Gaussian form, and from the results estimates of the mean and variance of the associated normal distribution could be made. In these cases where both the usual variable and the parameter be regarded as random variables we will denote the corresponding pdf as $f_R(x; \theta)$.

In Bayesian estimation, the emphasis is not on satisfying the requirements of 'good' point estimators as discussed in Section 5.1.2, but rather on minimizing 'information loss', expressed through a so-called *loss function* $l(\hat{\theta}; \theta)$. Expressed loosely, the latter gives the loss of information incurred by using the estimate $\hat{\theta}$ instead of the true value θ. In practice it is difficult to know what form to assume for the loss function, but a simple, common sense, form that suggests itself is

$$l(\hat{\theta}; \theta) = (\hat{\theta} - \theta)^2. \tag{8.82}$$

(A loss function that is bounded by zero, as in (8.82), is an example of a more general function found in decision theory, called a *risk function*.) The other quantities we need follow directly from work of previous chapters. Thus

$$j(x_1, x_2, \ldots, x_n; \theta) = f(x_1, x_2, \ldots, x_n | \theta) f_{\text{prior}}(\theta), \tag{8.83}$$

is the joint density of $x_1, x_2, \ldots, x_n$ and θ and

$$m(x_1, x_2, \ldots, x_n) = \int_{-\infty}^{\infty} j(x_1, x_2, \ldots, x_n, \theta) d\theta \tag{8.84}$$

is the marginal distribution of the x's. From equation (3.23) it then follows that the conditional distribution of θ given $x_1, x_2, \ldots, x_n$ is

$$c(\theta | x_1, x_2, \ldots, x_n) = \frac{j(x_1, x_2, \ldots, x_n; \theta)}{m(x_1, x_2, \ldots, x_n)}$$
$$= \frac{f(x_1, x_2, \ldots, x_n | \theta) f_{\text{prior}}(\theta)}{m(x_1, x_2, \ldots, x_n)}. \tag{8.85}$$

This is the posterior density $f_{\text{post}}(\theta | x_1, x_2, \ldots, x_n)$. We can now define a Bayes' estimator.

8.4. OTHER METHODS

Let $x_1, x_2, \ldots, x_n$ be a random sample of size n drawn from a density $f_R(x; \theta)$; and let $f_{\text{prior}}(\theta)$ be the prior density of θ and $f(x_1, x_2, \ldots, x_n|\theta)$ be the conditional density of the set x_i given θ. Furthermore, let $f_{\text{post}}(\theta|x)$ be the posterior density of θ given the set x_i, and let $l(\hat{\theta}, \theta)$ be the loss function. Then the *Bayes' estimator* of θ is that function defined by

$$\hat{\theta} = d(x_1, x_2, \ldots, x_n)$$

which minimizes the quantity

$$\mathbf{B}(\hat{\theta}; x_1, x_2, \ldots, x_n) = \int_{-\infty}^{\infty} l(\hat{\theta}, \theta) f_{\text{post}}(\theta|x_1, x_2, \ldots, x_n) d\theta. \qquad (8.86)$$

The disadvantage in using (8.86) is the necessity of assuming a form for both $f_{\text{prior}}(\theta)$ and $l(\hat{\theta}; \theta)$. The following example illustrates the use of the method.

EXAMPLE 8.8

Let $x_1, x_2, \ldots, x_n$ be an independent random sample of size n drawn from a normal density $f_R(x; \theta, a)$ with unknown mean θ and unit variance $a^2 = 1$. If θ is assumed to be normally distributed with known mean μ and unit variance $b^2 = 1$, find the Bayes' estimator for θ, using a loss function of the form $l(\hat{\theta}, \theta) = (\hat{\theta} - \theta)^2$.

From the above, setting $a = 1$,

$$f_R(x; \theta, a) = (2\pi)^{-1/2} \exp\left[-\frac{1}{2}(x-\theta)^2\right],$$

and hence

$$f(x_1, x_2, \ldots, x_n|\theta) = \frac{1}{(2\pi)^{n/2}} \exp\left[-\frac{1}{2}\left(\sum_{i=1}^{n} x_i^2 - 2\theta \sum_{i=1}^{n} x_i + n\theta^2\right)\right].$$

Also, setting $b = 1$,

$$f_{\text{prior}}(\theta) = (2\pi)^{-1/2} \exp[-(\theta-\mu)^2/2],$$

so that from (8.83)

$$j(x_1, x_2, \ldots, x_n, \theta) = \frac{1}{(2\pi)^{(n+1)/2}} \exp\left[-\frac{1}{2}\left(\sum_{i=1}^{n} x_i^2 + \mu^2\right)\right] \exp\left[-\frac{1}{2}(n+1)\theta^2 + (n\bar{x}+\mu)\theta\right],$$

and from (8.84)

$$m(x_1, x_2, \ldots, x_n) = (2\pi)^{-(n+1)/2} \exp\left[-\frac{1}{2}\left(\sum x_i^2 + \mu^2\right)\right] \int_{-\infty}^{\infty} \exp\left[\theta(n\bar{x}+\mu) - \frac{1}{2}(n+1)\theta^2\right] d\theta$$

$$= \frac{1}{(n+1)^{1/2}(2\pi)^{n/2}} \exp\left[-\frac{1}{2}\left(\sum x_i^2 + \mu^2\right) + \frac{1}{2}\frac{(n\bar{x}+\mu)^2}{n+1}\right].$$

Then using these in (8.85) gives

$$f_{\text{post}}(\theta|x_1, x_2, \ldots, x_n) = \left(\frac{n+1}{2\pi}\right)^{1/2} \exp\left\{-\frac{(n+1)}{2}\left[\theta - \frac{n\bar{x}+\mu}{n+1}\right]^2\right\},$$

and, using $l(\hat{\theta},\theta) = (\hat{\theta}-\theta)^2$ and this expression for $f_{\text{post}}(\theta|x_1, x_2, \ldots, x_n)$ in (8.86), we find, after some algebra,

$$B(\hat{\theta}; x_1, x_2, \ldots, x_n) = \left(\frac{n+1}{2\pi}\right)^{1/2} \int_{-\infty}^{\infty} (\hat{\theta}-\theta)^2 \exp\left\{-\frac{(n+1)}{2}\left[\theta - \frac{n\bar{x}+\mu}{n+1}\right]^2\right\} d\theta$$

$$= \hat{\theta}^2 - \frac{2\hat{\theta}(\bar{x}n+\mu)}{n+1} + \frac{1}{n+1} + \left(\frac{\bar{x}n+\mu}{n+1}\right)^2.$$

Finally, to minimize B we set

$$\frac{\partial B}{\partial \hat{\theta}}(\hat{\theta}; x_1, x_2, \ldots, x_n) = 0,$$

giving

$$\hat{\theta} = \frac{\mu + n\bar{x}}{n+1},$$

which is the Bayes' estimator for θ. It can be seen that $\hat{\theta}$ is the weighted average of the sample mean $\bar{x}$ and the prior mean μ.

If we extend the case studied in Example 8.8 to the situation where the variances a and b are not zero, then a useful general result is as follows, which is given without proof, but may be obtained by repeating the step in Example 8.8. If $\bar{x}$ is the mean of a random sample of size n from a normal population with known variance a^2, and the prior distribution of the population mean is a normal distribution with mean μ and variance b^2, then the posterior distribution of the population mean is also a normal distribution and the Bayes' estimators for the mean and variance are

$$\mu_1 = \frac{a^2\mu + nb^2\bar{x}}{a^2 + nb^2} \tag{8.87a}$$

and

$$\sigma_1^2 = \frac{a^2 b^2}{a^2 + nb^2}. \tag{8.87b}$$

If the prior was uniform, the posterior density is also normal, although in this case with $\mu_1 = \bar{x}$ and $\sigma_1^2 = a^2/n$, which are the limits of (8.87) as $n \to \infty$. In fact for *large* samples, equations (8.87) hold for an independent random sample of size n drawn form *any* distribution with a finite variance. This is the Bayesian statement of the central limit theorem.

Under very general conditions it can be shown that Bayes' estimators, independent of the assumed prior distribution $f_{\text{prior}}(\theta)$, are efficient, consistent, and a function of sufficient estimators.

It is useful to consider the relation between Bayes' estimators and those obtained from the maximum likelihood method. Using Bayes' theorem, the posterior pdf of (8.85) may be written in terms of the likelihood (which is not a pdf) as

$$f_{\text{post}}(\theta|\mathbf{x}) = \frac{L(\mathbf{x}|\theta)f_{\text{prior}}(\theta)}{\int L(\mathbf{x}|\theta')f_{\text{prior}}(\theta')d\theta'}. \tag{8.88}$$

In the absence of any prior information, it is common to take $f_{prior}(\theta)$ to be a constant and in this case the posterior pdf is proportional to the likelihood and the two methods are very similar. However, a uniform prior has potential problems. First, if the parameter can take on any values, $f_{prior}(\theta)$ cannot be normalized, although in practice this is not usually a difficulty because in the denominator it appears multiplied by the likelihood function. But a second problem is that one could take the prior to be uniform in a function of θ rather than the parameter itself and this would lead to a different posterior pdf and hence a different estimate. Thus Bayes' estimators with a uniform prior do not have the useful invariance property that ML estimators have. In practice, the distinction between different methods of estimation lessens as the sample size increases (because of the central limit theorem) and in particular Bayes' estimators depend less on the assume prior density.

PROBLEMS 8

8.1 Figure 8.3 shows some data fitted with polynomials of order 1, 2, and 3. Assuming the data are normally distributed, the χ^2 values for the fits are 13.9, 12.0, and 5.1, respectively. Comment on these results.

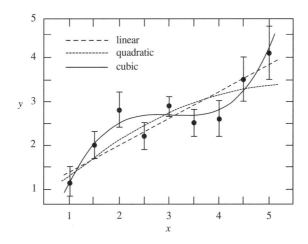

FIGURE 8.3 Data fitted with linear, quadratic, and cubic polynomials.

8.2 The table below shows the values of a quantity y, assumed to be normally distributed, and their associated errors σ, measured at six values of x.

i	1	2	3	4	5	6
x_i	1	2	3	4	5	6
y_i	2.5	3.0	6.0	9.0	10.5	10.5
σ_i	1	1	1	1	1	1

By successively fitting polynomials of increasing order, deduce the lowest order polynomial that gives an acceptable fit to the data and justify your answer. Find the coefficients of the polynomials corresponding to the best fit and their errors and plot the resulting best-fit curve.

8.3 An experiment determines two parameters λ_1 and λ_2 and finds values $y_1^{(1)} = 1.0$ and $y_2^{(1)} = -1.0$ with a variance matrix

$$\mathbf{V}^{(1)} = \begin{pmatrix} 2.0 & -1.0 \\ -1.0 & 1.5 \end{pmatrix} \times 10^{-2}.$$

A second experiment finds a new value of λ_2 to be $y_2^{(2)} = -1.1$ with a variance 10^{-2}. Find the least-squares estimates for λ_1 and λ_2 and their associated error matrix.

8.4 Rework Problem 8.3, but now with the constraint $\lambda_1 + \lambda_2 = 0$.

8.5 Measurements are made of the lengths $x_i (i = 1, 2, 3)$ of the sides of a right-angled triangle and the values $h_i (i = 1, 2, 3)$ found. If these are assumed to be normally distributed with equal variances σ^2, find the minimum chi-squared estimates for $x_i (i = 1, 2, 3)$.

8.6 Two determinations are made of the parameters of a straight line $y = ax + b$. The first is $a_1 = 4$, $b_1 = 12$ and the second is $a_2 = 3$, $b_2 = 14$. The associated variance matrices are

$$\mathbf{V}_1 = \begin{pmatrix} 1 & -1 \\ -1 & 2 \end{pmatrix} \quad \text{and} \quad \mathbf{V}_2 = \begin{pmatrix} 1 & -1 \\ -1 & 3 \end{pmatrix}.$$

Find the best estimate for a and b and the associated error matrix.

8.7 Let $r_1, r_2, \ldots, r_n$ be an independent random sample of size n drawn from a binomial density $f_R(r; p, n)$ with unknown parameter p. If p is assumed to be uniformly distributed in the interval $(0, 1)$, find the Bayes' estimator for p, using a loss function of the form $l(\hat{p}, p) = (\hat{p} - p)^2$. Compare your solution with that obtained by using the maximum likelihood method (Problem 7.7). Note the integral

$$\int_0^1 x^n (1-x)^m dx = \frac{n! m!}{(n+m+1)!}.$$

8.8 Use the method of moments to find an estimator for the parameter α in the two-parameter distribution:

$$f(x; \alpha, \beta) = \alpha \exp[-\alpha(x - \beta)], \quad \alpha, \beta > 0, \; x > 0$$

in terms of the first two sample moments.

CHAPTER 9

Interval Estimation

OUTLINE

9.1 Confidence Intervals: Basic Ideas 174
9.2 Confidence Intervals: General Method 177
9.3 Normal Distribution 179
   9.3.1 Confidence Intervals for the Mean 180
   9.3.2 Confidence Intervals for the Variance 182
   9.3.3 Confidence Regions for the Mean and Variance 183
9.4 Poisson Distribution 184
9.5 Large Samples 186
9.6 Confidence Intervals Near Boundaries 187
9.7 Bayesian Confidence Intervals 189

In Chapters 7 and 8, we discussed point estimation — the estimation of the value of a parameter. In practice, point estimation alone is not enough. It is also necessary to supply a statement about the error on the estimate. In those chapters, we did this by calculating the variance on the estimator and taking its square root, the standard deviation, as a 'standard error' to define error bars. In practice, because of the central limit theorem, most density functions lead to a normal form for the sampling density of the estimate in the case of large samples. In cases where this is not true, we could still use the standard deviation as a measure of uncertainty, but in these situations it is more usual to consider a generalization called *interval estimation*, based on the concept of a *confidence interval*, which is an interval constructed in such a way that a predetermined percentage of them will contain the true value of the parameter. This chapter will describe these ideas and their application, including the problematic case where an estimate leads to a value of a parameter that is close to its physical boundary.

9.1. CONFIDENCE INTERVALS: BASIC IDEAS

We have already encountered the idea of a confidence interval in Chapter 1, although it was not called that there. In Section 1.3.4, we noted that the distribution of the observations on a random variable x for large samples often, indeed usually, had a density $n(x)$ of approximately normal form about the sample mean $\bar{x}$ with variance σ^2. In that case, we could find values

$$C = \int_{x_L}^{x_U} n(x) dx$$

for any values x_L and x_U. The quantity C is called the *confidence coefficient* and is usually written $C = (1 - 2\alpha)$. (The reason for using the quantity $(1 - 2\alpha)$ will become clear later.) We also refer to $100C\% = 100(1 - 2\alpha)\%$ as the *confidence level*. The confidence coefficient corresponds to a random interval (x_L, x_U), called the *confidence interval* $x_L \leq x \leq x_U$, which depends only on the observed data. For example, from tables of the normal density, we know that $C = 0.683$, i.e., a confidence level of 68.3%, for a confidence interval $\mu - \sigma \leq x \leq \mu + \sigma$. In general, if the confidence coefficient is $C = (1 - 2\alpha)$, then $100(1 - 2\alpha)\%$ of the corresponding confidence intervals computed will include the true value of the parameter being estimated. Figure 9.1 shows an example of a confidence interval for a 90% confidence level of a normal distribution. Note that, in this case, the shaded areas both contain $\frac{1}{2}(100 - 90)\% = 5\%$ of the area of the distribution.

Confidence intervals are not uniquely defined by the value of the confidence level. In addition to the choice used in Fig. 9.1, where the probabilities above and below the interval are equal, called a *central interval*, we could, for example, have chosen a symmetric interval about the mean, so that $(x_U - \mu)$ and $(\mu - x_L)$ were equal, or values of x_L and x_U that minimize $(x_U - x_L)$, although, in practice, the construction of confidence intervals that are shortest for a given confidence coefficient is difficult, or may not even be possible. For symmetric distributions like the normal distribution, all three choices produce the same confidence intervals, but this is not true in general for asymmetric probability densities. The usual choice is the central interval.

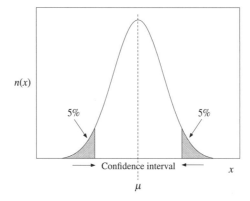

FIGURE 9.1 Central confidence interval corresponding to a 90% confidence level for a normal distribution.

9.1. CONFIDENCE INTERVALS: BASIC IDEAS

Suppose we are interested in estimating a single parameter θ from an experiment that consists of n observations of a random variable x drawn from a probability density $f(x; \theta)$. The sample $x_1, x_2, \ldots, x_n$ is used to construct an estimator $\hat{\theta}(x_1, x_2, \ldots, x_n)$ for θ, for example, by one of the methods discussed in earlier chapters. If $\hat{\theta}_e$ is the value of the estimator observed in the experiment, and $\hat{\sigma}_e$ is the estimate of its standard deviation, then the measurement would be given as $\theta = \hat{\theta}_e \pm \hat{\sigma}_e$. The interpretation of this is that if repeated estimates, all based on n observations of the random variable x, are made, they will be distributed according to the same sampling distribution $g(\hat{\theta}; \theta)$ centered around the true value θ and with a true standard deviation σ_θ that are estimated to be $\hat{\theta}_e$ and $\hat{\sigma}_e$. For most practical cases, $g(\hat{\theta}; \theta)$ will be approximately normal for large samples. Our aim is to find intervals about the estimator $\hat{\theta}$ such that we may make probabilistic statements concerning the probability of the true value θ being within the intervals.

One method that is applicable in many cases is the following. One finds, if possible, a function of the sample data and the parameter to be estimated, say u, which has a distribution independent of the parameter. Then a probability statement of the form

$$P[u_1 \leq u \leq u_2] = p$$

is constructed and converted into a probability statement about the parameter to be estimated. It is not always possible to find such a function, and in these cases more general methods (to be described in Section 9.2) must be used. For the present, we will illustrate this method by an example.

EXAMPLE 9.1

A sample of size 100 is drawn from a population with unit variance, but unknown mean μ. If $\hat{\mu}$ is estimated from the sample to be $\hat{\mu}_e = 1.0$, find a random interval for a confidence coefficient of 0.95.

The quantity

$$u = \left(\frac{\hat{\mu}_e - \mu}{\sigma/\sqrt{n}} \right) = 10(\hat{\mu}_e - \mu),$$

is, in general, normally distributed with mean zero and unit variance, and so has a density function

$$f(u) = \frac{1}{\sqrt{2\pi}} \exp\left(-\frac{u^2}{2}\right),$$

which is independent of μ. The probability that u lies between any two arbitrary values u_1 and u_2 is thus

$$P[u_1 \leq u \leq u_2] = \int_{u_1}^{u_2} f(t)dt.$$

Then, from Table C.1, we can find values $u_1 = -u_2 = -1.96$ such that

$$P[-1.96 \leq u \leq 1.96] = \int_{-1.96}^{1.96} f(t)dt = 0.95.$$

Transforming back to the variable μ, this becomes

$$P[\hat{\mu}_e - 0.196 \leq \mu \leq \hat{\mu}_e + 0.196] = 0.95,$$

and since $\hat{\mu}$ is estimated from the sample to be $\hat{\mu}_e = 1.0$, we have

$$P[0.804 \leq \mu \leq 1.196] = 0.95.$$

This is the required confidence interval. The interpretation of this is that if samples of size 100 were repeatedly drawn from the population, and if random intervals were computed as above for each sample, then 95% of those intervals would be expected to contain the true mean.

For obvious reasons, the intervals discussed above are called *two-tailed confidence intervals*. *One-tailed confidence intervals* are also commonly used. In these cases, the confidence coefficients are defined by

$$C_U = P[x < x_U] = \int_{-\infty}^{x_U} f(x) dx$$

if one is only interested in the upper limit of the variable, or

$$C_L = P[x > x_L] = \int_{x_L}^{\infty} f(x) dx$$

if one is only interested in its lower limit. It is worth emphasizing that a central interval corresponding to a confidence level C is not the same as a one-tailed limit corresponding to the same value of C. For example, for a normal distribution, the upper limit of a 90% two-tailed central confidence interval has 95% of the distribution below it and 5% above, whereas for a one-tailed confidence interval, a 90% upper limit has 90% of the distribution below it and 10% above.

EXAMPLE 9.2

Out of 1000 decays of an unstable particle, 9 are observed to be of type E. What can be said about the upper limit for the probability of a decay of this type?

The Poisson distribution is applicable here and we have $\mu = \sigma^2 = 9$. However, we also know that for $\mu \geq 9$, the Poisson distribution is well approximated by a normal distribution. Thus, the quantity

$$u = (x - \mu)/\sigma = (x - 9)/3,$$

is a standard normal variate. So, for example, from Table C.1,

$$P[u \leq 1.645] = 0.95,$$

and, hence, $x \leq 13.9$. Hence, the upper limit for the probability of this type of decay is $P \leq 0.014$ with 95% confidence.

The concept of interval estimation for a single parameter may be extended in a straightforward way to include simultaneous estimation of several parameters. Thus, a $100(1 - 2\alpha)\%$ *confidence region* is a region constructed from the sample such that, for repeatedly drawn samples, $100(1 - 2\alpha)\%$ of the regions would be expected to contain the set of parameters under estimation.

It should be remarked immediately that confidence intervals and regions are essentially arbitrary, because they depend on what function of the observations is chosen to be an estimator. This is easily illustrated by reference to the normal distribution of Example 9.1. If we use the sample mean as an estimator of the population mean, then for a confidence coefficient of 0.95,

$$P\left[\bar{x} - \frac{1.96\sigma}{\sqrt{n}} \leq \mu \leq \bar{x} + \frac{1.96\sigma}{\sqrt{n}}\right] = 0.95. \tag{9.1}$$

and the length of the interval is $2 \times 1.96\sigma/\sqrt{n}$. However, we could also use any given single observation to be an estimator, in which case, the confidence interval would be $\sqrt{n}$ times as long. An important property of ML estimators is that, for large samples, they provide confidence intervals and regions that, on average, are smaller than intervals and regions determined by any other method of estimation of the parameters.

9.2. CONFIDENCE INTERVALS: GENERAL METHOD

The method used in Section 9.1 requires the existence of functions of the sample and parameters that are distributed independently of the parameters. This is its disadvantage, for in many cases such functions do not exist. However, for these cases, there exists a more general method that we now describe.

Let $g(\hat{\theta}; \theta)$ be the sampling pdf of $\hat{\theta}$, the estimator for samples of size n drawn from a population density $f(x; \theta)$ containing a parameter θ. Figure 9.2 shows a plot of $g(\hat{\theta}; \theta)$ as a function of $\hat{\theta}$ for a given value of the true parameter θ. Also shown are two shaded regions that give the values of $\hat{\theta}$ for which

$$P[\hat{\theta} \geq h_\alpha(\theta)] = \int_{h_\alpha(\theta)}^{\infty} g(\hat{\theta}; \theta) d\hat{\theta} = 1 - G(h_\alpha; \theta) = \alpha \tag{9.2}$$

and

$$P[\hat{\theta} \leq h_\beta(\theta)] = \int_{-\infty}^{h_\beta(\theta)} g(\hat{\theta}; \theta) d\hat{\theta} = G(h_\beta; \theta) = \beta, \tag{9.3}$$

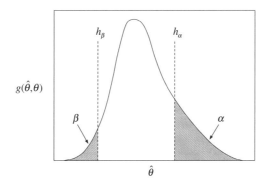

FIGURE 9.2 The density function of $g(\hat{\theta}; \theta)$ for a given value of the true parameter θ.

where G is the distribution function corresponding to the density $g(\hat{\theta}; \theta)$. Thus, for a fixed value of θ, a $100(1 - \alpha - \beta)\%$ confidence interval for $\hat{\theta}$ is

$$P[h_\alpha(\theta) \leq \hat{\theta} \leq h_\beta(\theta)] = \int_{h_\alpha(\theta)}^{h_\beta(\theta)} g(\hat{\theta}, \theta) d\hat{\theta} = 1 - \alpha - \beta. \tag{9.4}$$

Equations (9.2) and (9.3) determine the functions $h_\alpha(\theta)$ and $h_\beta(\theta)$. If the equations $\hat{\theta} = h_\alpha(\theta)$ and $\hat{\theta} = h_\beta(\theta)$ are plotted as a function of the true parameter θ, a diagram such as that shown in Fig. 9.3 would result. The region between the two curves is called the *confidence belt*. A vertical line through any value of θ, say $\bar{\theta}$, intersects $h_\alpha(\theta)$ and $h_\beta(\theta)$ at the values $\hat{\theta} = h_\alpha(\bar{\theta})$ and $\hat{\theta} = h_\beta(\bar{\theta})$, which determine the $100(1 - \alpha - \beta)\%$ confidence limits. Thus, (9.4) gives the probability for the estimator to be within the belt, *regardless of the value of θ*.

A horizontal line through some experimental value of $\hat{\theta} = \hat{\theta}_e$, corresponding to an estimate based on a sample of size n, cuts the curves at values $\theta_\alpha(\hat{\theta}_e)$ and $\theta_\beta(\hat{\theta}_e)$, where θ_α and θ_β are the values of the inverse functions $h_\alpha^{-1}(\hat{\theta})$ and $h_\beta^{-1}(\hat{\theta})$, respectively. Since the inequalities

$$\hat{\theta} \geq h_\alpha(\theta) \quad \text{and} \quad \hat{\theta} \leq h_\beta(\theta)$$

imply

$$\theta_\alpha \geq \theta \quad \text{and} \quad \theta_\beta \leq \theta,$$

respectively, (9.2) and (9.3) become

$$P[\theta_\alpha \geq \theta] = \alpha \quad \text{and} \quad P[\theta_\beta \leq \theta] = \beta$$

or, equivalently

$$P[\theta_\alpha(\hat{\theta}_e) \leq \theta \leq \theta_\beta(\hat{\theta}_e)] = 1 - \alpha - \beta.$$

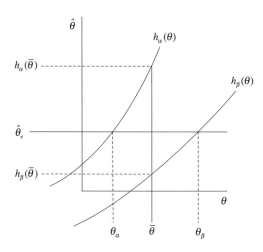

FIGURE 9.3 General method to construct a confidence interval.

Thus, to construct a confidence interval for θ, we first calculate an estimate $\hat{\theta}_e$ from a sample of size n. Then, we draw a horizontal line through $\hat{\theta}_e$ to cut the curves at values $\theta_\alpha(\hat{\theta}_e)$ and $\theta_\beta(\hat{\theta}_e)$, as shown in Fig. 9.3, so that, by construction, the required confidence limit is

$$P[\theta_\alpha(\hat{\theta}_e) \leq \theta \leq \theta_\beta(\hat{\theta}_e)] = 1 - \alpha - \beta. \tag{9.5}$$

A confidence interval is often expressed by asymmetric error bars in the same way as the use of a standard deviation. Thus, if we set $a = \theta_\alpha(\hat{\theta}_e)$ and $b = \theta_\beta(\hat{\theta}_e)$, then the result of the measurement would be written $\theta = \hat{\theta}_e {}^{+d}_{-c}$, where $c = \hat{\theta}_e - a$ and $d = b - \hat{\theta}_e$. If we are only interested in one-sided confidence intervals, then θ_α represents a lower limit on θ, i.e., $P[a \leq \theta] = 1 - \alpha$, and, similarly, θ_β represents an upper limit with $P[\theta \leq b] = 1 - \beta$.

To find the curves $h_\alpha(\theta)$ and $h_\beta(\theta)$ may be a lengthy procedure. However, in some cases, the values a and b may be obtained without knowing these curves. From (9.2) and (9.3), a and b are solutions of the equations

$$\alpha = \int_{\hat{\theta}_e}^{\infty} g(\hat{\theta}, a) d\hat{\theta} = 1 - G(\hat{\theta}_e; a), \tag{9.6a}$$

and

$$\beta = \int_{-\infty}^{\hat{\theta}_e} g(\hat{\theta}, b) d\hat{\theta} = G(\hat{\theta}_e; b), \tag{9.6b}$$

So, if these equations can be solved (possibly numerically), the confidence interval results directly.

The general method given above can be extended to the case of confidence regions for the p parameters of the population $f(x; \theta_1, \theta_2, \ldots, \theta_p)$, i.e., that region R in the parameter space, such that

$$P[\hat{\theta}_1, \hat{\theta}_2, \ldots, \hat{\theta}_p \text{ are contained in } R]$$
$$= \int_R \cdots \int g(\hat{\theta}_1, \hat{\theta}_2, \ldots, \hat{\theta}_n; \theta_1, \theta_2, \ldots, \theta_p) \prod_{i=1}^{p} d\hat{\theta}_i$$
$$= 1 - \alpha - \beta. \tag{9.7}$$

This can be done assuming that the sampling distribution of the estimators is a multivariate normal distribution with a given covariance matrix. This will not be pursued further here, except to say that the confidence region for two variables is approximately an ellipse and, for n variates, is an n-dimensional ellipsoid.

Finally, we note that the method cannot be used to obtain confidence regions for a subset r of the p parameters in the density $f(x; \theta_1, \theta_2, \ldots, \theta_p)$, except for the case of large samples. This is discussed in Section 9.5 below.

9.3. NORMAL DISTRIBUTION

Because the normal distribution is very widely used in physical sciences, we will obtain specific confidence intervals for its parameters.

9.3.1. Confidence Intervals for the Mean

From (9.1), it is clear that a confidence interval for the mean μ cannot be calculated unless the variance σ^2 is known, and so we will initially assume that this is the case. We will also assume, as usual, that the distribution of $\bar{x}$, the sample mean, is approximately normal with mean μ and standard deviation σ, i.e., its sampling distribution function is

$$G(\bar{x}; \mu, \sigma) = \frac{1}{\sqrt{2\pi}\sigma} \int_{-\infty}^{\bar{x}} \exp\left[-\frac{1}{2}\left(\frac{\bar{x}' - \mu}{\sigma}\right)^2\right] d\bar{x}'.$$

Then a confidence interval $[a, b]$ may be constructed if equations (9.6) can be solved. These are

$$\alpha = 1 - N(\bar{x}; a, \sigma)$$

and

$$\beta = N(\bar{x}; b, \sigma),$$

where N is the standardized form of the normal distribution function. The solutions for a and b are

$$a = \bar{x} - \sigma N^{-1}(1 - \alpha) \tag{9.8a}$$

and

$$b = \bar{x} + \sigma N^{-1}(1 - \beta), \tag{9.8b}$$

where N^{-1} is the inverse function of N, i.e., the quantile of the standardized normal distribution function, and we have taken $N^{-1}(\beta) = -N^{-1}(1 - \beta)$ for symmetry. The relationship between the inverse function and the confidence level is illustrated in Fig. 9.4.

If we consider a central confidence interval so that $\alpha = \beta = \gamma/2$, a common choice for the interval is to use values such that $N^{-1}(1 - \gamma/2) = 1, 2, \ldots$ Similarly, for a one-sided interval we could choose $N^{-1}(1 - \alpha) = 1, 2, \ldots$ Tables of the inverse function N^{-1} are published, but

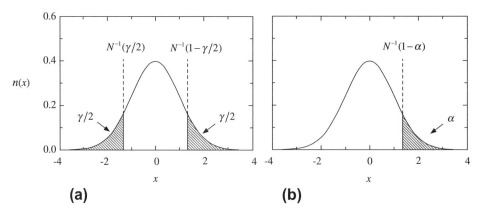

FIGURE 9.4 The standardized normal density $n(x)$ and the relationship between the inverse function N^{-1} and the confidence level for (a) a two-tailed central confidence level and (b) a one-tailed confidence level.

TABLE 9.1(a) Values of the confidence level for different values of the inverse of the standardized normal distribution N^{-1}: (A) for a central confidence interval with confidence level $(1 - \gamma)$ (see Fig. 9.4(a)); and (B) a one-tailed confidence interval with confidence level $(1 - \alpha)$ (see Fig. 9.4(b))

(A) Central two-tailed		(B) One-tailed	
$N^{-1}(1-\gamma/2)$	$1-\gamma$	$N^{-1}(1-\alpha)$	$1-\alpha$
1	0.6827	1	0.8413
2	0.9544	2	0.9772
3	0.9973	3	0.9987

in practice only a few values are commonly used. The resulting confidence levels for these values are shown in Table 9.1(a). The conventional 68.3% central confidence interval has $\alpha = \beta = \gamma/2$ with $N^{-1}(1 - \gamma/2) = 1$ and corresponds to 'one σ' errors bars. Alternatively, we could choose a convenient number for the confidence level itself and find the corresponding values of N^{-1}. Again, the commonly used values are shown in Table 9.1(b).

If σ^2 is not known, then, for large samples, we could use an estimate $\hat{\sigma}^2$ for this quantity without significant loss of precision, but, for small samples, this procedure is not satisfactory. The solution is to use the quantity

$$t = \frac{\bar{x} - \mu}{(s^2/n)^{1/2}} = (\bar{x} - \mu) \left[\frac{1}{n(n-1)} \sum_{i=1}^{n}(x_i - \bar{x})^2 \right]^{-1/2}, \quad (9.9)$$

which we have seen in Section 6.2 has a Student's t distribution with $(n-1)$ degrees of freedom, and only involves μ. Thus, we can find a number t_α such that

$$P[-t_\alpha \le t \le t_\alpha] = \int_{-t_\alpha}^{t_\alpha} f(t; n-1) dt = (1 - 2\alpha). \quad (9.10)$$

As in Example 9.1, we may now transform the inequality in (9.10) to give

$$P[\bar{x} - T_\alpha \le \mu \le \bar{x} + T_\alpha] = (1 - 2\alpha), \quad (9.11)$$

TABLE 9.1(b) Values of the inverse of the standardized normal distribution N^{-1} for different values of the confidence level: (A) for a central confidence interval and $N^{-1}(1 - \gamma/2)$ (see Fig. 9.4(a)); and (B) a one-tailed confidence interval with $N^{-1}(1 - \alpha)$ (see Fig. 9.4(b))

(A) Central two-tailed		(B) One-tailed	
$1-\gamma$	$N^{-1}(1-\gamma/2)$	$1-\alpha$	$N^{-1}(1-\alpha)$
0.90	1.645	0.90	1.282
0.95	1.960	0.95	1.645
0.99	2.576	0.99	2.326

where

$$T_\alpha = t_\alpha \left[\frac{1}{n(n-1)} \sum_{i=1}^{n} (x_i - \bar{x})^2 \right]^{1/2}.$$

The width of the interval is then $2T_\alpha$. The number t_α is called the $100\alpha\%$ *level of t*, and gives the point that cuts off $100\alpha\%$ of the area under the curve $f(t)$ on the upper tail.

EXAMPLE 9.3

Find: (a) a 95% central confidence interval for the mean of a normal distribution with unknown variance, given that the sample mean and sample variance are $\bar{x} = 5$ and $s^2 = 6$, respectively, using a sample size of 60; (b) an exact 95% central confidence interval using the Student's t distribution for the same statistics. Repeat the calculations for a sample size of 8 and comment on your results.

(a) For $n = 60$, we can use the normal approximation and the interval $[a, b]$ is then given by (9.8). Using Table 9.1(b), a 95% confidence interval is

$$\left[\left(\bar{x} - \frac{1.96 \times s}{\sqrt{n}} \right), \left(\bar{x} + \frac{1.96 \times s}{\sqrt{n}} \right) \right] = [4.38, 5.62]$$

and has a length 1.24.

(b) For an exact confidence level, we use the t distribution and (9.11). Then, using Table C.4, a 95% confidence interval is

$$\left[\left(\bar{x} - \frac{1.67 \times s}{\sqrt{n}} \right), \left(\bar{x} + \frac{1.67 \times s}{\sqrt{n}} \right) \right] = [4.47, 5.53]$$

which has a length 1.06, i.e., a reduction of 15%. Repeating the calculation using a sample size of 8, the confidence intervals are [3.30, 6.70], with a length 4.40, in the normal approximation, and [3.55, 6.45], with a length 2.90, using the exact form from the t distribution, a reduction of 34%. These differences are greater for $n = 8$ than for $n = 60$, because in the former, the small sample size means that the normal approximation is poor.

9.3.2. Confidence Intervals for the Variance

To find confidence intervals for the variance, we use the χ^2 distribution. We know that the quantity

$$\chi^2 = \frac{1}{\sigma^2} \sum_{i=1}^{n} (x_i - \bar{x})^2 \tag{9.12}$$

has a χ^2 distribution with $(n-1)$ degrees of freedom, and so we can use it to find numbers χ_1^2 and χ_2^2 such that

$$P[\chi_1^2 \leq \chi^2 \leq \chi_2^2] = \int_{\chi_1^2}^{\chi_2^2} f(\chi^2; n-1) d\chi^2 = 1 - 2\alpha,$$

or, equivalently,

$$P\left[\frac{1}{\chi_2^2}\sum_{i=1}^{n}(x_i-\bar{x})^2 \leq \sigma^2 \leq \frac{1}{\chi_1^2}\sum_{i=1}^{n}(x_i-\bar{x})^2\right] = 1-2\alpha. \quad (9.13)$$

Since the χ^2 distribution is not symmetric, the shortest confidence interval cannot be simply obtained for a given α. However, provided the number of degrees of freedom is not too small, a good approximation is to choose χ_1^2 and χ_2^2 such that $100\alpha\%$ of the area of $f(\chi^2)$ is cut off from each tail, i.e., such that

$$\int_{\chi_1^2}^{\infty} f(\chi^2; n-1)d\chi^2 = 1-\alpha,$$

and

$$\int_{\chi_2^2}^{\infty} f(\chi^2; n-1)d\chi^2 = \alpha.$$

Such numbers can easily be obtained from tables of the χ^2 distribution function.

EXAMPLE 9.4

The following random sample was drawn from a normal distribution with variance σ^2:

10 11 13 13 12 13 10 14 12 12

Find an approximate 99% central confidence interval for σ^2.

This is found by using (9.13). First, we find the sample mean $\bar{x} = 12$, and hence

$$\sum_{i=1}^{10}(x_i-\bar{x})^2 = 16.$$

For an approximate central confidence interval, we need to find values of χ_1^2 and χ_2^2 such that equal areas are cut off from the upper and lower tails of the chi-squared distribution function. So, for a 99% confidence level,

$$\int_{-\infty}^{\chi_1^2} f(\chi^2, n-1)d\chi^2 = 0.005 \quad \text{and} \quad \int_{\chi_2^2}^{\infty} f(\chi^2, n-1)d\chi^2 = 1 - \int_{-\infty}^{\chi_2^2} f(\chi^2, n-1)d\chi^2 = 0.005,$$

where $f(\chi^2, n-1)$ is the chi-squared distribution for $n-1$ degrees of freedom. Using Table C.4, for $n=10$, gives $\chi_1^2 = 1.73$ and $\chi_2^2 = 23.6$. Hence, from (9.13) the interval is [0.68, 9.25], with width 8.57.

9.3.3. Confidence Regions for the Mean and Variance

In constructing a confidence region for the mean and variance simultaneously, we cannot use the region bounded by the limits of the confidence intervals obtained separately for μ and

σ^2 (a rectangle in the (μ, σ^2) plane), because the quantities t of (9.9) and μ are not independently distributed, and hence the joint probability that the two intervals contain the true parameter values is not equal to the product of the separate probabilities. However, the distributions of $\bar{x}$ and $\sum(x_i - \bar{x})^2$ *are* independent and may be used to construct the required confidence region. Thus, for a $100(1 - 2\alpha)\%$ confidence region, we may find numbers $a_i (i = 1, 4)$ such that

$$P\left[-a_1 \leq \left(\frac{\bar{x} - \mu}{\sigma/\sqrt{n}}\right) \leq a_2\right] = (1 - 2\alpha)^{1/2}, \tag{9.14a}$$

and

$$P\left[-a_3 \leq \left(\frac{\sum(x_i - \bar{x})^2}{\sigma^2}\right) \leq a_4\right] = (1 - 2\alpha)^{1/2}. \tag{9.14b}$$

The joint probability is then $100(1 - 2\alpha)$ by virtue of the independence of the variables. The region defined by (9.14) will not, in general, be the smallest possible, but will not differ much from the minimum (which is roughly elliptical) unless the sample size is very small.

9.4. POISSON DISTRIBUTION

Another important distribution commonly met in physical science is the Poisson that we discussed in Section 4.8. Recall that the probability of observing k events is given by the Poisson density (4.47),

$$f(k;\lambda) = \frac{\lambda^k}{k!}\exp(-\lambda), \qquad \lambda > 0, \ k = 0, 1, 2, \ldots \tag{9.15}$$

and λ is the mean of the distribution, i.e., $\lambda = E[k]$. The aim is to a construct a confidence interval for a single measurement $\hat{\lambda}_e = k_e$. For values of $k_e \geq 9$, we can use the normal approximation to the Poisson, as we did in Example 9.2, but for smaller values, we must use the exact form of the distribution. The general technique given in Section 9.2 is not directly applicable because, for a discrete parameter, the functions h_α and h_β that define the confidence corridor do not exist for all values of the parameter. For example, in the present case, we would need to find values of h_α and h_β satisfying the conditions

$$P[\hat{\lambda} \geq h_\alpha(\lambda)] = \alpha \quad \text{and} \quad P[\hat{\lambda} \leq h_\beta(\lambda)] = \beta$$

for all values of λ. But, if α and β have fixed values, then because $\hat{\lambda}_e$ only takes on the discrete values k_e, these inequalities hold only for particular values of λ. However, we can still construct a confidence interval $[a, b]$ by using equations (9.6). For discrete variables, these become

$$\alpha = P[\hat{\lambda} \geq \hat{\lambda}_e; a] \quad \text{and} \quad \beta = P[\hat{\lambda} \leq \hat{\lambda}_e; b] \tag{9.16a}$$

and, for the case of a Poisson variable, using (9.15), they take the forms

$$\alpha = \sum_{k=k_e}^{\infty} f(k;a) = 1 - \sum_{k=0}^{k_e-1} f(k;a) = 1 - \sum_{k=0}^{k_e-1} \frac{a^k}{k!} e^{-a} \qquad (9.16b)$$

and

$$\beta = \sum_{k=0}^{k_e} f(n;b) = \sum_{k=0}^{k_e} \frac{b^k}{k!} e^{-b}. \qquad (9.16c)$$

For a given estimate $\hat{\lambda} = k_e$, these equations may be solved numerically by iteration to yield values for a and b. Some values of the upper and lower limits obtained for a range of values of k_e are given in Table 9.2. Note that a lower limit is not obtainable if $k_e = 0$.

The interpretation of equations (9.16a) is that if $\lambda = a$, the probability of observing a value greater than or equal to the one actually observed is α. Likewise, if $\lambda = b$, the probability of observing a value less than or equal to the one actually observed is β. The confidence intervals for the mean are

$$P[\lambda \geq a] \geq 1 - \alpha, \quad P[\lambda \leq b] \geq 1 - \beta$$

and

$$P[a \leq \lambda \leq b] \geq 1 - \alpha - \beta.$$

An important case is when $k_e = 0$, i.e., no events are observed. In this case, (9.16c) becomes $\beta = \exp(-b)$, or $b = -\ln \beta$.

TABLE 9.2 Lower and upper limits for a Poisson variable for various observed values k_e.

k_e	Lower limit a			Upper limit b		
	$\alpha = 0.1$	$\alpha = 0.05$	$\alpha = 0.01$	$\beta = 0.1$	$\beta = 0.05$	$\beta = 0.01$
0	—	—	—	2.30	3.00	4.61
1	0.11	0.05	0.01	3.89	4.74	6.64
2	0.53	0.36	0.15	5.32	6.30	8.41
3	1.10	0.82	0.44	6.68	7.75	10.04
4	1.74	1.37	0.82	7.99	9.15	11.60
5	2.43	1.97	1.28	9.27	10.51	13.11
6	3.15	2.61	1.79	10.53	11.84	14.57
7	3.89	3.29	2.33	11.77	13.15	16.00
8	4.66	3.98	2.91	12.99	14.43	17.40
9	5.43	4.70	3.51	14.21	15.71	18.78
10	6.22	5.43	4.13	15.41	16.96	20.14

EXAMPLE 9.5

How does the probability calculated in Example 9.2 change if no events were observed?

With less than about 9 events, the normal approximation used in Example 9.2 is not appropriate and we have to use the Poisson distribution. If we still work at a confidence level of 95%, so that $\beta = 0.05$, the upper limit obtained from (9.16c) is $b = -\ln(0.05) \simeq 3$, as shown in Table 9.2. Thus, if the number of occurrences of a rare event follows a Poisson distribution with mean λ and no such event is observed, the 95% upper limit for the mean is 3; that is, if the true mean were 3, then the probability of observing zero events is 5%. So, if no events were seen, the probability of the occurrence of a type E event is $P \leq 0.003$ with 95% confidence.

9.5. LARGE SAMPLES

In Chapter 7, we have seen that the large-sample distribution of the ML estimator $\hat{\theta}$ of a parameter θ in the density function $f(x; \theta)$ is approximately normal about θ as mean. In this situation, approximate confidence intervals may be simply constructed. The method is, by analogy with Example 9.1, to convert an inequality of the form

$$P\left[-u_\alpha \leq \frac{\hat{\theta} - \theta}{(\operatorname{var}\hat{\theta})^{1/2}} \leq u_\alpha\right] \simeq 1 - 2\alpha \qquad (9.17)$$

for the distribution of $\hat{\theta}$ expressed in standard measure, to an inequality for θ itself. Recall that α is defined by

$$\frac{1}{\sqrt{2\pi}} \int_{-u_\alpha}^{u_\alpha} \exp\left(-\frac{u^2}{2}\right) du = 1 - 2\alpha. \qquad (9.18)$$

This will be illustrated by applying the method to the binomial distribution.

EXAMPLE 9.6

Find an approximate 95% confidence interval for p, the parameter of the binomial distribution.

If we apply equation (7.14) to the binomial distribution of equation (4.29), we find (see Example 7.6)

$$\operatorname{var}(\hat{\theta}) \equiv \hat{\sigma}^2 = \frac{p(1-p)}{n}. \qquad (9.19)$$

An approximate $(1 - 2\alpha)$ confidence interval is then obtained from (9.17) by considering the statement

$$P\left[-u_\alpha \leq \frac{\hat{p} - p}{[p(1-p)/n]^{1/2}} \leq u_\alpha\right] \simeq 1 - 2\alpha, \qquad (9.20)$$

which, if we neglect terms of order $1/\sqrt{n}$, may be written as

$$P\left[\hat{p} - u_\alpha \left\{\frac{\hat{p}(1-\hat{p})}{n}\right\}^{1/2} \leq p \leq \hat{p} + u_\alpha \left\{\frac{\hat{p}(1-\hat{p})}{n}\right\}^{1/2}\right] \simeq 1 - 2\alpha. \quad (9.21)$$

So, a 95% confidence interval for p is defined by

$$P\left[\hat{p} - 1.96\left(\frac{\hat{p}(1-\hat{p})}{n}\right)^{1/2} \leq p \leq \hat{p} + 1.96\left(\frac{\hat{p}(1-\hat{p})}{n}\right)^{1/2}\right] \simeq 0.95.$$

The above method may be extended to confidence regions. In terms of the matrix M_{ij}, defined in equation (7.34), we know that

$$\chi^2 = \sum_{i=1}^{p}\sum_{j=1}^{p}(\hat{\theta}_i - \theta)_i M_{ij}(\hat{\theta}_j - \theta_j), \quad (9.22)$$

is approximately distributed as χ^2, with p degrees of freedom. So, just as we used (9.15) for the normal distribution, we can use the α percentage points of the χ^2 distribution to set up a confidence region for the parameters θ_i. It is an ellipsoid with the center at $(\theta_1, \theta_2, ..., \theta_p)$.

At the end of Section 9.2, we remarked that it was not possible, in general, to obtain a confidence region for a subset of the p parameters for samples of arbitrary size. However, for large samples, this *is* possible. If we wish to construct a region for a subset of r parameters ($r < p$), then the elements of the matrix M'_{ij} analogous to M_{ij} above are given by

$$M'_{ij} = (V')_{ij}^{-1}, \quad (9.23)$$

where the matrix V' is obtained by removing the last $(p - r)$ rows and columns in V_{ij}. The quadratic form

$$\chi'^2 = \sum_{i=1}^{r}\sum_{j=1}^{r}(\hat{\theta}_i - \theta_i)M'_{ij}(\hat{\theta}_j - \theta_j), \quad (9.24)$$

is then approximately distributed as χ^2 with r degrees of freedom, and will define an ellipsoid in the $\theta_i(1, 2, ..., r)$ space.

9.6. CONFIDENCE INTERVALS NEAR BOUNDARIES

In the discussion of point estimation in Chapters 7 and 8, it was implicitly assumed that an individual measurement could take on any value.[1] However, this assumption is not always true. An example often cited is that of the mass of a body, which cannot be negative. If the mass of a body is obtained by a direct measurement, for example by weighing it, then this

---

[1]The constraints discussed in the context of the least-squares method were on combinations of data or the parameters used in the fitting procedure.

condition is automatically satisfied. But a direct measurement may not always be possible. For example, in the case of a sub-atomic particle, it is usual to measure its energy E and momentum p. The mass m is then found from the general expression $m^2c^4 = E^2 - p^2c^2$, where c is the speed of light. But, in this situation, E and p are both random variables with associated uncertainties, so that even though both will be positive, the resulting experimental value of the squared mass, being the difference of two terms, can be, and sometimes is, found to be negative. In general, a measurement of any quantity θ, that is known to be positive for physical reasons, when found from the differences of random variables, can result in a negative value for a specific measurement of its estimator $\hat{\theta}$. In these circumstances, the construction and interpretation of confidence intervals must be treated with care if one is not to end up making misleading probability statements. Similar difficulties occur whenever a parameter has a physical boundary, but the method of estimation allows its estimator to take values in unphysical regions. The example below will illustrate this.

Consider the simple case where an estimator $\hat{\theta}$ of a parameter θ, known for physical reasons to be non-negative, is given in terms of two independent random variables, x and y, by $\hat{\theta} = x - y$. If x and y are both normally distributed with means μ_x and μ_y and variances σ_x^2 and σ_y^2, respectively, we know from the work of previous chapters that $\hat{\theta}$ is also normally distributed, with mean $\theta = \mu_x - \mu_y$ and variance $\sigma_\theta^2 = \sigma_x^2 + \sigma_y^2$. If, now, an experiment gives a value $\hat{\theta}_e$ for $\hat{\theta}$, then the upper limit for θ at a confidence level of $(1 - \beta)$ is obtained from (9.8b) and is

$$\theta_{up} = \hat{\theta}_e + \sigma_\theta N^{-1}(1 - \beta). \tag{9.25a}$$

For example, if $\hat{\theta}_e$ is measured to be -3.0 with $\sigma_\theta = 1.5$, where the latter is either known or estimated from the data, and we use a 95% confidence interval, with $N^{-1}(0.95) = 1.645$ (see Table 9.1b), then from (9.25a), $\theta_{up} = -0.532$. The interval, $[-\infty, -0.532]$, therefore will contain, by construction, the true value θ with a probability of 95%, regardless of the actual value of θ. Although this may look odd at first sight, there is nothing intrinsically wrong with this. If the true value of θ were zero, half of such estimates would be expected to be negative. But the upper limit is also in the unphysical region. Again, we would expect this for 5% of similar experiments if θ really were zero. There is nothing incorrect with the procedure; we have simply encountered an experiment that does not lie within the interval constructed by applying the frequency definition of probability. So, unless there are other compelling reasons for doing so, the data should certainly not be discarded as being 'wrong'. Nevertheless, since we know that θ cannot be negative, the measurement has not added to our prior knowledge in any significant way, so the question arises as to whether this single estimate can be better used.

Unfortunately, there is no unique answer to this question. The first possibility is to do nothing except to report the measurement. Other experiments will produce different values of $\hat{\theta}_e$ and by combining them (for example, by combining the likelihood functions for each experiment, as mentioned in Chapter 7, or by using the least-squares method as in Section 8.1.6), a more precise overall estimate $\hat{\theta}$ may be found. A second possibility is to increase the confidence limit until the upper limit enters the physical region. Using the example above, if the confidence limit is increased to 99%, with $N^{-1}(0.99) = 2.326$ from Table 9.1b, then from (9.25a), $\theta_{up} = 0.489$. Although this is comfortably greater than zero, it could be smaller than the precision of the experiment as measured by σ_θ, and in this example it is because we took $\sigma_\theta = 1.5$. An extreme example of the difficulty that this strategy can lead to is if a confidence level is deliberately chosen so that θ_{up} is only just positive. Thus, if we choose a confidence level of 97.725%, we would quote the value $\theta_{up} = 1.5 \times 10^{-5}$ at this confidence level, which is clearly absurd. A third possibility is to move a negative value of $\hat\theta_e$ to zero, before using (9.25a) to calculate θ_{up}, so that

$$\theta_{up} = \max(\hat\theta_e, 0) + \sigma_\theta N^{-1}(1-\beta). \tag{9.25b}$$

For the example above, this gives $\theta_{up} = 2.468$, which is both in the physical region and compatible with the precision of the experiment. The drawback with this method is that we can no longer interpret the computed interval as a range that will include the true value with a probability $(1-\beta)$. The actual probability will always be greater, because the value of θ_{up} from (9.25b) will always be greater than the value calculated from (9.25a).

9.7. BAYESIAN CONFIDENCE INTERVALS

The strategies outlined in Section 9.6 to handle the problem of confidence intervals when the estimated value of a parameter is close to a boundary all use the frequency interpretation of probability. A final possibility is to incorporate our prior knowledge, including the fact that the parameter cannot have a negative value, by using subjective probability, leading to so-called Bayesian intervals. Although such intervals may look similar to the confidence intervals discussed previously, because they are based on posterior probability densities their interpretation is very different and, for this reason, they are usually called *probability intervals* or *credible intervals*, to distinguish them from confidence intervals constructed using the frequency interpretation of probability. In the latter, x is a random variable and gives rise to a random interval that has a specific probability of containing the fixed, but unknown, value of the parameter θ. In the Bayesian approach, the parameter is random in the sense that we have a prior belief about its value and the interval can be thought of as fixed, once this information is available. Only if the prior distribution of the unknown parameter is chosen to be a uniform distribution are the two intervals equivalent.

The starting point for constructing credible intervals is Bayes' theorem, which we first introduced for discrete variables in equation (2.11) and generalized to the case of continuous variables in equation (3.22c). In the present context, it is more convenient to write Bayes' theorem in terms of the likelihood function, as was done in Section 8.4.3. Thus, for a single variable θ, if $L(\mathbf{x};\theta)$ is the likelihood function of the set of n variables $\mathbf{x}(x_1, x_2, \ldots, x_n)$ for a given value of θ, i.e.,

$$L(\mathbf{x}|\theta) = \prod_{i=1}^{n} f(x_i; \theta),$$

where $f(x;\theta)$ is the density function of the variables $\mathbf{x}$, then rewriting (8.88), the posterior probability density $f_{post}(\theta|\mathbf{x})$ is given by

$$f_{post}(\theta|\mathbf{x}) = \frac{L(\mathbf{x}|\theta)f_{prior}(\theta)}{\int L(\mathbf{x}|\theta')f_{prior}(\theta')d\theta'}, \tag{9.26}$$

where $f_{\text{prior}}(\theta)$ is the prior probability density for θ. The density $f_{\text{post}}(\theta|\mathbf{x})$ replaces the distributions (usually the normal) assumed in previous sections and can be used to construct an interval $[a, b]$, such that for any given probabilities α and β,

$$\alpha = \int_{-\infty}^{a} f_{\text{post}}(\theta|\mathbf{x}) d\theta \quad \text{and} \quad \beta = \int_{b}^{\infty} f_{\text{post}}(\theta|\mathbf{x}) d\theta.$$

Thus, $\alpha = \beta$ gives a central interval with a predetermined probability $(1 - \alpha - \beta)$; alternatively, one could choose $f_{\text{post}}(a|\mathbf{x}) = f_{\text{post}}(b|\mathbf{x})$, which leads to the shortest interval and is what is usually used in practice. The advantage of the subjective approach over the frequency approach is that, in principle, prior knowledge can be incorporated via the density $f_{\text{prior}}(\theta)$. The credible interval then contains a fraction of one's total belief about the parameter, in the sense that one would be prepared to bet, with well-defined odds that depend on α and β, that the true value of θ lies in the interval. The qualifier 'in principle' is necessary because, as in all applications of subjective probability, the problem arises in choosing a form for $f_{\text{prior}}(\theta)$. An example of the construction of a credible interval is given in Problem 9.7.

For the case discussed in Section 9.6, where $\theta > 0$, we can certainly set $f_{\text{prior}}(\theta) = 0$ for $\theta \leq 0$. Then, using (9.26), the upper limit is given by

$$1 - \beta = \int_{-\infty}^{\theta_{\text{up}}} f_{\text{post}}(\theta|\mathbf{x}) d\theta$$

$$= \int_{-\infty}^{\theta_{\text{up}}} L(\mathbf{x}|\theta) f_{\text{prior}}(\theta) d\theta \left[\int_{-\infty}^{\infty} L(\mathbf{x}|\theta) f_{\text{prior}}(\theta) d\theta \right]^{-1}.$$

But, for $\theta > 0$, it is not so clear what to do. If we invoke Bayes' postulate, we would choose

$$f_{\text{prior}}(\theta) = \begin{cases} 0 & \theta \leq 0 \\ 1 & \theta > 0 \end{cases} \tag{9.27}$$

While this has the advantage of simplicity, it also has a serious problem. Continuing with our example of identifying θ with a mass, usually one would have some knowledge of at least its order of magnitude from physical principles. If the body were an atom, for example, we would expect to find values of the order of 10^{-22} grams, and it would be unrealistic to assume that the probabilities of obtaining *any* positive value were all equal. Other forms have been suggested for $f_{\text{prior}}(\theta)$, but none are without their difficulties. Moreover, Bayes' postulate applied to different *functions* of θ result in different credible intervals, that is, the method is not invariant with respect to a nonlinear transformation of the parameter. Despite these limitations, in practice, the simple form (9.27) is often used.

PROBLEMS 9

9.1 Potential voters in an election are asked whether or not they will vote for candidate X. If 700 out of a sample of 2000 indicate that they would, find a 95% confidence interval for the fraction of voters who intend to vote for candidate X.

PROBLEMS 9

9.2 A signal of strength S is sent between two points. En route, it is subject to random noise N, known to be normally distributed with mean zero and variance 5, so that the received signal R has strength $(S + N)$. If the signal is sent 10 times with the results:

$$R_i \quad 6 \quad 7 \quad 11 \quad 15 \quad 12 \quad 8 \quad 9 \quad 14 \quad 5 \quad 13$$

construct (a) a 95% central confidence interval and (b) 95% upper and lower bounds, for S.

9.3 Rework Problem 9.2(a) for the case where σ is unknown.

9.4 In the study of very many decays of an elementary particle, 8 events of a rare decay mode are observed. If the expected value of events due to other random processes is 2, what is the 90% confidence limit for the actual number of decays?

9.5 Electrical components are manufactured consecutively and their effective lifetimes τ are found to be given by the exponential distribution of (4.26) with a common parameter λ. Use the fact that the sample mean lifetime $\bar{\tau}$ of n components is related to the χ^2 variable with $2n$ degrees of freedom by $2\lambda n \bar{\tau} \simeq \chi^2_{2n}$, to construct a 90% confidence interval for the population lifetime, given that the sample mean for the first 15 components is 200 hrs.

9.6 Extend the work of Section 9.3.1 to the case of two normal populations $N(\mu_1, \sigma_1^2)$ and $N(\mu_2, \sigma_2^2)$, to derive a confidence interval for the difference $(\mu_1 - \mu_2)$. If a national physics examination is taken by a sample of 40 women and 80 men, and produces average marks of 60 and 70, with standard deviations of 8 and 10, respectively, use your result to construct a 95% confidence interval for the difference in the marks for all men and women eligible to take the examination.

9.7 Electrical components are manufactured with lifetimes that are approximately normally distributed with a standard deviation of 15. Prior experience suggests that the lifetime is a normal random variable with a mean of 800 hrs and a standard deviation of 10 hrs. If a random sample of 25 components has an average lifetime of 780 hrs, use the results given in Section 9.6 to find a 95% credible interval for the mean and compare it with a 95% confidence interval.

9.8 All school leavers applying for a place at university take a standard national English language test. At a particular school, the test was taken by 16 boys and 13 girls. The boys' average mark was 75 with a standard deviation of 8, and the corresponding numbers for the girls were 80 and 6. Assuming a normal distribution, find a 95% confidence interval for the ratio σ_b^2/σ_g^2, where $\sigma_{b,g}^2$ are, respectively, the variances of all boys and girls who take the test nationally.

CHAPTER 10

Hypothesis Testing I: Parameters

OUTLINE

10.1 Statistical Hypotheses — 194	10.3 Normal Distribution — 204
10.2 General Hypotheses: Likelihood Ratios — 198	10.3.1 Basic Ideas — 204
10.2.1 Simple Hypothesis: One Simple Alternative — 198	10.3.2 Specific Tests — 206
10.2.2 Composite Hypotheses — 201	10.4 Other Distributions — 214
	10.5 Analysis of Variance — 215

    In earlier chapters, we have discussed one of the two main branches of statistical inference as applied to physical science: estimation. We now turn to the other main branch: hypothesis testing. This is a large topic and so, for convenience, it has been split into two parts. In this chapter we consider situations where the parent distribution is known, usually a normal distribution, either exact or approximate, and the aim is to test hypotheses about its parameters, for example, whether they do, or do not, have certain values, rather than to estimate the values of the parameters. This topic was touched upon in previous chapters, particularly Chapter 6, where we discussed the use of the χ^2, t, and F distributions, and much of the preliminary work has been done in Chapter 9, when confidence intervals were constructed. The aim of the present chapter is to bring together and extend those ideas to discuss hypothesis testing on parameters in a systematic way. The Bayesian approach to hypothesis testing will not be discussed. It is very similar, from a calculational viewpoint, to the frequency approach, but uses the appropriate posterior probability distributions, as defined in Chapter 9, and hence, by analogy with Bayesian confidence intervals, the interpretation is different.

    In the following chapter, we consider other types of hypotheses, such as whether a sample of data is compatible with an assumed distribution, or whether a sample is really random. We also discuss hypotheses that can be applied to situations that are occasionally met in physical science where the data are non-numeric.

10.1. STATISTICAL HYPOTHESES

Consider a set of random variables $x_1, x_2, \ldots, x_n$ defining a sample space S of n dimensions. If we denote a general point in the sample space by E, then if R is a region in S, any hypothesis concerning the probability that E falls in R, i.e., $P[E \in R]$, is called a *statistical hypothesis*. Furthermore, if the hypothesis determines $P[E \in R]$ completely, then it is called *simple*; otherwise, it is called *composite*. For example, when testing the significance of the mean of a sample, it is a statistical hypothesis that the parent population is normal. Furthermore, if the parent population is postulated to have mean μ and variance σ^2, then the hypothesis is simple, because the density function is then completely determined.

The hypothesis under test is called the *null* hypothesis and denoted H_0. The general procedure for testing H_0 is as follows: Assuming the hypothesis to be true, we can find a region S_c in the sample space S such that the probability of E falling in S_c is any pre-assigned value α, called the *significance level*. The region $S_0 = (S - S_c)$ is called the *region of acceptance*, and S_c is called the *critical region*, or *region of rejection*. If the observed event E falls in S_c, we reject H_0; otherwise, we accept it. It is worth being clear about the meaning of the phrase 'accept the hypothesis H_0'. It does *not* mean that H_0 is definitely true; rather, it means that the data are *not inconsistent* with the hypothesis, in the sense that the observed value would be expected at least $\alpha\%$ of the time if the null hypothesis were true. In practice, as we shall discuss below, the critical region is determined by a statistic, the nature of which depends upon the hypothesis to be tested.

Just as there are many confidence intervals for a given confidence level, so there are many possible acceptance regions for a given hypothesis at a given significance level α. For all of them, the hypothesis will be rejected, although true, in some cases. Such 'false negatives' are called *type I errors* and their probability, denoted by $P[I]$, is equal to the significance level of the test. The value of α is arbitrary, and the choice of a suitable value depends on how important the consequence of rejecting H_0 is. Thus, if its rejection could have serious consequences, such as a substantial loss, either of money, time etc., or even lives in extreme cases, then one would tend to be conservative and choose a value $\alpha = 0.05, 0.01$, or even smaller. Values commonly used in physical science are 0.1 and 0.05. It is also possible that even though the hypothesis is false, we fail to reject it. These 'false positives' are called *type II errors*. We are led to the following definition of *error probabilities*.

Consider a parameter θ and two hypotheses, the null hypothesis $H_0 : \theta \in R_0$ and the alternative $H_a : \theta \in R_a$, where R_o and R_a are two mutually exclusive and exhaustive regions of the parameter space. Further, let S_0 and S_a be the acceptance and critical regions of the sample space S associated with the event $E \equiv (x_1, x_2, \ldots, x_n)$, assuming H_0 to be true. Then, the probability of a type I error is

$$P[I] = P[E \in S_a | H_0 : \theta \in R_0], \tag{10.1}$$

and, if H_0 is false, but H_a is true, the probability of a type II error is

$$P[II] = P[E \in S_0 | H_a : \theta \in R_a]. \tag{10.2}$$

The two types of error are inversely related; for a fixed sample size, one can only be reduced at the expense of increasing the other. From (10.1) and (10.2), we can define a useful quantity, called the *power*, that can be used to compare the relative merits of the two tests. The *power* of a statistical test is defined as

$$\beta(\theta) \equiv P[E \in S_a | H_a : \theta \in R_a] = 1 - P[E \in S_0 | H_a : \theta \in R_a] \quad (10.3)$$

and so is the probability of rejecting the hypothesis when it is false. Clearly, an acceptable test should result in a power that is large. From (10.1) and (10.2), it follows that

$$\beta(\theta) = \begin{cases} P[\text{I}], & \theta \in R_0 \\ 1 - P[\text{II}], & \theta \in R_a. \end{cases} \quad (10.4)$$

To see how these definitions are used in practice, we will look at a simple example concerning a normal population with unknown mean μ and known variance σ^2. We will test the null hypothesis $H_0: \mu = \mu_0$ against the alternative $H_a: \mu \neq \mu_0$, where μ_0 is a constant, using a sample of size n. Since the arithmetic mean is an unbiased estimator for μ, it is reasonable to accept the null hypothesis if $\bar{x}$ is not too different from μ_0, so the critical region for the test is determined by the condition

$$P[|\bar{x} - \mu_0| > c : H_0] = \alpha,$$

where H_0 indicates that the probability is calculated under the assumption that H_0 is true, i.e., that the mean is μ_0. In this case, we know that $\bar{x}$ is normal distribution with mean μ_0 and variance σ^2/n, so that the quantity $z = (\bar{x} - \mu_0)/(\sigma/\sqrt{n})$ has a standard normal distribution and thus

$$P[z > c\sqrt{n}/\sigma] = \alpha/2.$$

However, we know that

$$P[z > z_{\alpha/2}] = \alpha/2$$

and so $c = z_{\alpha/2} \sigma/\sqrt{n}$. Thus, at the significance level α, we conclude that

$$\frac{\sqrt{n}}{\sigma}|\bar{x} - \mu_0| > z_{\alpha/2} \Rightarrow H_0 \text{ is rejected} \quad (10.5a)$$

and

$$\frac{\sqrt{n}}{\sigma}|\bar{x} - \mu_0| \leq z_{\alpha/2} \Rightarrow H_0 \text{ is accepted.} \quad (10.5b)$$

This is shown in Fig. 10.1 and, for obvious reasons, this is called a *two-tailed test*.

The decision of whether or not to accept a null hypothesis using a test with a fixed value of α often depends on very small changes in the value of the test statistic. It is therefore useful to have a way of reporting the result of a significance test that does not depend on having to choose a value of the significance level beforehand. This may be done by calculating the so-called *p-value* of the test. It may be defined in several ways. For example, the smallest level of significance which would lead to rejection of the null hypothesis using the observed sample, or the probability of observing a value of the test statistic that contradicts the null hypothesis at least as much as that computed from the sample. Thus H_0 will be accepted (rejected) if the significance level α is less than (greater than or equal to) the *p*-value.

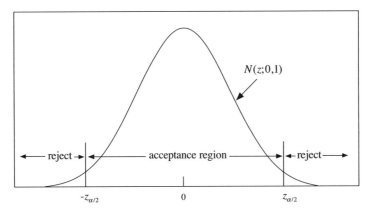

FIGURE 10.1 Two-tailed test for $H_0: \mu = \mu_0$ against $H_a: \mu \neq \mu_0$.

EXAMPLE 10.1

A manufacturer aims to produce power packs with an output O of 10 watts and a standard deviation of 1 watt. To control the quality of its output, 20 randomly chosen power packs are tested and found to have an average value of 9.5 watts. Test the hypothesis $H_0: O = 10$ against the alternative $H_a: O \neq 10$ at a 5% significance level. What is the lowest significance level at which H_0 would be accepted?

If we assume we can use the normal distribution, then the test statistic is

$$z = \frac{|\bar{x} - O|}{\sigma/\sqrt{n}} = 2.236,$$

if H_0 is true. This value is to be compared with $z_{0.025}$ because we are using a two-tailed test. From Table C.1, $z_{0.025} = 1.96$ and since $z > z_{0.025}$, the null hypothesis must be rejected at the 5% significance level, and the critical region from (10.5a) is

$$\bar{x} < 9.56 \text{ and } \bar{x} > 10.44.$$

Notice that even though the lower of these values is very close to the observed mean, we are forced to reject H_0 because we have set the value of α before the test. The *p*-value for the sample is given by

$$p = P[|z| > 2.236] = 2P[z > 2.236] = 0.0254,$$

again using Table C.1. Thus, for $\alpha > 0.0254$, H_0 would still be rejected, but for values lower than this, it would not be rejected.

We also have to consider type II errors, because if these turn out to be large, the test may not be useful. In principle, this is not an easy question to address. In the present example, each value of $\mu \neq \mu_0$ leads to a different sampling distribution, and, hence, a different value of the type II error. In practice, however, what can be done is to find a curve that, for a given value of α, gives the probability of a type II error for any value of $(\mu - \mu_0)$.

10.1. STATISTICAL HYPOTHESES

This is called the *operating characteristic (OC) curve*. From the definition (10.2), and using the fact that

$$z = \frac{\bar{x} - \mu}{\sigma/\sqrt{n}}$$

is approximately distributed as standard normal distribution, it is straightforward to show that the OC curve is given by

$$P[\text{II};\mu] = N\left(\frac{\mu_0 - \mu}{\sigma/\sqrt{n}} + z_{\alpha/2}\right) - N\left(\frac{\mu_0 - \mu}{\sigma/\sqrt{n}} - z_{\alpha/2}\right) \tag{10.6}$$

and is symmetric about the point where $(\mu - \mu_0)$ equals zero.

Figure 10.2(a) shows a plot of (10.6) as a function of μ for the data of Example 10.1. For this example, the maximum of the OC curve is at $\mu = 10$, where $(1 - \alpha) = 0.95$. The curve shows that the probability of making a type II error is larger when the true mean is close to the value μ_0 and decreases as the difference becomes greater, but if $\mu \approx \mu_0$, a type II error will presumably have smaller consequences. The associated power curve $1 - P[\text{II};\mu]$ is given in Fig. 10.2(b). This shows that the power of the test increases to unity as the true mean gets further away from the H_0 value, which is a statement of the fact that it is easier to reject an hypothesis as it gets further from the truth. Decreasing α, with a fixed sample size, reduces the power of the test, whereas increasing the sample size produces a power curve with a sharper minimum and will increase the power of the test, except at $\mu = \mu_0$. Once α and the sample size n are chosen, the size of the type II error is determined, but we can also use the OC curve to calculate the sample size for a fixed α that gives a specific value for the type II error. From (10.6), this is equivalent to

$$P[\text{II};\mu_n] = N\left(\frac{\mu_0 - \mu_n}{\sigma/\sqrt{n}} + z_{\alpha/2}\right) - N\left(\frac{\mu_0 - \mu_n}{\sigma/\sqrt{n}} - z_{\alpha/2}\right),$$

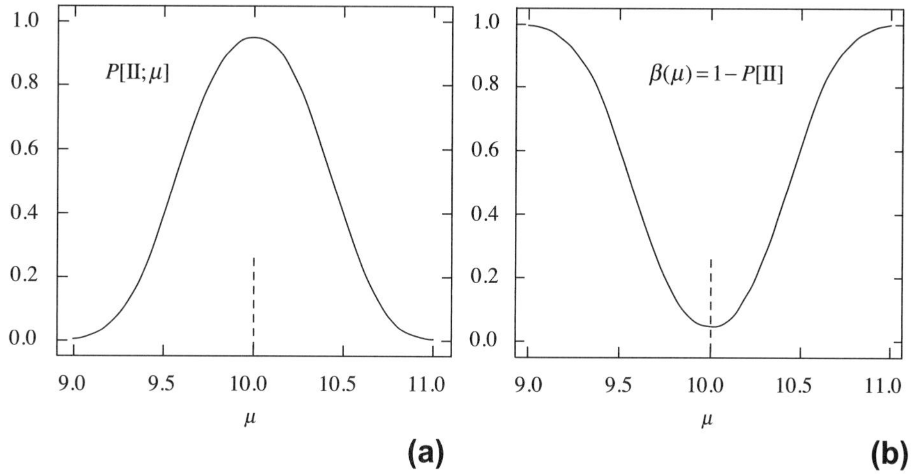

FIGURE 10.2 (a) Operating characteristic curve as a function of; μ and (b) power curve, both for the data for Example 10.1.

where $P[\text{II}; \mu_n]$ is the probability of accepting H_0 when the true mean is μ_n. Given $P[\text{II}; \mu_n]$ and α, a solution for n may be obtained from this relation numerically.

10.2. GENERAL HYPOTHESES: LIKELIHOOD RATIOS

The example in Section 10.1 exhibits all the steps necessary for testing an hypothesis. In this section, we turn to tests derived from a consideration of likelihood ratios, starting with the simplest of all possible situations, that of a simple hypothesis and one simple alternative. This case is not very useful in practice, but it will serve as an introduction to the method.

10.2.1. Simple Hypothesis: One Simple Alternative

The likelihood ratio λ for a sample $x_i (i = 1, 2, \ldots, n)$ of size n having a density $f(x, \theta)$ is defined by

$$\lambda \equiv \frac{\prod_{i=1}^{n} f(x_i; \theta_0)}{\prod_{i=1}^{n} f(x_i; \theta_a)} = \frac{L(\theta_0)}{L(\theta_a)}. \tag{10.7a}$$

Then, for a fixed $k > 0$, the *likelihood ratio test* for deciding between a simple null hypothesis $H_0: \theta = \theta_0$ and the simple alternative $H_a: \theta = \theta_a$ is

$$\begin{aligned} &\text{for } \lambda > k, \quad H_0 \text{ is accepted} \\ &\text{for } \lambda < k, \quad H_0 \text{ is rejected} \\ &\text{for } \lambda = k, \quad \text{either action is taken.} \end{aligned} \tag{10.7b}$$

The inequality $\lambda > k$ determines the acceptance and critical region S_0 and S_a, respectively, as illustrated by the following example.

EXAMPLE 10.2

If x is a random sample of size one, drawn from a normal distribution with mean and variance both equal to 1, find the acceptance and critical regions for testing the null hypothesis $H_0: \mu = -1$ against the alternative $H_a: \mu = 0$ for $k = e^{1/2}$.

The normal density with unit variance is

$$f(x; \theta) = \frac{1}{\sqrt{2\pi}} \exp\left[-\frac{(x-\mu)^2}{2}\right],$$

and from (10.7a), the likelihood ratio is

$$\lambda = \exp\left[-\frac{1}{2}(x+1)^2\right] \exp\left[\frac{1}{2}x^2\right] = \exp\left[-\frac{1}{2}(2x+1)\right].$$

So, with $k = e^{1/2}$, the inequality $\lambda > k$ implies $e^{-x} > e$, which is true for $-\infty < x < -1$ and determines the acceptance region for H_0. Likewise, $\infty > x \geq -1$ determines the critical region.

For a fixed sample size, the method of testing described in Section 10.1 concentrates on controlling only type I errors, and type II errors are calculated *a posteriori*. A better test would be one that for a null hypothesis $H_0: \theta \in R_0$, with an alternative $H_a: \theta \in R_a$, gives

$$P[\text{I}] \leq \alpha, \quad \text{for } \theta \in R_0,$$

and maximizes the power

$$\beta(\theta) = 1 - P[\text{II}], \quad \text{for } \theta \in R_a.$$

For the case of a simple null hypothesis and a simple alternative, such a test exists, and is defined as follows.

The critical region R_k, which, for a fixed significance level α, maximizes the power of the test of the null hypothesis $H_0: \theta = \theta_0$ against the alternative $H_a: \theta = \theta_a$, where $x_1, x_2, \ldots, x_n$ is a sample of size n from a density $f(x;\theta)$, is that region for which the likelihood ratio

$$\lambda = \frac{L(\theta_0)}{L(\theta_a)} < k, \tag{10.8a}$$

for a fixed number k, and

$$\int_{R_k} \cdots \int \prod_{i=1}^{n} f(x_i;\theta_0) dx_i = \alpha. \tag{10.8b}$$

This result is known as the *Neyman–Pearson lemma*. The proof is as follows.

The object is to find the region R that maximizes the power

$$\beta = \int_R L(\theta_a) d\mathbf{x},$$

subject to the condition implied by equation (10.8b), i.e.,

$$\int_R L(\theta_0) d\mathbf{x} = \alpha.$$

Consider the region R_k defined to be that where the likelihood ratio

$$\lambda = \frac{L(\theta_0)}{L(\theta_a)} < k.$$

In R_k, it follows that

$$\int_{R_k} L(\theta_a) d\mathbf{x} > \frac{1}{k} \int_{R_k} L(\theta_0) d\mathbf{x}.$$

But, for *all* regions, equation (10.8b) must hold, and so we have, for *any* region R,

$$\int_{R_k} L(\theta_a) d\mathbf{x} > \frac{1}{k} \int_R L(\theta_0) d\mathbf{x}. \tag{10.9a}$$

Now for a region R outside R_k,

$$\lambda = \frac{L(\theta_0)}{L(\theta_a)} > k,$$

and hence,

$$\frac{1}{k}\int_R L(\theta_0)d\mathbf{x} > \int_R L(\theta_a)d\mathbf{x}. \qquad (10.9b)$$

Combining the two inequalities (10.9a) and (10.9b) gives

$$\int_{R_k} L(\theta_a)d\mathbf{x} > \int_R L(\theta_a)d\mathbf{x}, \qquad (10.9c)$$

which is true for any R, and all R_k such that $\lambda < k$. Thus R_k is the required critical region. Once λ is chosen, the values of type I and type II errors, and hence the power, are determined.

We will illustrate the use of the Neyman–Pearson lemma by an example involving the normal distribution.

EXAMPLE 10.3

If θ is the mean of a normal population with unit variance, test the null hypothesis $H_0:\theta = 2$ against the alternative $H_a:\theta = 0$, given a sample of size n.

Using the normal density, the likelihood ratio is

$$\lambda = \exp\left[-\frac{1}{2}\sum_{i=1}^n (x_i - 2)^2\right]\left\{\exp\left[-\frac{1}{2}\sum_{i=1}^n x_i^2\right]\right\}^{-1}$$

$$= \exp\left[\sum_{i=1}^n (2x_i - 2)\right] = \exp[2n\bar{x} - 2n],$$

and thus, from (10.7b), H_0 is accepted if $\lambda > k$, i.e., if

$$\bar{x} > c = \frac{\ln k}{2n} + 1, \qquad (10.10)$$

and rejected if $\bar{x} < c$.

The error probabilities for Example 10.3 are given by the shaded areas in Fig. 10.3. To find the point for which $P[I]$ is a given value for a fixed value of n, we note that when $\theta = 2$,

$$P[I] = P[\bar{x} < c | \theta = 2] = \alpha,$$

so, for $\alpha = 0.05$ and $n = 4$ (say), using Table C.1 gives

$$c = \theta - \frac{1.645}{\sqrt{4}} = 1.1775,$$

FIGURE 10.3 Error probabilities for Example 10.3.

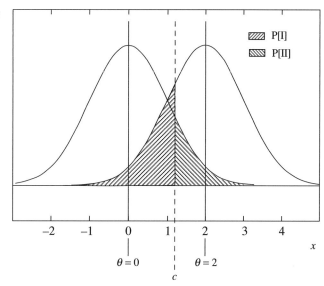

and for this value of c,

$$P[\text{II}] = P[\bar{x} > 1.1775 | \theta = 0] = 0.009.$$

It is also possible to find the sample size necessary to control the values of $P[\text{I}]$ and $P[\text{II}]$. Thus, for example, if we want the error probabilities to be $P[\text{I}] = 0.03$ and $P[\text{II}] = 0.01$, then using approximate numbers from Table C.1, we require that

$$c = \theta_0 - \frac{1.88}{\sqrt{n}} = 2 - \frac{1.88}{\sqrt{n}},$$

and

$$c = \theta_a + \frac{2.33}{\sqrt{n}} = \frac{2.33}{\sqrt{n}},$$

simultaneously. This gives $n = 4.43$, and so a sample size of 5 would suffice.

10.2.2. Composite Hypotheses

The case considered in Section 10.2.1 is really only useful for illustrative purposes. More realistic situations usually involve composite hypotheses. The first that we will consider is when the null hypothesis is simple and the alternative is composite, but may be regarded as an aggregate of simple hypotheses. If the alternative is H_a, then for each of the simple hypotheses in H_a, say H'_a, we may construct, for a given α, a region R for testing H_0 against H'_a. However, R will vary from one hypothesis H'_a to the next and we are therefore faced with the problem of determining the best critical region for the totality of hypotheses H'_a. Such a region is called the *uniformly most powerful (UMP)* and a UMP test is defined as follows.

A test of the null hypothesis $H_0 : \theta \in R_0$ against the alternative $H_a : \theta \in R_a$ is a *uniformly most powerful (UMP) test* at the significance level α if the critical region of the test is such that

$$P[\mathrm{I}] \leq \alpha \quad \text{for all} \quad \theta \in R_0$$

and

$$\beta(\theta) = 1 - P[\mathrm{II}] \quad \text{is a maximum for each} \quad \theta \in R_a.$$

The following simple example will illustrate how such a UMP test may be constructed.

EXAMPLE 10.4

Test the null hypothesis $H_0 : \mu = \mu_0$ against the alternative $H_a : \mu > \mu_0$, for a normal distribution with unit variance, using a sample of size n.

The hypothesis H_a may be regarded as an aggregate of hypotheses H'_a of the form $H'_a : \mu = \mu_a$ where $\mu_a > \mu_0$. The likelihood ratio for testing H_0 against H'_a is, from (10.7a),

$$\lambda = \exp\left\{-\frac{1}{2}[2n\bar{x}(\mu_a - \mu_0) + n(\mu_0^2 - \mu_a^2)]\right\}.$$

The Neyman–Pearson lemma may now be applied for a given k and gives the critical region

$$\bar{x} > c = \frac{-\ln k}{n(\mu_a - \mu_0)} + \frac{1}{2}(\mu_0 + \mu_a).$$

Thus the critical region is of the form $\bar{x} > c$, regardless of the value of μ_a, provided $\mu_a > \mu_0$. Therefore, to reject H_0 if $\bar{x} > c$ tests H_0 against $H_a : \mu > \mu_0$. The number c may be found from

$$P[\mathrm{I}] = \alpha = \left(\frac{n}{2\pi}\right)^{1/2} \int_c^\infty \exp\left[-\frac{n}{2}(\bar{x} - \mu_0)^2\right] d\bar{x}.$$

The integral is evaluated by substituting $u = \sqrt{n}(\bar{x} - \mu_0)$ and using Table C.1. For example, choosing $\alpha = 0.025$ gives $c = \mu_0 + 1.96/\sqrt{n}$.

A more complicated situation that can occur is testing one composite hypothesis against another, for example, testing the null hypothesis $H_0 : \theta_1 < \theta < \theta_2$ against $H_a : \theta < \theta_1, \theta > \theta_2$. In such cases, a UMP test does not exist, and other tests must be devised whose power is not too inferior to the maximum power tests. A useful method is to construct a test having desirable large-sample properties and hope that it is still reasonable for small samples. One such test is the *generalized likelihood ratio* described below.

Let $x_1, x_2, \ldots, x_n$ be a sample of size n from a population density $f(x; \theta_1, \theta_2, \ldots, \theta_p)$, where S is the parameter space. Let the null hypothesis be $H_0 : (\theta_1, \theta_2, \ldots, \theta_p) \in R_0$, and the alternative be $H_a : (\theta_1, \theta_2, \ldots, \theta_p) \in (S - R_0)$. Then, if the likelihood of the sample is denoted by $L(S)$ and its maximum value with respect to the parameters in the region S denoted by $L(\hat{S})$, the *generalized likelihood ratio* is given by

$$\lambda = \frac{L(\hat{R}_0)}{L(\hat{S})}, \qquad (10.11)$$

and $0 < \lambda < 1$. Furthermore, if $P[I] = \alpha$, then the critical region for the generalized likelihood ratio test is $0 < \lambda < A$ where

$$\int_0^A g(\lambda|H_0)d\lambda = \alpha,$$

and $g(\lambda|H_0)$ is the density of λ when the null hypothesis H_0 is true. Again, we will illustrate the method by an example.

EXAMPLE 10.5

Use the generalized likelihood ratio to test the null hypothesis $H_0 : \mu = 2$ against $H_a : \mu \neq 2$, for a normal density with unit variance.

In this example, the region R_U is a single point $\mu = 2$, and $(S - R_U)$ is the rest of the real axis. The likelihood is

$$L = \left(\frac{1}{2\pi}\right)^{n/2} \exp\left[-\frac{1}{2}\sum_{i=1}^n (x_i - \mu)^2\right]$$

$$= \left(\frac{1}{2\pi}\right)^{n/2} \exp\left[-\frac{1}{2}\sum_{i=1}^n (x_i - \bar{x})^2 - \frac{n}{2}(\bar{x} - \mu)^2\right],$$

and the maximum value of $L(S)$ is obtained when $\mu = \bar{x}$, i.e.

$$L(\hat{S}) = \left(\frac{1}{2\pi}\right)^{n/2} \exp\left[-\frac{1}{2}\sum_{i=1}^n (x_i - \bar{x})^2\right]. \tag{10.12}$$

Similarly,

$$L(\hat{R}_0) = \left(\frac{1}{2\pi}\right)^{n/2} \exp\left[-\frac{1}{2}\sum_{i=1}^n (x_i - \bar{x})^2 - \frac{n}{2}(\bar{x} - 2)^2\right], \tag{10.13}$$

and so the generalized likelihood ratio is

$$\lambda = \exp\left[-\frac{n}{2}(\bar{x} - 2)^2\right]. \tag{10.14}$$

If we use $\alpha = 0.025$, the critical region for the test is given by $0 < \lambda < A$, where

$$\int_0^A g(\lambda|H_0)d\lambda = 0.025.$$

Now, if H_0 is true, $\bar{x}$ is normally distributed with mean 2 and variance $1/n$. Then, $n(\bar{x} - 2)^2$ is distributed as chi-square with one degree of freedom. Taking the logarithm of (10.14), it follows that $(-2\ln\lambda)$ is also distributed as chi-square with one degree of freedom. Setting $\chi^2 = -2\ln\lambda$, and using Table C.4 gives

$$0.025 = \int_0^A g(\lambda|H_0)d\lambda = \int_{-2\ln\lambda}^\infty f(\chi^2;1)d\chi^2$$

$$= \int_{5.02}^\infty f(\chi^2;1)d\chi^2.$$

Thus, the critical region is defined by $-2\ln\lambda > 5.02$, i.e., $n(\bar{x}-2)^2 > 5.02$, or

$$\bar{x} > 2 + \frac{2.24}{\sqrt{n}}; \quad \bar{x} < 2 - \frac{2.24}{\sqrt{n}}. \tag{10.15}$$

The generalized likelihood ratio test has useful large-sample properties. These can be stated as follows. Let $x_1, x_2, \ldots, x_n$ be a random sample of size n drawn from a density $f(x;\theta_1,\theta_2,\ldots,\theta_p)$, and let the null hypothesis be

$$H_0: \theta_i = \bar{\theta}_i, \quad i = 1, 2, \ldots, k < p,$$

with the alternative

$$H_a: \theta_i \neq \bar{\theta}_i.$$

Then, when H_0 is true, $-2\ln\lambda$ is approximately distributed as chi-square with k degrees of freedom, if n is large.

To use this result to test the null hypothesis H_0 with $P[\text{I}] = \alpha$, we need to only compute $-2\ln\lambda$ from the sample, and compare it with the α level of the chi-square distribution. If $-2\ln\lambda$ exceeds the α level, H_0 is rejected; if not H_0 is accepted.

10.3. NORMAL DISTRIBUTION

Because of the great importance of the normal distribution, in this section we shall give some more details concerning tests involving this distribution.

10.3.1. Basic Ideas

In Section 10.1, we discussed a two-tailed test of the hypothesis $H_0: \mu = \mu_0$ against the alternative $H_a: \mu \neq \mu_0$ for the case where the population variance is known $n(x:\mu,\sigma^2)$. To recap the result obtained there, the null hypothesis is rejected if the quantity $W_0 = \sqrt{n}(\bar{x}-\mu_0)/\sigma$ is greater than W_γ in modulus, i.e., if $|W_0| > W_\gamma$, where

$$P[|W| \geq W_\gamma] = \left(\frac{1}{2\pi}\right)^{1/2}\left\{\int_{-\infty}^{-W_\gamma} e^{-t^2/2}dt + \int_{W_\gamma}^{\infty} e^{-t^2/2}dt\right\}$$

$$= 2[1 - N(W_\gamma;0,1)] = 2\gamma.$$

If the alternative hypothesis is $H_a: \mu > \mu_0$, then $P[\text{I}]$ is the area under only *one* of the tails of the distribution, and the significance level of the test is thus

$$P[\text{I}] = \alpha = \gamma.$$

Such a test is called a *one-tailed test*. We also showed how to find the probability of a type II error and, from this, the power of the test. If, for definiteness, the alternative is taken to be $H_a: \mu = \mu_a$, then the power is given by

$$\beta = 1 - P\left[-W_{\alpha/2} - \left(\frac{\mu_a-\mu_0}{\sigma/\sqrt{n}}\right) \leq W_a \leq -W_{\alpha/2} + \left(\frac{\mu_a-\mu_0}{\sigma/\sqrt{n}}\right)\right].$$

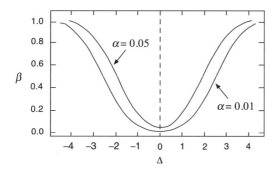

FIGURE 10.4 The power of a test comparing two means for a normal population with known variance.

If $\mu_a - \mu_0$ is small, then

$$P[\text{II}] \simeq 1 - P[\text{I}],$$

and hence

$$\beta \simeq \alpha.$$

Thus, the power of the test will be very low. This situation can only be improved by making $(\mu_a - \mu_0)$ large, or by having n large. This is in accord with the common-sense view that it is difficult to distinguish between two close alternatives without a large quantity of data. The situation is illustrated in Fig. 10.4, which shows the power β as a function of the parameter $\Delta = \sqrt{n}(\mu_a - \mu_0)/\sigma$ for two sample values of α, the significance level. This is a generalization of the specific case shown in Fig. 10.2(b).

We are now in a position to review the general procedure followed to test a hypothesis:

1. State the null hypothesis H_0, and its alternative H_a;
2. Specify $P[\text{I}]$ and $P[\text{II}]$, the probabilities for errors of types I and II, respectively, and compute the necessary sample size n.[1] In practice, $P[\text{I}] = \alpha$ and n are commonly given. However, since even a relatively small $P[\text{II}]$ is usually of importance, a check should always be made to ensure that the values of α and n used lead to a suitable $P[\text{II}]$.
3. Choose a test statistic and determine the critical region for the test. Alternatively, calculate the p-value and then choose a suitable value of α *a posteriori*.
4. Accept or reject the null hypothesis H_0, depending on whether or not the value obtained for the sample statistic falls inside or outside the critical region.

A graphical interpretation of the above scheme is shown in Fig. 10.5. The curve $f_0(\theta|H_0)$ is the density function of the test statistic θ if H_0 is true and $f_a(\theta|H_a)$ is its density function if H_a is true. The hypothesis H_0 is rejected if $\theta > \theta_\alpha$, and H_a is rejected if $\theta < \theta_\alpha$. The probabilities of the errors of types I and II are also shown. It is perhaps worth repeating that failure to reject a hypothesis does not necessarily mean that the hypothesis is true. However, if we can reject the hypothesis on the basis of the test, then we *can* say that there is experimental evidence against it.

---

[1] Tables for this purpose applying to some of the tests we will consider are given in O.L. Davies *Design and Analysis of Industrial Experiments, Research Vol 1*, Oliver and Boyd Ltd (1948).

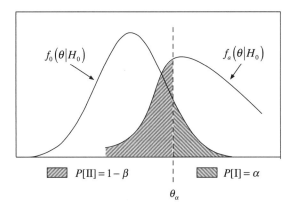

FIGURE 10.5 Graphical representation of a general hypothesis test.

10.3.2. Specific Tests

We shall now turn to more practical cases, where one of the parameters of the distribution is unknown, and use the general procedure given above to establish some commonly used tests, deriving them from the likelihood ratio.

(a) *Test of whether the mean is different from some specified value.*

The null hypothesis in this case is

$$H_0: \mu = \mu_0, \ 0 < \sigma^2 < \infty,$$

and the alternative is

$$H_0: \mu \neq \mu_0, \ 0 < \sigma^2 < \infty,$$

since the variance is unknown. The parameter space is

$$S = \{-\infty < \mu < \infty; 0 < \sigma^2 < \infty\},$$

and the acceptance region associated with the null hypothesis is

$$R_0 = \{\mu = \mu_0 : 0 < \sigma^2 < \infty\}.$$

In this case, the null hypothesis is not simple because it does not specify the value of σ^2. For large samples we could use an estimate s^2 for σ^2 and then take over the results of Section 10.3.1. However, for small samples, this procedure could lead to errors and so we must devise a test where σ^2 is not explicitly used. Such a test is based on the use of the Student's t-distribution. Its derivation is as follows.

The likelihood function for a sample of size n drawn from the population is given by

$$L = \frac{1}{(2\pi)^{n/2}} \frac{1}{\sigma^n} \exp\left[-\frac{1}{2} \sum_{i=1}^{n} \left(\frac{x_i - \mu}{\sigma}\right)^2\right], \quad (10.16)$$

10.3. NORMAL DISTRIBUTION

and we have seen in Chapter 7 that the ML estimators of μ and σ^2 are

$$\hat{\mu} = \frac{1}{n}\sum_{i=1}^{n} x_i \quad \text{and} \quad \hat{\sigma}^2 = \frac{1}{n}\sum_{i=1}^{n}(x_i - \bar{x})^2. \tag{10.17}$$

Using (10.17) in (10.16) gives

$$L(\hat{S}) = \left[\frac{n}{2\pi\sum(x_i - \bar{x})^2}\right]^{n/2} e^{-n/2}. \tag{10.18}$$

To maximize L in R_0, we set $\mu = \mu_0$, giving

$$L' = \frac{1}{(2\pi)^{n/2}}\frac{1}{\sigma^n}\exp\left[-\frac{1}{2}\sum_{i=1}^{n}\left(\frac{x_i - \mu_0}{\sigma}\right)^2\right].$$

Then, the value of σ^2 that maximizes L' is

$$\hat{\sigma}^2 = \frac{1}{n}\sum_{i=1}^{n}(x_i - \mu_0)^2,$$

and hence

$$L(\hat{R}_0) = \left[\frac{n}{2\pi\sum(x_i - \mu_0)^2}\right]^{n/2} e^{-n/2}. \tag{10.19}$$

So, from (10.18) and (10.19), the generalized likelihood ratio is

$$\lambda = \left[\frac{\sum(x_i - \bar{x})^2}{\sum(x_i - \mu_0)^2}\right]^{n/2}. \tag{10.20}$$

We must now find the distribution of λ if H_0 is true. Rewriting (10.20) gives

$$\lambda = \left[1 + \frac{n(\bar{x} - \mu_0)^2}{\sum(x_i - \bar{x})^2}\right]^{-n/2} = \left(1 + \frac{t^2}{n-1}\right)^{-n/2}, \tag{10.21}$$

where

$$t = \left[\frac{n(n-1)}{\sum(x_i - \bar{x})^2}\right]^{1/2}(\bar{x} - \mu_0)$$

is distributed as the t-distribution with $(n-1)$ degrees of freedom. From (10.21), a critical region of the form $0 < \lambda < A$ is equivalent to the region $t^2 > F(A)$. Thus, a significance level of α corresponds to the pair of intervals

$$t < -t_{\alpha/2} \quad \text{and} \quad t > t_{\alpha/2},$$

where

$$\int_{t_{\alpha/2}}^{\infty} f(t; n-1)\,dt = \alpha/2, \tag{10.22}$$

and $f(t; n-1)$ is the t-distribution with $(n-1)$ degrees of freedom. If t lies between $-t_{\alpha/2}$ and $t_{\alpha/2}$, then H_0 is accepted; otherwise, it is rejected. This is a typical example of a two-tailed test, and is exactly equivalent to constructing a $100(1-2\alpha)\%$ confidence interval for μ, and accepting H_0 if μ lies within it. The test is summarized as follows:

Observations	n values of x
Significance level	α
Null hypothesis	$H_0: \mu = \mu_0, \quad 0 < \sigma^2 < \infty$
Alternative hypothesis	$H_0: \mu \neq \mu_0, \quad 0 < \sigma^2 < \infty$
Test statistic	$t = \left[\dfrac{n(n-1)}{\sum(x_i - \bar{x})^2}\right]^{1/2} (\bar{x} - \mu_0) = \dfrac{(\bar{x} - \mu_0)}{s/\sqrt{n}}$ (10.23)

Decision criterion. The test statistics obeys a t-distribution with $(n-1)$ degrees of freedom if null hypothesis is true; so if the observed value of t lies between $-t_{\alpha/2}$ and $t_{\alpha/2}$, where the latter is defined by (10.22), the null hypothesis is accepted; otherwise, it is rejected. Alternatively, it is accepted (rejected) if the calculated p-value is larger (smaller) than a specified value of α.

This may be generalized in an obvious way to test the null hypothesis against the alternatives $H_a: \mu > \mu_0$ and $H_a: \mu < \mu_0$. The test statistic is the same, but the critical regions are now $t > t_\alpha$ and $t < -t_\alpha$, respectively.

The above procedure has controlled type I errors by specifying the significance level. We must now consider the power of the test. This is no longer a simple problem, because if H_0 is not true, then the statistic t no longer has a Student's t-distribution. If the alternative hypothesis is

$$H_a: \mu = \mu_a, \quad 0 < \sigma^2 < \infty$$

and H_a is true, it can be shown that t obeys a *noncentral t-distribution* of the form

$$f^{nc}(t) = \frac{2^{-(\nu-1)/2}}{\Gamma(\nu/2)(\nu\pi)^{1/2}} \left(1 + \frac{t^2}{\nu}\right)^{-(\nu+1)/2} \exp\left[-\frac{1}{2}\left(\frac{\delta^2}{1 + t^2/\nu}\right)\right]$$

$$\times \int_0^\infty x^\nu \exp\left[-\frac{1}{2}\left(x - \frac{t\nu}{(t^2 + \nu)^{1/2}}\right)\right] dx,$$

where $\nu = n - 1$ and $\delta = \sqrt{n}|\mu_a - \mu_0|/\sigma$. Unfortunately, this distribution, apart from being exceedingly complex, contains the population variance σ^2 in the *noncentrality parameter* δ. An estimate of the power of the test may be obtained by replacing σ^2 by the sample variance s^2 in the noncentral distribution and then using tables of the distribution.

Noncentral distributions typically arise if we wish to consider the power of a test, and are generally functions of a noncentrality parameter that itself is a function of the alternative hypothesis and a population parameter that is unknown.

EXAMPLE 10.6

An experiment finds the mean lifetime of a rare nucleus to be $\tau = 125$ μs. A second experiment finds 20 examples of the same decay with lifetimes given below (in μs):

$$\begin{array}{cccccccccc}
120 & 113 & 120 & 140 & 136 & 117 & 140 & 136 & 110 & 119 \\
123 & 119 & 118 & 120 & 134 & 121 & 137 & 129 & 137 & 140
\end{array}$$

Test whether the data from the second experiment are compatible with those from the first experiment at the 10% level.

Formally, we are testing the null hypothesis $H_0 : \tau = 125$ with the alternative $H_a : \tau \neq 125$, using test (a) above. Firstly, using the data given, we find the mean to be $\bar{x} = 126.45$ and the standard deviation $\sigma = 10.02$, and using these, then calculate the value of t as 0.647. Next, we find $t_{\alpha/2}$ from (10.23) for $\alpha = 0.10$ and 19 degrees of freedom. Using Table C.5, this gives $t_{\alpha/2} = 1.73$. As this value is greater than the observed value of t, the null hypothesis is accepted, that is, the two experiments are compatible at this significance level.

Another use of the Student's t-distribution is contained in the following test, which we will state without proof.

(b) *Test of whether the means of two populations having the same, but unknown, variance differ.*

Observations	m values of x_1, n values of x_2
Significance level	α
Null hypothesis	$H_0 : \mu_1 = \mu_2$, $-\infty < \sigma_{1,2}^2 < \infty$ $(\sigma_1^2 = \sigma_2^2)$
Alternative hypothesis	$H_a : \mu_1 \neq \mu_2$, $-\infty < \sigma_{1,2}^2 < \infty$ $(\sigma_1^2 = \sigma_2^2)$
Test statistic	$t = \dfrac{(\bar{x}_1 - \bar{x}_2)}{s_p \sqrt{(1/m) + (1/n)}}$, (10.24a)

where the pooled sample variance, s_p^2, is given by

$$s_p^2 = \frac{(m-1)s_1^2 + (m-1)s_2^2}{m + m - 2}. \tag{10.24b}$$

Decision criterion: The test statistic obeys a t-distribution with $(m + n - 2)$ degrees of freedom if the null hypothesis is true; so, if the observed value of t lies in the range $t_{-\alpha/2} < t < t_{\alpha/2}$, where

$$\int_{t_{\alpha/2}}^{\infty} f(t; m + n - 2) dt = \alpha/2, \tag{10.25}$$

the null hypothesis is accepted; otherwise, it is rejected.

Again, this test may be extended to the cases $\mu_1 > \mu_2$ and $\mu_1 < \mu_2$, with critical regions $t > t_\alpha$ and $t < -t_\alpha$, respectively. It may also be extended in a straightforward way to test the null hypothesis $H_0: \mu_1 - \mu_2 = d$, where d is a specified constant, and to cases where the two variances are both unknown and unequal.

We will now consider two tests associated with the variance of a normal population, and, by analogy with the discussion of tests involving the mean, we start with a test of whether the variance is equal to some specific value.

(c) Test of whether the variance is equal to some specific value

The null hypothesis in this case is

$$H_0: \sigma^2 = \sigma_0^2, \quad -\infty < \mu < \infty,$$

and the alternative is

$$H_a: \sigma^2 \neq \sigma_0^2, \quad -\infty < \mu < \infty,$$

since the mean is unknown. The parameter space is

$$S = \{-\infty < \mu < \infty; 0 < \sigma^2 < \infty\},$$

and the acceptance region associated with the null hypothesis is

$$R_0 = \{-\infty < \mu < \infty; \sigma^2 = \sigma_0^2\}.$$

The test will involve the use of the χ^2 distribution and will again be derived by the method of likelihood ratios.

As before, the likelihood function for a sample of size n drawn from the population is given by

$$L = \left(\frac{1}{2\pi}\right)^{n/2} \frac{1}{\sigma^n} \exp\left[-\frac{1}{2}\sum_{i=1}^{n}\left(\frac{x_i - \mu}{\sigma}\right)^2\right]$$

and in the acceptance region R_0

$$L = \left(\frac{1}{2\pi}\right)^{n/2} \frac{1}{\sigma_0^n} \exp\left[-\frac{1}{2}\sum_{i=1}^{n}\left(\frac{x_i - \mu}{\sigma_0}\right)^2\right].$$

This expression is a maximum when the summation is a minimum, i.e., when $\bar{x} = \mu$. Thus,

$$L(\hat{R}_0) = \left(\frac{1}{2\pi}\right)^{n/2} \frac{1}{\sigma_0^n} \exp\left[-\frac{1}{2}\sum_{i=1}^{n}\left(\frac{x_i - \bar{x}}{\sigma_0}\right)^2\right] = \left(\frac{1}{2\pi}\right)^{n/2} \frac{1}{\sigma_0^n} \exp\left[-\frac{(n-1)s^2}{2\sigma_0^2}\right],$$

where s^2 is the sample variance. To maximize L in S, we have to solve the maximum likelihood equations. The solutions have been given in (10.18) and, hence,

$$L(\hat{S}) = \left(\frac{1}{2\pi}\right)^{n/2} \frac{1}{s^n} \left(\frac{n}{n-1}\right)^{n/2} \exp\left(-\frac{n}{2}\right).$$

10.3. NORMAL DISTRIBUTION

We may now form the generalized likelihood ratio

$$\lambda = \frac{L(\hat{R}_0)}{L(\hat{S})} = \left[\frac{(n-1)s^2}{n\sigma_0^2}\right]^{n/2} \exp\left[\frac{n}{2} - \frac{(n-1)s^2}{2\sigma_0^2}\right],$$

from which we see that a critical region of the form $\lambda < k$ is equivalent to the region

$$k_1 < \frac{s^2}{\sigma_0^2} < k_2,$$

where k_1 and k_2 are constants depending on n and α, the significance level of the test. If H_0 is true, then $(n-1)s^2/\sigma_0^2$ obeys a χ^2 distribution with $(n-1)$ degrees of freedom and so, in principle, the required values of k_1 and k_2 could be found. A good approximation is to choose values of k_1 and k_2 using equal right and left tails of the chi-squared distribution. Thus, we are led to the following test procedure.

Observations	n values of x
Significance level	α
Null hypothesis	$H_0: \sigma^2 = \sigma_0^2$
Alternative hypothesis	$H_a: \sigma^2 \neq \sigma_0^2$
Test statistic	$\chi^2 = \sum_{i=1}^{n}\left(\frac{x_i - \bar{x}}{\sigma_0}\right)^2 = \frac{s^2}{\sigma_0^2}(n-1).$ (10.26)

Decision criterion. The test statistics obeys a χ^2 distribution with $(n-1)$ degrees of freedom if the null hypothesis is true; so if the observed value of χ^2 lies in the interval $\chi^2_{1-\alpha/2} < t < \chi^2_{\alpha/2}$, where

$$\int_{t_{\alpha/2}}^{\infty} f(\chi^2; n-1)dt = \alpha/2, \tag{10.27}$$

the null hypothesis is accepted; otherwise, it is rejected.

As previously, we now have to examine the question: what is the probability of a type II error in this test? If the alternative hypothesis is

$$H_a: \sigma^2 = \sigma_a^2,$$

and if H_a is true, then the quantity

$$\chi_a^2 = \frac{s^2}{\sigma_a^2}(n-1)$$

will be distributed as χ^2 with $(n-1)$ degrees of freedom. Thus, from the definition of the power function, we have

$$\beta = 1 - P\left[\chi^2_{\alpha/2}(n-1) \geq (n-1)\frac{s^2}{\sigma_0^2} \geq \chi^2_{1-\alpha/2}(n-1)\right],$$

and therefore,

$$\beta = 1 - P\left[\chi^2_{\alpha/2}(n-1)\frac{\sigma_0^2}{\sigma_a^2} \geq (n-1)\frac{s^2}{\sigma_a^2} \geq \chi^2_{1-\alpha/2}(n-1)\frac{\sigma_0^2}{\sigma_a^2}\right]. \tag{10.28}$$

Having fixed the significance level α and the values of σ_0 and σ_a, we can read off from tables the probability that a chi-square variate with $(n-1)$ degrees of freedom lies between the two limits in the square brackets.

Again, this test may be simply adapted to deal with the hypotheses $H_a: \sigma^2 > \sigma_0^2$ and $H_a: \sigma^2 < \sigma_0^2$. The critical regions are $\chi^2 > \chi^2_{\alpha}$ and $\chi^2 < \chi^2_{1-\alpha}$, respectively.

EXAMPLE 10.7

Steel rods of notional standard lengths are produced by a machine whose specifications states that, regardless of the length of the rods, their standard deviation will not exceed 2 (in centimeter units). During commissioning, a check is made on a random sample of 20 rods, whose lengths (in centimeters) were found to be

$$\begin{array}{cccccccccc}
105 & 104 & 103 & 98 & 100 & 102 & 103 & 97 & 99 & 106 \\
105 & 102 & 99 & 100 & 98 & 97 & 102 & 101 & 100 & 99
\end{array}$$

Test at the 10% level whether the machine is performing according to its specification.

Here, we are testing the null hypothesis $H_0: \sigma^2 \leq \sigma_0^2$ against the alternative $H_a: \sigma^2 > \sigma_0^2$, where $\sigma_0^2 = 4$. That is, we are using a one-tailed test, which is an adaptation of test (c) above. From the data, $\bar{x} = 101$ and $s^2 = 7.474$, so the test statistic is

$$\chi^2 = s^2(n-1)/\sigma_0^2 = 35.5.$$

Now, from Table C.4, we find that for $\alpha = 0.10$ and 19 degrees of freedom, $\chi^2_{0.1} = 27.2$. As this value is less than χ^2, we conclude that the machine is not performing according to its specification.

The final test concerns the equality of the variances of two normal populations, which we quote without proof.

(d) Test of whether the variances of two populations with different, but unknown, means differ.

Observations	m values of x_1, n values of x_2
Significance level	α
Null hypothesis	$H_0: \sigma_1^2 = \sigma_2^2$
Alternative hypothesis	$H_a: \sigma_1^2 \neq \sigma_2^2$
Test statistic	$F = \dfrac{s_1^2}{s_2^2} = \dfrac{(n-1)\sum_i(x_{1i}-\bar{x}_1)^2}{(m-1)\sum_j(x_{2i}-\bar{x}_2)^2}$, (10.29)

10.3. NORMAL DISTRIBUTION

Decision criterion. The test statistic obeys the F distribution with $(m-1)$ and $(n-1)$ degrees of freedom if null hypothesis is true; so, if the observed value of F lies in the interval

$$\left[F_{\alpha/2}(n-1,m-1)\right]^{-1} < F < F_{\alpha/2}(m-1,n-1),$$

where

$$\int_{t_{\alpha/2}}^{\infty} f(F; m-1, n-1) dF = \alpha/2, \qquad (10.30)$$

the null hypothesis is accepted; otherwise, it is rejected.

To calculate the power of the test, we note that $P[\text{II}]$ depends on the value of σ_1^2/σ_2^2. If the true value of this ratio is δ then, since $(m-1)s_1^2/\sigma_1^2$ for a sample from a normal population is distributed as χ_{m-1}^2, we find s_1^2/s_2^2 is distributed as

$$\frac{\sigma_1^2}{\sigma_2^2} F(m-1, n-1) = \delta F(m-1, n-1).$$

Thus,

$$\beta = 1 - P\left[F_{1-\alpha/2}(m-1, n-1) \leq \frac{s_1^2}{s_2^2} \leq F_{\alpha/2}(m-1, n-1)\right]$$

is equivalent to

$$\beta = 1 - P\left[\frac{F_{1-\alpha/2}(m-1, n-1)}{\delta} \leq F \leq \frac{F_{\alpha/2}(m-1, n-1)}{\delta}\right]. \qquad (10.31)$$

TABLE 10.1 Summary of hypothesis tests on a normal distribution

H_0	Test statistic	H_a	Critical region
$\mu = \mu_0$ σ^2 unknown	$t = \dfrac{\bar{x} - \mu_0}{s/\sqrt{n}}$ distributed as t with $n-1$ degrees of freedom	$\mu \neq \mu_0$ $\mu > \mu_0$ $\mu < \mu_0$	$t > t_{\alpha/2}$ and $t < -t_{\alpha/2}$ $t > t_\alpha$ $t < -t_\alpha$
$\mu_1 = \mu_2$ $\sigma_1^2 = \sigma_2^2$ unknown	$t = \dfrac{(\bar{x}_1 - \bar{x}_2)}{s_p\sqrt{(1/m)+(1/n)}}$, where $s_p^2 = \dfrac{(m-1)s_1^2 + (n-1)s_2^2}{m+n-2}$, distributed as t with $(m+n-2)$ degrees of freedom	$\mu_1 \neq \mu_2$ $\mu_1 > \mu_2$ $\mu_1 < \mu_2$	$t > t_{\alpha/2}$ and $t < -t_{\alpha/2}$ $t > t_\alpha$ $t < -t_\alpha$
$\sigma^2 = \sigma_0^2$ μ unknown	$\chi^2 = s^2(n-1)/\sigma_0^2$ distributed as χ^2 with $(n-1)$ degrees of freedom	$\sigma^2 \neq \sigma_0^2$ $\sigma^2 > \sigma_0^2$ $\sigma^2 < \sigma_0^2$	$\chi^2 < \chi^2_{1-\alpha/2}$ and $\chi^2 > \chi^2_{\alpha/2}$ $\chi^2 > \chi^2_\alpha$ $\chi^2 < \chi^2_{1-\alpha}$
$\sigma_1^2 = \sigma_2^2$ $\mu_1 \neq \mu_2$ unknown	$F = s_1^2/s_2^2$ distributed as F with $(m-1)$ and $(n-1)$ degrees of freedom	$\sigma_1^2 \neq \sigma_2^2$ $\sigma_1^2 > \sigma_2^2$ $\sigma_1^2 < \sigma_2^2$	$F < F_{1-\alpha/2}$ and $F > F_{\alpha/2}$ $F > F_\alpha$ $F < F_{1-\alpha}$

For any given value of δ, these limits may be found from tables of the F distribution. It can be shown, by consulting these tables, that the power of the F test is rather small unless the ratio of variances is large, a result that is in accordance with common sense.

As before, this test may be adapted to test the null hypotheses $\sigma_1^2 > \sigma_2^2$ and $\sigma_1^2 < \sigma_2^2$. The critical regions are $f > f_\alpha$ and $f < f_{1-\alpha}$, respectively.

A summary of some of the tests mentioned above is given in Table 10.1.

10.4. OTHER DISTRIBUTIONS

The ideas discussed in previous sections can be applied to other distributions and we will look briefly at two important examples, the binomial and Poisson.

Consider a situation where there are only two outcomes of a trial, for example, either satisfactory or unsatisfactory, with the latter outcome having a probability p. In this situation, the number of unacceptable outcomes in n independent random trials is given by the binomial distribution. As an example, we will test the null hypothesis $H_0 : p \leq p_0$ against the alternative $H_a : p > p_0$, where p_0 is some specified value. If the observed number of unacceptable outcomes is x, then from equation (4.34),

$$P[x \geq k] = \sum_{r=k}^{n} \binom{n}{r} p^r (1-p)^{n-r}.$$

It follows that when H_0 is true, that is, when $p \leq p_0$,

$$P[x \geq k] \leq \sum_{r=k}^{n} \binom{n}{r} p_0^r (1-p_0)^{n-r} \equiv B(k, p_0).$$

In practice, if we calculate $B(k, p_0)$, we accept the null hypothesis if $B(k, p_0) > \alpha$; otherwise reject it. Alternatively, from the value of $B(k, p_0)$, we can decide at what value of α the null hypothesis can be accepted and decide at that stage whether it is a reasonable value.

EXAMPLE 10.8

In the literature of a supplier of capacitors, it is stated that no more than 1% of its products are defective. A buyer checks this claim by testing a random sample of 100 capacitors at the 5% level and finds that 3 are defective. Does the buyer have a claim against the supplier?

We need to calculate the probability that in a random sample of 100 capacitors, there would be at least 3 that are defective if $p_0 = 0.01$. From the binomial distribution with $p = p_0$,

$$P[x \geq 3] = 1 - P[x < 3]$$
$$= 1 - \sum_{r=0}^{2} \binom{100}{r} (0.01)^r (0.99)^{100-r}.$$

This may be evaluated exactly by direct calculation, or approximately by using the normal approximation to the binomial for large n. The value of $P[x \geq 3]$ then corresponds approximately to

$$P\left[z \geq \frac{2.5 - \mu}{\sigma}\right],$$

where $\mu = np_0 = 1$ and $\sigma = [np_0(1-p_0)]^{1/2} = 1.005$. The direct calculation gives 0.0694 compared to the normal approximation of 0.0681, and as these are both greater than 0.05, the hypothesis cannot be rejected and no claim can be made against the supplier.

In the case of a random Poisson variable k distributed with parameter λ, as an example we could consider testing the null hypothesis $H_0: \lambda = \lambda_0$ against the alternative $H_a: \lambda \neq \lambda_0$. By analogy with the procedure followed for the binomial distribution, for a given significance level, we would calculate the probability that the observed value of k is greater than or less than the value predicted if the null hypothesis is true and compare this with the significance level. Alternatively, the p-value could be calculated and, from this, a value of the significance level found, above which the null hypothesis would be rejected. An example is given in Problem 10.9.

Just as for the normal distribution, these tests may be extended to other related cases, for example to test the equality of the parameters p_1 and p_2 in two Bernoulli populations, or the equality of the parameters of two Poisson distributions, but we will not pursue this further.

Finally, can we test a null hypothesis, such as $H_0: \mu = \mu_0$, against a suitable alternative when the population distribution is unknown? The answer is yes, but the methods are generally less powerful than those available when the distribution is known. Hypothesis testing in the former situation is an example of *nonparametric statistics* and is discussed in Chapter 11.

10.5. ANALYSIS OF VARIANCE

In Section 10.3, we discussed how to test the hypothesis that the means of two normal distributions with the same, but unknown, variance differ. It is natural to consider how to extend that discussion to the case of several means. The technique for doing this is called *analysis of variance*, usually abbreviated to *ANOVA*. It can be used, for example, to test the consistency of a series of measurements carried out under different conditions, or whether different manufacturers are producing a particular component to the same standard. ANOVA is an important technique in biological and social sciences, but is much less used in the physical sciences and so the discussion here will be very brief.

Consider the case of m groups of measurements, each leading to an average value $\mu_i (i = 1, 2, ..., m)$. We wish to test the null hypothesis

$$H_0: \mu_1 = \mu_2 = ... = \mu_m \tag{10.32a}$$

against the alternative

$$H_a: \mu_i \neq \mu_j \text{ for some } i \neq j, \tag{10.32b}$$

at some significance level. Thus, we are testing whether the various measurements all come from the same population and, hence, have the same variance. ANOVA is a method for splitting the total variation in the data into independent components that measure different sources of variation. A comparison is then made of these independent estimates of the common variance σ^2. In the simplest case, there are two independent estimates. The first of these, s_W^2, is obtained from measurements *within* groups, and their corresponding means. It is a valid estimator of σ^2, *independent* of whether H_0 is true or not. The second estimator, s_B^2, is obtained from the measurements *between* groups, and their corresponding means. It is *only* valid when H_0 is true, and moreover if H_0 is false, it will tend to exceed σ^2. The test is, therefore, to compare the values of these two estimators and to reject H_0 if the ratio s_B^2/s_W^2 is sufficiently large. This is done using the *F* distribution.

The above is an example of a *one-way analysis* because the observations are classified under a single classification, for example, the manufacturing process in the example above. ANVOA may be extended to deal with multi-way classifications. In our example, a second classification might be the country where the components were made. This situation is referred to as *multi-way analysis*, but we will discuss only the simpler, one-way analysis. In addition, we will assume that all samples are of the same size. This condition is preferable in practical work, but can be relaxed if necessary at the expense of somewhat more complicated equations.

The procedure starts with random samples, each of size n selected from m populations. Each sample is assumed to be normally and independently distributed with means $\mu_1 = \mu_2 = \ldots = \mu_m$ and a common, but unknown, variance σ^2. If x_{ij} is the *j*th sample value from the *i*th population, the sample mean of the *i*th population is

$$\bar{x}_i = \frac{1}{n}\sum_{j=1}^{n} x_{ij}, \quad i = 1, 2, \ldots, m \tag{10.33a}$$

and the mean of the whole sample is

$$\bar{x} = \frac{1}{mn}\sum_{i=1}^{r}\sum_{j=1}^{n} x_{ij} = \frac{1}{m}\sum_{i=1}^{m} \bar{x}_i \tag{10.33b}$$

The estimate of the variance from the entire sample is given by

$$s^2 = SST/(mn-1) \tag{10.34}$$

where *SST* is the *total sum of squares* given by

$$SST = \sum_{i=1}^{m}\sum_{j=1}^{n}(x_{ij} - \bar{x})^2. \tag{10.35}$$

By writing the right-hand side of this expression as

$$\sum_{i=1}^{m}\sum_{j=1}^{n}(x_{ij} - \bar{x}_i + \bar{x}_i - \bar{x})^2,$$

and noting that by virtue of (10.33a,b),

$$\sum_{i=1}^{m}\sum_{j=1}^{n}(x_{ij} - \bar{x}_i)(\bar{x}_i - \bar{x}) = 0,$$

the quantity *SST* may be written as the sum of two terms,

$$SST = SSB + SSW, \tag{10.36}$$

where

$$SSB = n \sum_{i=1}^{m} (\bar{x}_i - \bar{x})^2 \qquad (10.37)$$

is the *sum of squares between the groups*, and

$$SSW = \sum_{i=1}^{m} \sum_{j=1}^{n} (x_{ij} - \bar{x}_i)^2 \qquad (10.38)$$

is the *sum of squares within the groups*. Alternative forms for (10.35), (10.37) and (10.38), that are more convenient for calculations, are:

$$SST = \sum_{i=1}^{m} \sum_{j=1}^{n} x_{ij}^2 - nm\bar{x}^2, \qquad (10.39a)$$

$$SSB = n \sum_{i=1}^{m} (\bar{x}_i - \bar{x})^2 \qquad (10.39b)$$

and, using (10.36),

$$SSW = \sum_{i=1}^{m} \sum_{j=1}^{n} x_{ij}^2 - n \sum_{i=1}^{m} \bar{x}_i^2. \qquad (10.39c)$$

The two sums, SSW and SSB, form the basis of the two independent estimates of the variance we require. From previous work on the chi-squared distribution, in particular its additive property, we know that if H_0 is true, the quantity $s_W^2 = SSW/m(n-1)$ is distributed as χ^2 with $m(n-1)$ degrees of freedom; $s_B^2 = SSB/(m-1)$ is distributed as χ^2 with $(m-1)$ degrees of freedom; and $s_T^2 = SST/(nm-1)$ is distributed as χ^2 with $(nm-1)$ degrees of freedom. The quantities s_T^2, s_B^2 and s_w^2 are also called the *mean squares*. The test statistic is then $F = s_B^2/s_W^2$, which is distributed with $(m-1)$ and $m(n-1)$ degrees of freedom and is unbiased if H_0 is true. The variance of SSW is the same, whether H_0 is true or not, but the variance of SSB if H_0 is not true may be calculated and is greater if H_0 is false. Thus, since s_B^2 overestimates σ^2 when H_0 is false, the test is a one-tailed one, with the critical region such that H_0 is rejected at a significance level α when $F > F_\alpha[(m-1), m(n-1)]$. Alternatively, one could calculate the *p*-value for a value v of the test statistic, i.e.

$$p - \text{value} = P[F\{(m-1), m(n-1)\} \geq v]$$

and choose the significance level *a posteriori*. The test procedure is summarized in Table 10.2.

TABLE 10.2 Analysis of variance for a one-way classification

Source of variation	Sums of squares	Degrees of freedom	Mean square	Test statistic
Between groups	SSB	$m-1$	$s_B^2 = SSB/(m-1)$	$F = s_B^2/s_W^2$
Within groups	SSW	$m(n-1)$	$s_W^2 = SSW/[m(n-1)]$	
Total	SST	$nm-1$	$s^2 = SST/(nm-1)$	

EXAMPLE 10.9

Three students each use five identical balances to weigh a number of objects to find their average weight, with the results as show below.

$$
\begin{array}{llllll}
S1 & 22 & 17 & 15 & 20 & 16 \\
S2 & 29 & 23 & 22 & 20 & 28 \\
S3 & 29 & 31 & 27 & 30 & 32
\end{array}
$$

Test, at a 5% significance level, the hypothesis that the average weights obtained do not depend on which student made the measurements.

This can be solved using a simple spreadsheet. Intermediate values are:

$$\bar{x}_1 = 18.0, \; \bar{x}_2 = 24.4, \; \bar{x}_3 = 29.8 \text{ and } \bar{x} = 24.07.$$

Then, from (10.39)

$$SSB = 209.4, \; SSW = 249.6$$

and so

$$F = s_B^2 / s_W^2 = 2.10.$$

From Table C.5, $F_{0.05}(4, 10) = 3.48$ and since $F < 3.48$, the hypothesis cannot be rejected at this significance level.

PROBLEMS 10

10.1 A signal with constant strength S is sent from location A to location B. En route, it is subject to random noise distributed as $N(0, 8)$. The signal is sent 10 times and the strength of the signals received at B are:

$$14 \quad 15 \quad 13 \quad 16 \quad 14 \quad 14 \quad 17 \quad 15 \quad 14 \quad 18$$

Test, at the 5% level, the hypothesis that the signal was transmitted with strength 13.

10.2 A factory claims to produce ball bearings with an overall mean weight of $W = 250$g and a standard deviation of 5g. A quality control check of 100 bearing finds an average weight of $\overline{W} = 248$g. Test, at a 5% significance level, the hypothesis $H_0: W = 250$g against the alternative $H_a: W \neq 250$g. What is the critical region for $\overline{W}$ and what is the probability of accepting H_0, if the true value of the overall mean weight is 248g?

10.3 Nails are sold in packets with an average weight of 100 g and the seller has priced the packets in the belief that 95% of them are within 2g of the mean. A sample of 20 packets are weighed with the results:

$$
\begin{array}{cccccccccc}
100 & 97 & 89 & 93 & 103 & 105 & 93 & 110 & 101 & 102 \\
98 & 99 & 105 & 106 & 89 & 103 & 90 & 93 & 92 & 106
\end{array}
$$

Assuming an approximate normal distribution, test this hypothesis at a 10% significance level against the alternative that the variance is greater than expected.

10.4 A supplier sells resistors of nominal value 5 ohms in packs of 10 and charges a premium by claiming that the average value of the resistors in a pack is not less than 5 ohms. A buyer tests this claim by measuring the values of the resistors in two packs with the results:

5.3 5.4 4.9 5.1 5.0 4.8 5.1 5.2 4.7 4.9
5.2 5.5 4.7 4.6 5.5 5.4 5.0 5.0 4.8 5.4

Test the supplier's claim at a 10% significance level.

10.5 New production technique M1 has been developed for the abrasive material of car brakes and it is claimed that it reduces the spread in the lifetimes of the product. It is tested against the existing technique M2, by measuring samples of the effective lifetimes of random samples of each production type. Use the lifetime data below to test the claim at a 5% significance level.

M1   98 132 109 116 131 124 117 120 116 99 109 113
M2 100 120 134   99 130 113 106 124 118

10.6 Two groups of students study for a physics examination. Group 1 of 15 students study full-time at college and achieve an average mark of 75% with a standard deviation of 3. Group 2 of 6 students study part-time at home and achieve an average mark of 70% with a standard deviation of 5. If the two populations are assumed to be normally distributed with equal variances, test at a 5% level of significance the claim that full-time study produces better results.

10.7 Six samples of the same radioisotope are obtained from four different sources and their activities in kBq measure and found to be:

S1	S2	S3	S4
91	129	119	100
100	127	141	93
99	100	123	89
89	98	137	110
110	97	132	132
96	113	124	116

Test, at a 10% significance level, the hypothesis that the mean activity does not depend on the source of the supply.

10.8 A manufacturer of an electrical device states that if it is stress tested with a high voltage, on average no more than 4% will fail. A buyer checks this by stress testing a random sample of 1000 units and finds that 50 fail. What can be said about the manufacturer's statement at a 10% significance level?

10.9 The supplier in Example 10.8 claims that a defective rate of 1% means that no more than 10 unacceptable capacitors are distributed on any day. This is checked by daily testing batches of capacitors. A five-day run of such tests resulted in 13, 12, 15, 12, 11 defective capacitors. Is the supplier's claim supported at the 5% significance level?

CHAPTER 11

Hypothesis Testing II: Other Tests

OUTLINE

11.1 Goodness-of-Fit Tests	221	11.3.1 Sign Test	233
11.1.1 Discrete Distributions	222	11.3.2 Signed-Rank Test	234
11.1.2 Continuous Distributions	225	11.3.3 Rank-Sum Test	236
11.1.3 Linear Hypotheses	228	11.3.4 Runs Test	237
11.2 Tests for Independence	231	11.3.5 Rank Correlation Coefficient	239
11.3 Nonparametric Tests	233		

In Chapter 10 we discussed how to test a statistical hypothesis H_0 about a single population parameter, such as whether its mean μ was equal to a specific value μ_0, against a definite alternative H_a, for example that μ was greater than μ_0, and gave a very brief explanation about how this may be extended to many parameters by the method known as analysis of variance. In this chapter we will discuss a range of other tests. Some of these address questions about the population as a whole, without always referring to a specific alternative, which is left as implied. Examples are those that examine whether a set of observations is described by a specific probability density, or whether a sample of observations is random, or whether two sets of observations are compatible. We will also discuss some tests that are applicable to non-numeric data. We will start by looking at the first of these questions.

11.1. GOODNESS-OF-FIT TESTS

In Section 8.4.1 we introduced the method of estimation known as 'minimum chi-square', and at the beginning of that chapter we briefly discussed how the same technique could be used to test the compatibility of repeated measurements. In the latter applications we are testing the statistical hypothesis that estimates produced by the measurement process all come from the same population. Such procedures, for obvious reasons, are known as *goodness-of-fit tests* and are widely used in physical science.

11.1.1. Discrete Distributions

We will start by considering the case of a discrete random variable x that can take on a finite number of values $x_i (i = 1, 2, ..., k)$ with corresponding probabilities $p_i (i = 1, 2, ..., k)$. We will test the null hypothesis

$$H_0 : p_i = \pi_i, \quad i = 1, 2, ..., k, \tag{11.1a}$$

against the alternative

$$H_a : p_i \neq \pi_i, \tag{11.1b}$$

where π_i are specified fixed values and

$$\sum_{i=1}^{k} \pi_i = \sum_{i=1}^{k} p_i = 1.$$

To do this we will use the method of likelihood ratios that was developed in Chapter 10. The likelihood function for a sample of size n is

$$L(\mathbf{p}) = \prod_{i=1}^{k} p_i^{f_i},$$

where $f_i = f_i(\mathbf{x})$ is the observed frequency of the value x_i. The maximum value of $L(\mathbf{p})$ if H_0 is true is

$$\max L(\mathbf{p}) = L(\boldsymbol{\pi}) = \prod_{i=1}^{k} \pi_i^{f_i}. \tag{11.2}$$

To find the maximum value of $L(\mathbf{p})$ if H_a is true, we need to know the ML estimator of $\mathbf{p}$, i.e., $\hat{\mathbf{p}}$. Thus we have to maximize

$$\ln L(\mathbf{P}) = \sum_{i=1}^{k} f_i \ln p_i, \tag{11.3a}$$

subject to the constraint

$$\sum_{i=1}^{k} p_i = 1 \tag{11.3b}$$

Introducing the Lagrange multiplier Λ, the variation function is

$$P = \ln L(\mathbf{P}) - \Lambda \left[\sum_{i=1}^{k} p_i - 1 \right],$$

11.1. GOODNESS-OF-FIT TESTS

and setting $\partial P / \partial p_i = 0$, gives $p_j = f_i / \Lambda$. Now since

$$\sum_{j=1}^{k} p_j = 1 = \frac{1}{\Lambda} \sum_{j=1}^{k} f_j = \frac{n}{\Lambda}.$$

the required ML estimator is $p_j = f_i / n$. Thus the maximum value of $L(\mathbf{P})$ if H_a is true is

$$L(\hat{\mathbf{p}}) = \prod_{i=1}^{k} \left(\frac{f_i}{n}\right)^{f_i}. \qquad (11.4)$$

The likelihood ratio is therefore, from (11.2) and (11.4),

$$\lambda = \frac{L(\boldsymbol{\pi})}{L(\hat{\mathbf{p}})} = n^n \prod_{i=1}^{k} \left(\frac{\pi_i}{f_i}\right)^{f_i}. \qquad (11.5)$$

Finally, H_0 is accepted if $\lambda < \lambda_c$, where λ_c is a given fixed value of λ that depends on the confidence level of the test, and H_0 is rejected if $\lambda > \lambda_c$.

EXAMPLE 11.1

A die is thrown 60 times and the resulting frequencies of the faces are as shown in the table.

Face	1	2	3	4	5	6
Frequency	9	8	12	11	6	14

Test whether the die is 'true' at a 10% significance level.
From equation (11.5),

$$-2 \ln \lambda = -2 \left[n \ln n + \sum_{i=1}^{k} f_i \ln \left(\frac{\pi_i}{f_i}\right) \right], \qquad (11.6)$$

where

$$\pi_i = 1/6; \quad n = 60; \quad \text{and} \quad k = 6.$$

Thus

$$-2 \ln \lambda = -2[60 \ln 60 - 60 \ln 6 - 9 \ln 9 \ldots - 14 \ln 14] \simeq 4.3.$$

We showed in Chapter 10 that $-2 \ln \lambda$ is approximately distributed as χ^2, with $(k-1)$, in this case 5, degrees of freedom. (There are only 5 degrees of freedom, not 6, because of the constraint (11.3b).) From Table C.4 we find

$$\chi^2_{0.1}(5) = 9.1,$$

and since $\chi^2 < 9.1$, we can accept the hypothesis that the die is true, i.e.,

$$H_0 : p_i = 1/6, \quad i = 1, 2, \ldots, 6,$$

at a 10% significance level. Alternatively, we could calculate the *p*-value for a χ^2 value of 4.3. This is approximately 0.5, so that the hypothesis would be accepted for all significance levels $\alpha < 0.5$.

An alternative goodness-of-fit test is due to Pearson. He considered the statistic

$$X^2 = \sum_{i=1}^{k} \frac{(f_i - e_i)^2}{e_i} = \sum_{i=1}^{k} \frac{(f_i - n\pi_i)^2}{n\pi_i}, \qquad (11.7)$$

where f_i are the observed frequencies and e_i are the expected frequencies under the null hypothesis. Pearson showed that X^2 is approximately distributed as χ^2 with $(k-1)$ degrees of freedom for large values of n. At first sight the two statistics X^2 and $-2\ln\lambda$ appear to be unrelated, but in fact they can be shown to be equivalent asymptotically. To see this, we define

$$\Delta_i \equiv (f_i - n\pi_i)/n\pi_i,$$

and write (11.6) as

$$-2\ln\lambda = 2\sum_{i=1}^{k} f_i \ln(1 + \Delta_i).$$

For small Δ_i we can expand the logarithm as

$$\ln(1+\Delta_i) = \Delta_i - \Delta_i^2/2 + \ldots,$$

which gives

$$-2\ln\lambda = 2\sum_{i=1}^{k}\left[(f_i - n\pi_i) + n\pi_i\right]\left[\Delta_i - \frac{1}{2}\Delta_i^2 + O\!\left(n^{-3/2}\right)\right]$$

$$= 2\sum_{i=1}^{k}\left[(f_i - n\pi_i)\Delta_i + n\pi_i\Delta_i - \frac{1}{2}n\pi_i\Delta_i^2 + O\!\left(n^{-1/2}\right)\right].$$

Using the definition of Δ_i and the fact that

$$\sum_{i=1}^{k}\Delta_i\pi_i = 0,$$

gives

$$-2\ln\lambda = \sum_{i=1}^{k}\left[n\pi_i\Delta_i^2 + O\!\left(n^{-1/2}\right)\right] = X^2\left[1 + O\!\left(n^{-1/2}\right)\right].$$

So X^2 and $-2\ln\lambda$ are asymptotically equivalent statistics, although they will differ for small samples. Using the data of Example 11.1, gives $X^2 = 4.20$, compared to $-2\ln\lambda = 4.30$. The Pearson statistic is easy to calculate and in practice is widely used. It is usually written as

$$\chi^2 = \sum_{i=1}^{k} \frac{(o_i - e_i)^2}{e_i}, \qquad (11.8)$$

where o_i are the observed frequencies and e_i are the expected frequencies under the null hypothesis.

EXAMPLE 11.2

A manufacturer produces an electrical component in four different and increasing qualities A, B, C, and D. These are produced independently with probabilities 0.20, 0.20, 0.25, and 0.35, respectively. A laboratory purchases a number of components and pays a lower price by agreeing to accept a sample chosen at random from A, B, C, and D. To test that the manufacturer is not in practice supplying more lower quality components than expected, the lab tests a random sample of 60 of them and finds 15, 7, 19, and 19 are of quality A, B, C, and D, respectively. At what significance level do the components not satisfy the manufacturer's claim?

We are testing the hypothesis

$$H_0 : p_i = \pi_i, \quad i = 1, 2, \ldots, k$$

where

$$\pi_i = 0.20, \ 0.20, \ 0.25, \ 0.35, \ \text{for } i = 1, 2, 3, 4, \ \text{respectively.}$$

We start by calculating the expected frequencies under H_0, i.e., $e_i = 60\pi_i$. Thus

$$e_i = 12, \ 12, \ 15, \ 21, \quad \text{for } i = 1, 2, 3, 4, \text{ respectively.}$$

Then from (11.8) we find $\chi^2 = 4.09$ for 3 degrees of freedom. The p-value of the test is therefore $p = P[\chi^2 \geq 4.09 : H_0]$ and from Table C.4 this is approximately 0.25. Thus H_0 would only be rejected at significance levels above about 0.25, i.e., in practice the manufacturer is fulfilling the contract.

11.1.2. Continuous Distributions

For the case of continuous distributions, the null hypothesis is usually that a population is described by a certain density function $f(x)$. This hypothesis may be tested by dividing the observations into k intervals and then comparing the observed interval frequencies $o_i (i = 1, 2, \ldots, k)$ with the expected values $e_i (i = 1, 2, \ldots, k)$ predicted by the postulated density function, using equation (11.8). For practical work a rule of thumb is that the number of observations in each bin should not be less than about 5. This may necessitate combining data from two or more bins until this criterion is satisfied.

If the expected frequencies π_i are unknown, but are estimated from the sample in terms of r parameters to be $\hat{\pi}_i$ then the statistic

$$\chi^2 = \sum_{i=1}^{k} \frac{(o_i - n\hat{\pi}_i)^2}{n\hat{\pi}_i}, \qquad (11.9)$$

is also distributed as χ^2 but now with $(k-1-r)$ degrees of freedom. In using the chi-square test of a continuous distribution with unknown parameters, one always has to be careful that the method of estimating the parameters still leads to an asymptotic χ^2 distribution. In general, this will not be true if the parameters are estimated either from the original data or from the grouped data. The correct procedure is to estimate the parameters $\boldsymbol{\theta}$ by the ML method using the likelihood function

$$L(\boldsymbol{\theta}) = \prod_{i=1}^{k} [p_i(\mathbf{x}, \boldsymbol{\theta})]^{f_i},$$

where p_i is the appropriate density function. Such estimates are usually difficult to obtain; however if one uses the simple estimates then one may be working at a higher significance level than intended. This will happen, for example, in the case of the normal distribution.

EXAMPLE 11.3

A quantity x is measured 100 times in the range 0–40 and the observed frequencies o in eight bins are given below.

x	0–5	5–10	10–15	15–20	20–25	25–30	30–35	35–40
o	5	9	15	26	18	14	8	5

Test the hypothesis that x has a normal distribution.

If the hypothesis is true, then the expected frequencies e have a normal distribution with the same mean and standard deviation as the observed data. The latter may be estimated from the sample and are $\bar{x} = 19.6$ and $s = 8.77$. Using these we can find the expected frequencies for each of the bins. This may be done either by finding the z value at each of the bin boundaries and using Tables C.1 of the normal distribution function or by direct integration of the normal pdf. Either way, the expected frequencies are

e	3.5	8.9	16.3	21.8	21.3	15.1	7.8	2.9

Since the first and last bins have entries less than 5, we should combine these with the neighboring bins, so that we finally get

o	14	15	26	18	14	13
e	12.4	16.3	21.8	21.3	15.1	10.7

Then from (11.9) we find $\chi^2 = 2.21$ and because we have estimated two parameters from the data, this is for 3 degrees of freedom. From Table C.4, there is a probability of approximately 50% of finding a value at least large as this for 3 degrees of freedom; so the hypothesis of normality is very consistent with the observations.

A disadvantage of Pearson's χ^2 test is that the data have to be binned, which could be a problem in cases where the number of observations is small. One method that does not require binning is the *Kolmogorov–Smirnov test*. In this test we start with a sample y_i of size n from a continuous distribution and test the hypothesis H_0 that the distribution function is $F(x)$. This is done by defining a piecewise continuous function $F_e(x)$ as the proportion of the observed values that are less than x, i.e., $(y_i \leq x)/n$. If H_0 is true, $F_e(x)$ will be close to

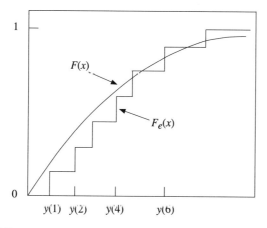

FIGURE 11.1 Construction of the Kolmogorov–Smirnov statistic.

$F(x)$ and a natural test statistic is the quantity $D \equiv \max|F_e(x) - F(x)|$, where the maximum is found by considering all values of x. This is called the Kolmogorov–Smirnov statistic.

To compute D, the sample is first ordered in increasing values and relabeled $y(1), y(2), \ldots, y(n)$. Then $F_e(x)$ consists of a series of steps with increases of $1/n$ at the points $y(1), y(2), \ldots, y(n)$ as shown in Fig. 11.1. Now because D is defined in terms of a modulus, it can be written as

$$D = \max\{\max[F_e(x) - F(x)], \max[F(x) - F_e(x)]\}$$
$$= \max\left\{\frac{j}{n} - F[y(j)], F[y(j)] - \frac{j-1}{n}\right\}, \quad (11.10)$$

where j takes all values from 1 to n. So if the data lead to a value of $D = d$, the p-value of the test is

$$p = P[D \geq d : H_0].$$

Although this probability in principle depends on whether H_0 is true, it can be shown that in practice it is independent of the form of $F(x)$ and, without proof, a test at significance level α can be found by considering the quantity

$$D^* \equiv (\sqrt{n} + 0.12 + 0.11/\sqrt{n})D, \quad (11.11)$$

which approximates to $\sqrt{n}\,D$ for large samples. Then if $P[D^* \geq d^*] = \alpha$, the critical values of d_α^* are

$$d_{0.1}^* = 1.224, \quad d_{0.05}^* = 1.358, \quad d_{0.025}^* = 1.480, \quad d_{0.01}^* = 1.626. \quad (11.12)$$

Finally, a test at a significance level α would reject H_0 if the observed value of D^* is at least as large as d_α^*.

EXAMPLE 11.4

A sample of size six is drawn from a population and the values in ascending order are:

1	2	3	4	5	6
0.20	0.54	0.71	1.21	1.85	2.45

Test the hypothesis that the sample comes from a population with the distribution function of Example 3.2. From the data and the distribution function of Example 3.2, we can construct the following table:

j	1	2	3	4	5	6
$F[y(j)]$	0.001	0.018	0.035	0.123	0.283	0.443
$j/n - F[y(j)]$	0.166	0.316	0.465	0.544	0.551	0.557
$F[y(j)] - (j-1)/n$	0.001	−0.149	−0.298	−0.377	−0.385	−0.390

From this table and equation (11.10) we have $D = 0.557$ and hence from (11.11) we have $D^* = 1.456$. Finally, using the critical values given in (11.12), we reject the null hypothesis at the 0.05 level of significance, but not at the 0.025 level.

11.1.3. Linear Hypotheses

In Section 8.1.3 we briefly mentioned the use of the χ^2 and F distributions as goodness-of-fit tests in connection with the use of the linear least-squares method of estimation. These applications were designed to test hypotheses concerning the quality of the approximation of the observations by some assumed expression linear in the parameters. We shall generalize that discussion now to consider some other hypothesis tests that can be performed using the least-squares results.

We have seen in Section 8.1 that the weighted sum of residuals $S = \mathbf{R}^T\mathbf{V}^{-1}\mathbf{R}$, where

$$\mathbf{R} = \mathbf{Y} - \mathbf{\Phi}\mathbf{\Theta},$$

and $\mathbf{V}$ is the variance matrix of the observations, is distributed as χ^2 with $(n-p)$ degrees of freedom, where n is the number of observations and p is the number of parameters $\theta_k (k = 1, 2, ..., p)$. We also saw (compare equation (8.31)) that

$$\mathbf{R}^T\mathbf{V}^{-1}\mathbf{R} = (\mathbf{Y} - \mathbf{Y}^0)^T\mathbf{V}^{-1}(\mathbf{Y} - \mathbf{Y}^0) - (\hat{\mathbf{\Theta}} - \mathbf{\Theta})^T\mathbf{M}^{-1}(\hat{\mathbf{\Theta}} - \mathbf{\Theta}),$$

where $\mathbf{M}$ is the variance matrix of the parameters. It follows from the additive property of χ^2 that since

$$(\mathbf{Y} - \mathbf{Y}^0)^T\mathbf{V}^{-1}(\mathbf{Y} - \mathbf{Y}^0)$$

is distributed as χ^2 with n degrees of freedom, the quantity

$$(\hat{\mathbf{\Theta}} - \mathbf{\Theta})^T\mathbf{M}^{-1}(\hat{\mathbf{\Theta}} - \mathbf{\Theta})$$

11.1. GOODNESS-OF-FIT TESTS

is distributed as χ^2 with p degrees of freedom. To test deviations from the least-squares estimates for the parameters we need to know the distribution of

$$(\hat{\Theta} - \Theta)^T E^{-1}(\hat{\Theta} - \Theta)$$

where E is the error matrix of equation (8.32). In the notation of Section 8.1.2,

$$E = \hat{w}(\Phi^T W \Phi)^{-1}, \qquad (11.13)$$

and so

$$(\hat{\Theta} - \Theta)^T E^{-1}(\hat{\Theta} - \Theta) = \frac{1}{\hat{w}}(\hat{\Theta} - \Theta)^T (\Phi^T W \Phi)(\hat{\Theta} - \Theta).$$

That is,

$$(\hat{\Theta} - \Theta)^T E^{-1}(\hat{\Theta} - \Theta) = (\hat{\Theta} - \Theta)^T M^{-1}(\hat{\Theta} - \Theta) \left(\frac{w}{\hat{w}}\right).$$

But we have seen above that

$$(\hat{\Theta} - \Theta)^T M^{-1}(\hat{\Theta} - \Theta)$$

is distributed as χ^2 with p degrees of freedom, and so

$$(\hat{\Theta} - \Theta)^T E^{-1}(\hat{\Theta} - \Theta)/p$$

is distributed as

$$\frac{\chi^2(p)/p}{\chi^2(n-p)/(n-p)} = F(p, n-p).$$

Thus to test the hypothesis $H_0 : \Theta = \Theta_0$, we compute the test statistic

$$F_0 = (\hat{\Theta} - \Theta)^T E^{-1}(\hat{\Theta} - \Theta)/p \qquad (11.14)$$

and reject the hypothesis at a significance level of α if $F_0 > F_\alpha(p, n-p)$.

The foregoing discussion is based on the work of Section 8.2, where we considered the least-squares method in the presence of linear constraints on the parameters. By analogy we will now generalize the discussion to include the general linear hypothesis

$$H_0 : C_{lp}\theta_p = Z_l, \quad l \leq p. \qquad (11.15)$$

This may be a hypothesis about all of the parameters, or any subset of them. The null hypothesis H_0 may be tested by comparing the least-squares solution for the weighted sum of residuals when H_0 is true, i.e., S_c, with the sum in the unconstrained situation, i.e., S. In the notation of Section 8.2, the additional sum of residuals $S_a = S_c - S$, which is present if the hypothesis H_0 is true, is distributed as χ^2 with l degrees of freedom, independently of S, which itself is distributed as χ^2 with $(n-p)$ degrees of freedom. Thus the ratio

$$F = \frac{S_a/l}{S/(n-p)}, \qquad (11.16)$$

is distributed as $F(l, n - p)$. Using the results of Section 8.2 we can then show that

$$F = (Z - C\hat{\Theta})^T (CEC^T)^{-1} (Z - C\hat{\Theta})/l. \tag{11.17}$$

Thus H_0 is rejected at the α significance level if $F > F_\alpha(l, n - p)$. (Compare the discussion at the end of Section 8.1.3.)

EXAMPLE 11.5

An experiment results in the following estimates for three parameters, based on ten measurements

$$\hat{\theta}_1 = 2; \quad \hat{\theta}_2 = 4; \quad \hat{\theta}_3 = 1,$$

with an associated error matrix

$$E = \begin{pmatrix} 1 & 0 & 0 \\ 0 & 2 & 1 \\ 0 & 1 & 1 \end{pmatrix}.$$

Test the hypothesis

$$H_0 : \theta_1 = 0; \quad \theta_2 = 0$$

at the 5% significance level.

For the above hypothesis,

$$\hat{\Theta} = \begin{pmatrix} 2 \\ 4 \\ 1 \end{pmatrix}; \quad C = \begin{pmatrix} 1 & 0 & 0 \\ 0 & 1 & 0 \end{pmatrix}; \quad Z = \begin{pmatrix} 0 \\ 0 \end{pmatrix},$$

and the calculated value of F from equation (11.17) is 6. From Table C.5 of the F distribution, we find that

$$F_\alpha(l, n - p) = F_{0.05}(2, 7) = 4.74,$$

and so we can reject the hypothesis at a 5% significance level.

Finally, we have to consider the power of the test of the general linear hypothesis, i.e., we have to find the distribution of F if H_0 is not true. Now $S/(n - p)$ is distributed as $\chi^2/(n - p)$ regardless of whether H_0 is true or false, but S_a/l is only distributed as χ^2/l if H_0 is true. If H_0 is false, then S_a/l will in general be distributed as *noncentral* χ^2, which has, for l degrees of freedom, a density function

$$f^{nc}(\chi^2; l) = \sum_{p=0}^{\infty} \left(\frac{e^{-\lambda} \lambda^p}{p!} \right) f(\chi^2; l + 2p),$$

where $f(\chi^2; l + 2p)$ is the density function for a χ^2 variable with $(l + 2p)$ degrees of freedom, and the noncentrality parameter is

$$\lambda = \frac{1}{2} (C\Theta - Z)^T (CMC^T)^{-1} (C\Theta - Z).$$

It follows that F is distributed as a *noncentral F distribution*. Tables of the latter distribution are available to construct the power curves. A feature of the noncentral F distribution is that the power of the test increases as λ increases.

11.2. TESTS FOR INDEPENDENCE

The χ^2 procedure of Section 11.1 can also be used to construct a test for the independence of variables. Suppose n observations have been made and the results are characterized by two random variables x and y that can take the discrete values $x_1, x_2, ..., x_r$ and $y_1, y_2, ..., y_c$, respectively. (Continuous variables can be accommodated by dividing the range into intervals, as described in Section 11.1.2) If the number of times the value x_i is observed together with y_j is n_{ij}, then the data can be summarized in the matrix form shown in Table 11.1, called an $r \times c$ *contingency table*.

If we denote by p_i the marginal probability for x_i, i.e., the probability for x to have the value x_i, independent of the value of y, and likewise denoted by q_j the marginal probability for y_j, then if the null hypothesis H_0 is that x and y are independent variables, the probability for observing x_i simultaneously with y_j is $p_i q_j$ ($i = 1, 2, ..., r; j = 1, 2, ..., c$). Since the marginal probabilities are not specified in the null hypothesis they must be estimated. Using the maximum likelihood method, this gives

$$\hat{p}_i = \frac{1}{n} \sum_{j=1}^{c} n_{ij}, \quad i = 1, 2, ..., r, \tag{11.18a}$$

and

$$\hat{q}_j = \frac{1}{n} \sum_{i=1}^{r} n_{ij}, \quad j = 1, 2, ..., c, \tag{11.18b}$$

and if H_0 is true, the expected values for the elements of the contingency table are $n\hat{p}_i\hat{q}_j$. So H_0 can be tested by calculating the quantity

$$\chi^2 = \sum_{i=1}^{r} \sum_{j=1}^{c} \frac{(n_{ij} - n\hat{p}_i\hat{q}_j)^2}{n\hat{p}_i\hat{q}_j} \tag{11.19}$$

TABLE 11.1 An $r \times c$ contingency table

	y_1	y_2	$\cdots$	y_c
x_1	n_{11}	n_{12}	$\cdots$	n_{1c}
x_2	n_{21}	n_{22}	$\cdots$	n_{2c}
$\vdots$	$\vdots$	$\vdots$		$\vdots$
x_r	n_{r1}	n_{r1}	$\cdots$	n_{rc}

and comparing it with χ^2_α at a significance level α. It remains to find the number of degrees of freedom for the statistic χ^2. First we note that because

$$\sum_{i=1}^{r} \hat{p}_i = \sum_{i=j}^{c} \hat{q}_j = 1, \qquad (11.20)$$

only $(r-1)$ parameters p_i and $(c-1)$ parameters q_i need to be estimated, i.e., $(r+c-2)$ in total. Therefore the number of degrees of freedom is

$$(\text{number of entries} - 1) - (r + c - 2) = (r-1)(c-1). \qquad (11.21)$$

EXAMPLE 11.6

A laboratory has three pieces of test apparatus of the same type, and they are used by four technicians on a one-month rota. A record is kept of the number of machine breakdowns for each month, and the average number is shown in the table according to which technician was using the machine.

Technician	Equipment			Total
	E1	E2	E3	
T1	3	6	1	10
T2	0	1	2	3
T3	3	2	4	9
T4	6	2	1	9
Total	12	11	8	31

Are the variables E, the equipment number, and T, the technician who used the equipment, independent random variables in determining the rate of breakdowns?

In the notation above, $r = 4$, $c = 3$ and $n = 31$. So from (11.18), using the data given in the table,

$$\hat{p}_1 = \frac{10}{31}, \quad \hat{p}_2 = \frac{3}{31}, \quad \hat{p}_3 = \frac{9}{31}, \quad \hat{p}_4 = \frac{9}{31},$$

and

$$\hat{q}_1 = \frac{12}{31}, \quad \hat{q}_2 = \frac{11}{31}, \quad \hat{q}_3 = \frac{8}{31}.$$

Then from (11.19)

$$\chi^2 = \sum_{i=1}^{c}\sum_{j=1}^{r} \frac{(n_{ij} - n\hat{p}_i\hat{q}_j)^2}{n\hat{p}_i\hat{q}_j} = 10.7,$$

and this is for $(r-1)(c-1) = 6$ degrees of freedom. The p-value is

$$p = P[\chi^2_6 \geq 5.66 : H_0] = \alpha$$

and from Table C.4, $\alpha \approx 0.1$; so the data are consistent at the 10% significance level with the independence of the two variables in determining the number of breakdowns.

11.3. NONPARAMETRIC TESTS

The tests that were discussed in Chapter 10 all assumed that the population distribution was known. However there are cases were this is not true, and to deal with these situations we can order the observations by rank and apply tests that do not rely on information about the underlying distribution. These are variously called *nonparametric tests*, or *distribution-free tests*. Not making assumptions about the population distribution is both a strength and a weakness of such tests; a strength because of their generality and a weakness because they do not use all the information contained in the data. An additional advantage is that such tests are usually quick and easy to implement, but because they are less powerful than more specific tests we have discussed, the latter should usually be used if there is a choice. In addition, although the data met in physical science are usually numeric, occasionally we have to deal with non-numeric data, for example, when testing a piece of equipment the outcome could 'pass' or 'fail'. Some nonparametric tests can also be applied to these situations. The discussion will be brief, and not all proofs of statements will be given.

11.3.1. Sign Test

In Section 10.5 we posed the question of whether it is possible to test hypotheses about the average of a population when its distribution is unknown. One simple test that can do this is the *sign test*, and as an example of its use we will test the null hypothesis $H_0 : \mu = \mu_0$ against some alternative, such as $H_a : \mu = \mu_a$ or $H_a : \mu > \mu_0$ using a random sample of size n in the case where n is small, so that the sampling distribution may not be normal. In general if we make no assumption about the form of the population distribution, then in the sign test and those that follow, μ refers to the median, but if we know that the population distribution is symmetric, then μ is the arithmetic mean. For simplicity the notation μ will be used for both cases.

We start by assigning a plus sign to all data values that exceed μ_0 and a minus sign to all those that are less than μ_0. We would expect the plus and minus signs to be approximately equal and any deviation would lead to rejection of the null hypothesis at some significance level. In principle, because we are dealing with a continuous distribution, no observation can in principle be exactly equal to μ_0, but in practice approximate equality will occur depending on the precision with which the measurements are made. In these cases the points of 'practical equality' are removed from the data set and the value of n reduced accordingly. The test statistic X is the number of plus signs in the sample (or equally we could use the number of minus signs). If H_0 is true, the probabilities of obtaining a plus or minus sign are equal to ½ and so X has a binomial distribution with $p = p_0 = 1/2$. Significance levels can thus be obtained from the binomial distribution for one-sided and two-sided tests at any given level α.

For example, if the alternative hypothesis is $H_a : \mu > \mu_0$, then the largest critical region of size not exceeding α is obtained from the inequality $x \geq k_\alpha$, where

$$\sum_{x=k_\alpha}^{n} B(x : n, p_0) \leq \alpha, \tag{11.22a}$$

and B is the binomial probability with $p_0 = p = \frac{1}{2}$ if H_0 is true. Similarly, if $H_a : \mu < \mu_0$, we form the inequality $x \leq k'_\alpha$, where k'_α is defined by

$$\sum_{x=0}^{k'_\alpha} B(x : n, p_0) \leq \alpha \qquad (11.22\text{b})$$

Finally, if $H_a : \mu \neq \mu_0$, i.e., we have a two-tailed test, then the largest critical region is defined by

$$x \leq k'_{\alpha/2} \quad \text{and} \quad x \geq k_{\alpha/2}. \qquad (11.22\text{c})$$

For sample sizes greater than about 10, the normal approximation to the binomial may be used with mean $\mu = np$ and $\sigma^2 = np(1-p)$.

EXAMPLE 11.7

A mobile phone battery needs to be regularly recharged even if no calls are made. Over 12 periods when charging was required, it was found that the intervals in hours between chargings were:

50 35 45 65 39 38 47 52 43 37 44 40

Use the sign test to test at a 10% significance level the hypothesis that the battery needs recharging on average every 45 hours.

We are testing the null hypothesis $H_0 : \mu_0 = 45$ against the alternative $H_a : \mu_0 \neq 45$. First we remove the data point with value 45, reducing n to 11, and then assign a plus sign to those measurements greater than 45 and a minus sign to those less than 45. This gives $x = 4$ as the number of plus signs. As this is a two-tailed test, we need to find the values of $k_{0.05}$ and $k'_{0.05}$ for $n = 11$. From Table C.2, these are $k'_{0.05} = 3$ and $k_{0.05} = 9$. Since $x = 4$ lies in the acceptance region, we accept the null hypothesis at this significance level.

The sign test can be extended in a straightforward way to two-sample cases, for example, to test the hypothesis that $\mu_1 = \mu_2$ using samples of size n drawn from two non-normal distributions. In this case the differences $d_i (i = 1, 2, ..., n)$ of each pair of observations is replaced by a plus or minus sign depending on whether d_i is greater than or less than zero, respectively. If the null hypothesis instead of being $\mu_1 - \mu_2 = 0$ is instead $\mu_1 - \mu_2 = d$, then the procedure is the same, but the quantity d is subtracted from each d_i before the test is made.

11.3.2. Signed-Rank Test

The sign test uses only the positive and negative signs of the differences between the observations and μ_0 in a one-sample case (or the signs of the differences d_i between observations in a paired sample case). Because it ignores the magnitudes of the differences, it can, for example, lead to conclusions that differ from those obtained using the t-distribution, which assumes the population distribution is normal. There is however another test, called the *Wilcoxon signed-rank test*, or simply the *signed-rank test*, that does take into account the magnitude of these differences. The test proceeds as follows.

We will assume that the null hypothesis is $H_0 : \mu = \mu_0$ and that the distribution is symmetric. If the latter is not true then the tests will refer to the median of the distribution. First, the differences $d_i = x_i - \mu_0 (i = 1, 2, \ldots, n)$ (or $d_i = x_i - y_i$) are found and any that are zero discarded. The absolute values of the remaining set of differences are then ranked in ascending order, i.e., rank 1 assigned to the smallest absolute value of d_i, rank 2 assigned to the second smallest absolute value of d_i, etc. If $|d_i|$ is equal for two or more values, their rank is assigned to be the average of the ranks they would have had if they had been slightly different and so distinguishable. For example, if two differences are equal and notionally have rank 5, both would be assigned a rank of 5.5 and the next term would be assigned a rank 7.

We now define w_+ and w_- to be the sum of the rank numbers corresponding to $d_i > 0$ and $d_i < 0$, respectively, and $w = \min\{w_+, w_-\}$. For different samples of the same size n, the values of these statistics will vary, and for example to test $H_0 : \mu = \mu_0$ against the alternative $H_a : \mu < \mu_0$, i.e., a one-tailed test, we would reject H_0 if w_+ is small and w_- large. (Similarly for testing $H_0 : \mu_1 = \mu_2$ against $H_a : \mu_1 < \mu_2$.) Likewise if w_+ is large and w_- small, we would accept $H_a : \mu > \mu_0$ (and similarly $H_a : \mu_1 > \mu_2$). For a two-tailed test with $H_a : \mu \neq \mu_0$, or $H_a : \mu_1 \neq \mu_2$, H_0 would be rejected if either w_+, w_-, and hence w, were sufficiently small. If the two-tailed case is generalized to $H_0 : \mu_1 - \mu_2 = d_0$, then the same test can be used by subtracting d_0 from each of the differences d_i, as in the sign test, and in this case the distribution need not be symmetric.

The final question is to decide what is 'large' and 'small' in this context. For small $n < 5$, it can be shown that provided $\alpha < 0.05$ for a one-tailed test, or less than 0.10 for a two-tailed test, any value of w will lead to acceptance of the null hypothesis. The test is summarized in Table 11.2.

For n greater than about 25, it can be shown that the sampling distribution of w_+, or w_-, tends to a normal distribution with

$$\text{mean } \mu = \frac{n(n+1)}{4} \text{ and variance } \sigma^2 = \frac{n(n+1)(2n+1)}{24} \quad (11.23)$$

so that the required probabilities may be found using the standardized variable $z_\pm = (w_\pm - \mu)/\sigma$. For the range $5 \leq n \leq 25$ it is necessary to calculate the explicit probabilities or obtain values from tabulations, such as given in Table C.7.

TABLE 11.2 Signed-rank test

H_0	H_a	Test statistic
$\mu = \mu_0$	$\mu \neq \mu_0$	w
	$\mu < \mu_0$	w_+
	$\mu > \mu_0$	w_-
$\mu_1 = \mu_2$	$\mu_1 \neq \mu_2$	w
	$\mu_1 < \mu_2$	w_+
	$\mu_1 > \mu_2$	w_-

EXAMPLE 11.8

Rework Example 11.7 using the rank-sign test.
Using the data of Example 11.7 with $\mu_0 = 45$, we have

x_i	50	35	45	65	39	38	47	52	43	37	44	40
d_i	5	−10	0	20	−6	−7	2	7	−2	−8	−1	−5
$|d_i|$	5	10		20	6	7	2	7	2	8	1	5
rank	4.5	10		11	6	7.5	2.5	7.5	2.5	9	1	4.5

where the point with $d_i = 0$ has been discarded and the rank of ties have been averaged. From the rank numbers we calculate $w_+ = 25.5$, $w_- = 40.5$, and so $w = \min\{w_+, w_-\} = 25.5$. From Table C.7, the critical region for $n = 11$ in a two-tailed test with $\alpha = 0.10$ is strictly $w \leq 13$ (although $w \leq 14$ is also very close), but since w is greater than either of these values we again accept the null hypothesis at this significance level.

11.3.3. Rank-Sum Test

The *rank-sum test*, which is also called the *Wilcoxon, or Mann–Whitney, rank-sum test*, is used to compare the means of two continuous distributions. When applied to the case of non-normal distributions, it is more powerful than the two-sample *t*-test discussed in Chapter 10.

We will use it to test the null hypothesis $H_0 : \mu_1 = \mu_2$ against some suitable alternative. First, samples of sizes n_1 and n_2 are selected from the two populations, and if $n_1 \neq n_2$ we assume that $n_2 > n_1$. Then the $n = n_1 + n_2$ observations are arranged in an ascending order and a rank number 1, 2, ..., n is assigned to each. In the case of ties, the rank number is taken to be the mean of the rank numbers that the observations would have had if they had been distinguishable, just as in the signed-rank test. We now proceed in a similar way to that used in the signed-rank test.

Let $w_{1,2}$ be the sums of the rank numbers corresponding to the $n_{1,2}$ sets of observations. With repeated samples, the values of $w_{1,2}$ will vary and may be viewed as random variables. So, just as for the signed-rank test, $H_0 : \mu_1 = \mu_2$ will be rejected in favor of $H_a : \mu_1 < \mu_2$ if w_1 is small and w_2 is large, and similarly it will rejected in favor of $H_a : \mu_1 > \mu_2$ if w_1 is large and w_2 is small. For a two-tailed test, H_0 is rejected in favor of $H_a : \mu_1 \neq \mu_2$ if $w = \min\{w_1, w_2\}$ is sufficiently small. It is common practice to work with the statistics

$$u_{1,2} = w_{1,2} - \frac{n_{1,2}(n_{1,2}+1)}{2}. \tag{11.24}$$

The test is summarized in Table 11.3.

For large sample sizes the variables $u_{1,2}$ are approximately normally distributed with

$$\text{mean } \mu = \frac{n_1 n_2}{2} \quad \text{and} \quad \text{variance } \sigma^2 = \frac{n_1 n_2 (n_1 + n_2 + 1)}{12}. \tag{11.25}$$

11.3. NONPARAMETRIC TESTS

TABLE 11.3 Rank-sum test

H_0	H_a	Test statistic
$\mu_1 = \mu_2$	$\mu_1 \neq \mu_2$	u
	$\mu_1 < \mu_2$	u_1
	$\mu_1 > \mu_2$	u_2

Conventionally this approximation is used for $n_1 \geq 10$ and $n_2 \geq 20$. Then, as usual, the statistics $z_{1,2} = (u_{1,2} - \mu)/\sigma$ are standardized normal variates and may be used to find suitable critical regions for the test. For smaller samples, critical values of u may be found from tables, such as Table C.8. The rank-sum test may be extended to accommodate many populations, when it is called the Kruskal–Wallis test.

EXAMPLE 11.9

A laboratory buys power packs from two sources S1 and S2. Both types have a nominal rating of 10 volts. The results of testing a sample of each type yield the following data for the actual voltage:

S1	9	8	7	8		
S2	10	9	8	11	7	9

Use the rank-sum test to test the hypothesis that the two types of pack supply the same average voltage at the 10% significance level.

The data are first ranked as follows:

S1	9		8		7		8			
S2		10		9		8		11	7	9
Rank	7	9	4	7	1.5	4	4	10	1.5	7

where we have averaged the ranking numbers for ties. From this table we calculate $w_1 = 16.5$, $w_2 = 38.5$ and hence, in the notations above, $u_1 = 6.5$ and $u_2 = 17.5$, so that $u = 6.5$. From Table C.8, the critical region u, for a two-tailed test at significance level 0.10 and sample sizes $n_1 = 4$ and $n_2 = 6$ is $u \leq 3$. Since the calculated value of u is greater than 3, we accept the hypothesis at the 10% significance level.

11.3.4. Runs Test

It is a basic assumption in much of statistics that a data set constitutes a simple random sample from some underlying population distribution. But we have noted in Section 5.2.1 that it is often difficult to ensure that randomness has been achieved and so a test for this is desirable. The *runs test* does this. In this test, the null hypothesis H_0 is that the observed data set is a random sample.

To derive the test, we shall consider a simple case where the elements of the sample $x_i (i - 1, 2, 3 \ldots, N)$ can only take the values A or B. (In a non-numeric situation these might be 'pass' and 'fail'.) A *run* is then defined as a consecutive sequence of either A's or B's, for example

$$AAABAABBBBBAABBBABBAA.$$

In this example, there are $n = 10$ A's, $m = 11$ B's, with $N = n + m = 21$, and $r = 9$ runs. The total number of permutations of the N points is ${}_N C_n = N!/(n!m!)$, and if H_0 were true, then each would be equally likely. We therefore have to find the probability mass function for r, the number of runs. This is

$$P[r = k : H_0] \equiv \frac{\text{probability of obtaining } k \text{ runs}}{{}_N C_n}. \tag{11.26}$$

The numerator can be found by considering only the A's and calculating for a given n how many places are there where a run can terminate and a new one begin. If k is even, there are $(k/2 - 1)$ places where a run ceases and a new one starts. There are $(n - 1)$ of these where the first run could cease and the second start, $(n - 2)$ where the second ceases and the third starts, and so on, giving ${}_{n-1}C_{k/2-1}$ distinct arrangements for the A's. There is a corresponding factor for the possible arrangements of the B's and the product of these two terms must be multiplied by a factor of two because the entire sequence could start with either an A or a B. Thus,

$$P[r = 2k : H_0] = 2 \frac{{}_{(n-1)}C_{(k-1)} \times {}_{(m-1)}C_{(k-1)}}{{}_N C_n}. \tag{11.27a}$$

An analogous argument for k odd gives

$$P[r = 2k+1 : H_0] = \frac{\left({}_{(n-1)}C_{(k-1)} \times {}_{(m-1)}C_k\right) + \left({}_{(n-1)}C_k \times {}_{(m-1)}C_{(k-1)}\right)}{{}_N C_n}. \tag{11.27b}$$

For small values of n and m these expressions can be used to calculate the required probabilities, but it is rather tedious. Usually one consults tabulations, an example of which is Table C.9. For example, for a two-tailed test at significance level α, the critical region is defined by the inequalities $r \leq a$ and $r \geq b$, where a is the largest value of r for which

$$P[r \leq a : H_0] \leq \alpha/2, \tag{11.28a}$$

and b is the smallest value of r_0 for which

$$P[r \geq b : H_0] \geq \alpha/2. \tag{11.28b}$$

The values of a and b are given in Table C.9, and may be used for a one-tailed test at a significance level of 0.05, or a two-tailed test at a significance level of 0.1.

For large n and m, it can be shown that r is approximately normally distributed with mean and variance given by

$$\mu = \frac{2nm}{n+m} + 1 \quad \text{and} \quad \sigma^2 = \frac{2nm(2nm - N)}{N^2(N-1)}, \tag{11.29}$$

so the variable $z \equiv (r - \mu)/\sigma$ will be approximately distributed as a standardized normal distribution. In this case the p-value is approximately

$$p\text{-value} = 2\min\{N(z), 1 - N(z)\}. \tag{11.30}$$

One use of the run test is to supplement the χ^2 goodness-of-fit test described in Section 11.1.1. An example of this is given below.

EXAMPLE 11.10

Figure 11.2 shows a linear least-squares fit to a set of 24 data points. The χ^2 value is 19.8 for 22 degrees of freedom, which is acceptable. Use the runs test to test at a 10% significance level whether the data are randomly distributed about the fitted line.

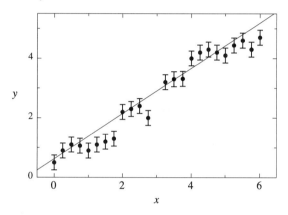

FIGURE 11.2   Linear fit to data.

The data have $n = 7$, $m = 16$ (one point is on the fitted line) and for a two-tailed test at a 10% significance level, the critical regions may be found using Table C.9. They are $r \leq 6$ and $r \geq 15$. Since the observed value of r is 9, the hypothesis that the data are randomly distributed about the best-fit line must be accepted.

11.3.5. Rank Correlation Coefficient

In previous chapters we have used the sample (Pearson's) correlation coefficient ρ to measure the correlation between two sets of continuous random variables x and y. If instead we replace their numerical values by their rankings, then we obtain the *rank correlation coefficient* (this is due to Spearman and so is also called the *Spearman rank correlation coefficient*) denoted ρ_R. It is given by

$$\rho_R = 1 - \frac{6}{n(n^2 - 1)} \sum_{i=1}^{n} d_i^2, \tag{11.31}$$

where n is the number of data points and d_i are the differences in the ranks of x_i and y_i. In practice, ties are treated as in the previous tests, i.e., the differences are averaged as if the ties could be distinguished. The rank correlation coefficient ρ_R is similar to ρ. It is a number between $+1$ and -1 with the extreme values indicating complete positive or negative correlation. For example, $\rho_R = 1$ implies that the ranking numbers of x_i and y_i are identical and a value close to zero indicates that the ranking numbers are uncorrelated. The advantages of using the Spearman correlation coefficient rather than Pearson's are the usual ones: no assumptions need to be made about the distribution of the x and y variables, and the test can be applied to non-numeric data.

The significance of the rank correlation coefficient is found by considering the distribution of ρ_R under the assumption that x and y are independent. In this case $\rho_R = 0$ and values of critical values can be calculated. An example is Table C.10. Note that the distribution of values of ρ_R is symmetric about $\rho_R = 0$, so left-tailed areas are equal to right-tailed areas for a given significance level α and for a two-tailed test the critical regions are equal in the two tails of the distribution. For larger values $n \geq 30$, the normal distribution may be used with mean zero and variance $(n-1)^{-1/2}$, so that $z = \rho_R \sqrt{n-1}$ is a standard normal variable.

EXAMPLE 11.11

A laboratory manager is doubtful whether regular preventive maintenance is leading to fewer breakdowns of his portfolio of equipment, so he records details of the annual rate of breakdowns (B) and the interval in months between services (I) for 10 similar machines (M), with the results shown below. Test the hypothesis that there is no correlation between B and I at a 5% significance level.

M	1	2	3	4	5	6	7	8	9	10
B	2	5	4	3	6	8	9	10	1	12
I	3	4	3	5	6	4	6	5	3	6

We start by rank ordering B and I to give B_R, I_R, and $d^2 \equiv (B_R - I_R)^2$:

M	B	I	B_R	I_R	$d^2 \equiv (B_R - I_R)^2$
1	2	3	2	2	0
2	5	4	5	4.5	2.25
3	4	3	4	2	4
4	3	5	3	6.5	12.25
5	6	6	6	8	4
6	8	4	7	4.5	6.25
7	9	6	8	8	0
8	10	5	9	6.5	6.25
9	1	3	1	2	1
10	12	6	10	8	4

Then from (11.31) we find $\rho_R = 0.76$. We will test the null hypothesis $H_0 : \rho_R = 0$ against the alternative $H_0 : \rho_R > 0$. From Table C.10, we find that we would reject H_0 at the 0.05 significance level because $\rho_R > 0.564$. Recall from the remarks in Section 1.3.3 that this does *not* mean that preventive maintenance *causes* fewer breakdowns, but simply means that there is a significant correlation between the two variables so that this assumption cannot be ruled out.

PROBLEMS 11

11.1 A radioactive source was observed for successive periods of 1 minute and the number of particles emitted of a specific type during 500 intervals recorded. The resulting observations o_i are shown below.

counts	0	1	2	3	4	5	6	7	8	9	10	11	12
o_i		1	8	38	67	75	85	89	66	39	15	10	5	2

Test the hypothesis that the number of particles emitted has a Poisson distribution with parameter $\lambda = 5$.

11.2 Two experiments give the following results for the value of a parameter (assumed to be normally distributed), 2.05 ± 0.01 and 2.09 ± 0.02, what can one say about their compatibility?

11.3 A sample of size ten is drawn from a population and the values are:.

1	2	3	4	5	6	7	8	9	10
48	55	65	77	94	118	135	150	167	190

Use the Kolmogorov–Smirnov technique to test at a 10% significance level the hypothesis that the sample comes from an exponential population with mean 100.

11.4 Four measurements of a quantity give values 1.12, 1.13, 1.10, and 1.09. If they all come from the same normal population with $\sigma^2 = 4 \times 10^{-4}$, test at a 5% significance level the hypothesis that the populations are identical and have a common mean $\mu_0 = 1.09$ against the alternative that $\mu_0 \neq 1.09$.

11.5 Three operatives (O) are given the task of testing identical electrical components for a fixed period of time in the morning (M), afternoon (A), and evening (E) and the numbers successfully tested are given below.

	M	A	E
O1	65	70	75
O2	95	89	70
O3	85	70	58

Test at the 10% significance level the hypothesis that the variables O and the time of day are independent variables in determining the number of components tested.

11.6 Use the normal approximation to the sign test to calculate the p-value for the hypothesis that the median of the following numbers is 35.

9	21	34	47	54	55	53
47	38	28	21	15	11	8

11.7 The sample of size 10 shown below was drawn from a symmetric population.

x	12	11	18	17	15	19	19	20	17	16

Use the signed-rank test to examine at a 5% significance level the hypothesis that the population mean is 15.5.

11.8 Two samples of weights are measured and give the following data:

S1	9	6	7	8	7		
S2	8	7	9	10	9	11	8

Use the rank-sum test to test at a 5% significance level the hypothesis that the mean value obtained from S1 is smaller than that from S2.

11.9 Test at a 10% significance level whether the following numbers are randomly distributed about their mean.

2	4	7	3	7	9	5	9	3	6
7	5	2	8	9	7	3	4	2	1

APPENDIX A

Miscellaneous Mathematics

OUTLINE

A.1 Matrix Algebra 243 A.2 Classical Theory of Minima 247

A.1. MATRIX ALGEBRA

A *matrix* **A** is a two-dimensional array of numbers, which is written as

$$\mathbf{A} = \begin{pmatrix} a_{11} & a_{12} & \cdots & a_{1n} \\ a_{21} & a_{22} & \cdots & a_{2n} \\ \vdots & \vdots & & \vdots \\ a_{m1} & a_{m2} & \cdots & a_{mn} \end{pmatrix}$$

where the general element in the ith row and jth column is denoted by a_{ij}. A matrix with m rows and n columns is said to be of *order* $(m \times n)$. For the cases $m = 1$ and $n = 1$, we have the *row* and *column* matrices

$$(a_1 \quad a_2 \quad \ldots \quad a_n) \quad \text{and} \quad \begin{pmatrix} a_1 \\ a_2 \\ \vdots \\ a_m \end{pmatrix},$$

respectively.

Matrices are frequently used in Chapter 8 to write the set of n linear equations in p unknowns:

$$\sum_{j=1}^{p} a_{ij} x_j = b_i, \quad i = 1, 2, \ldots, n$$

Statistics for Physical Sciences: An Introduction Copyright © 2012 Elsevier Inc. All rights reserved.

in the compact form

$$\mathbf{AX} = \mathbf{B},$$

where $\mathbf{A}$ is of order $(n \times p)$, $\mathbf{X}$ is an $(p \times 1)$ column vector, and $\mathbf{B}$ is a $(n \times 1)$ column vector. The set of n column vectors $\mathbf{X}_i (i = 1, 2, \ldots, n)$, all of the same order, are said to be *linearly dependent* if there exist n scalars $\alpha_i (i = 1, 2, \ldots, n)$, not all zero, such that

$$\sum_{i=1}^{n} \alpha_i \mathbf{X}_i = \mathbf{0},$$

where $\mathbf{0}$ is the *null matrix*, i.e., a matrix with all its elements zero. If no such set of scalars exists, then the set of vectors is said to be *linearly independent*.

The *transpose* of the matrix $\mathbf{A}$, denoted by $\mathbf{A}^T$, is obtained by interchanging the rows and columns of $\mathbf{A}$, so

$$\mathbf{A}^T = \begin{pmatrix} a_{11} & a_{21} & \cdots & a_{m1} \\ a_{12} & a_{22} & \cdots & a_{m2} \\ \vdots & \vdots & & \vdots \\ a_{1n} & a_{2n} & \cdots & a_{mn} \end{pmatrix}$$

is an $(n \times m)$ matrix.

A matrix with an equal number of rows and columns is called a *square* matrix, and if a square matrix $\mathbf{A}$ has elements such that $a_{ji} = a_{ij}$, it is said to be *symmetric*. A particular example of a symmetric matrix is the *unit-matrix* $\mathbf{1}$, with elements equal to unity for $i = j$, and zero otherwise. A symmetric matrix $\mathbf{A}$ is said to be *positive definite* if for any vector $\mathbf{V}$, (i) $\mathbf{V}^T \mathbf{A} \mathbf{V} \geq 0$ and (ii) $\mathbf{V}^T \mathbf{A} \mathbf{V} = 0$ implies $\mathbf{V} = \mathbf{0}$. A square matrix with elements $a_{ij} \neq 0$ only if $i = j$ is called *diagonal*; the unit matrix is an example of such a matrix. The line containing the elements $a_{11}, a_{22}, \ldots, a_{nn}$ is called the *principal*, or *main*, *diagonal* and the sum of its terms is the *trace* of the matrix, written as

$$\mathrm{Tr}\, \mathbf{A} = \sum_{i=1}^{n} a_{ii}.$$

The *determinant* of a square $(n \times n)$ matrix $\mathbf{A}$ is defined by

$$\det \mathbf{A} = |\mathbf{A}| \equiv \sum (\pm a_{1i} a_{2j} \ldots a_{nk}), \tag{A.1}$$

where the summation is taken over all permutations of $i, j, \ldots, k$, where these indices are the integers $1, 2, \ldots, n$. The positive sign is used for even permutations and the negative sign for odd permutations. The *minor* m_{ij} of the element a_{ij} is defined as the determinant obtained from $\mathbf{A}$ by deleting the ith row and the jth column, and the *cofactor* of a_{ij} is defined as $(-1)^{i+j}$ times the minor m_{ij}. The determinant of $\mathbf{A}$ may also be written in terms of its cofactors. For example, if

$$\mathbf{A} = \begin{pmatrix} 2 & 1 & 3 \\ 1 & 2 & 4 \\ 2 & 1 & 1 \end{pmatrix}, \tag{A.2}$$

then

$$\det \mathbf{A} = 2 \times \begin{vmatrix} 2 & 4 \\ 1 & 1 \end{vmatrix} - 1 \times \begin{vmatrix} 1 & 4 \\ 2 & 1 \end{vmatrix} + 3 \times \begin{vmatrix} 1 & 2 \\ 2 & 1 \end{vmatrix}$$

$$= 2 \times (2-4) - (1-8) + 3 \times (1-4) = -6.$$

The *adjoint* matrix is defined as the transposed matrix of cofactors and is denoted by $\mathbf{A}^\dagger$. Thus, the adjoint of the matrix in (A.2) is

$$\mathbf{A}^\dagger = \begin{pmatrix} -2 & 2 & -2 \\ 7 & -4 & -5 \\ -3 & 0 & 3 \end{pmatrix}. \tag{A.3}$$

A matrix $\mathbf{A}$ with real elements that satisfies the condition $\mathbf{A}^T = \mathbf{A}^{-1}$ is said to be *orthogonal* and its complex analog is a *unitary* matrix for which $\mathbf{A}^\dagger = \mathbf{A}^{-1}$. For any real symmetric matrix $\mathbf{A}$, a unitary matrix $\mathbf{U}$ may be found, that when multiplying $\mathbf{A}$ transforms it to diagonal form, i.e., the matrix $\mathbf{UA}$ has zeros everywhere except on the principal diagonal.

One particular determinant we have met is the *Jacobian J*. Consider n random variables $x_i (i = 1, 2, \ldots, n)$ which are themselves function of n other linearly independent random variables $y_i (i = 1, 2, \ldots, n)$, and assume that the relations can be inverted to give $x_i(y_1, y_2, \cdots, y_n)$. If the partial derivatives $\partial y_i / \partial x_j$ are continuous for all i and j, then J is defined by (some authors use the term Jacobian to mean the determinant of J, i.e., $|J|$)

$$J \equiv \frac{\partial(y_1, y_2, \cdots y_n)}{\partial(x_1, x_2, \cdots x_n)} \equiv \begin{pmatrix} \frac{\partial y_1}{\partial x_1} & \frac{\partial y_2}{\partial x_1} & \cdots & \frac{\partial y_n}{\partial x_1} \\ \frac{\partial y_1}{\partial x_2} & \frac{\partial y_2}{\partial x_2} & \cdots & \frac{\partial y_n}{\partial x_2} \\ \vdots & \vdots & & \vdots \\ \frac{\partial y_1}{\partial x_n} & \frac{\partial y_2}{\partial x_n} & \cdots & \frac{\partial y_n}{\partial x_n} \end{pmatrix}.$$

The Jacobian has been used, for example, in Section 3.4, when discussing functions of a random variable. In this case, if n random variables $x_i (i = 1, 2, \ldots, n)$ have a joint probability density $f(x_1, x_2, \ldots, x_n)$, then the joint probability density $g(y_1, y_2, \ldots, y_n)$ of a new set of variates y_i, which are themselves function of the n variables $x_i (i = 1, 2, \ldots, n)$ defined by $y_i = y_i(x_1, x_2, \ldots, x_n)$, is given by

$$g(y_1, y_2, \ldots, y_n) = f(x_1, x_2, \ldots, x_n)|J|.$$

Two matrices may be added and subtracted if they contain the same number of rows and columns, and such addition is both commutative and associative. The (inner) product $\mathbf{A} = \mathbf{BC}$ of two matrices $\mathbf{B}$ and $\mathbf{C}$ has elements given by

$$a_{ij} = \sum_k b_{ik} c_{kj}$$

and so is defined only if the number of columns in the first matrix **B** is equal to the number of rows in the second matrix **C**. Matrix multiplication is not, in general, commutative, but is associative.

Division of matrices is more complicated and needs some preliminary definitions. If we form all possible square submatrices of the matrix **A** (not necessarily square), and find that at least one determinant of order r is nonzero, but all determinants of order $(r+1)$ are zero, then the matrix is said to be of *rank r*. A square matrix of order n with rank $r < n$ has det **A** $= 0$ and is said to be *singular*. The rank of a matrix may thus be expressed as the greatest number of linearly independent rows or columns existing in the matrix, and so, for example, a nonsingular square matrix of order $(n \times n)$ must have rank n. Conversely, if a square matrix **A**, of order $(n \times n)$, has rank $r = n$, then it is nonsingular and there exists a matrix $\mathbf{A}^{-1}$, known as the *inverse* matrix, such that

$$\mathbf{A}\mathbf{A}^{-1} = \mathbf{A}^{-1}\mathbf{A} = \mathbf{1}, \tag{A.4}$$

where **1** is a diagonal matrix with elements of unity along the principle diagonal and zeros elsewhere. This is the analogous process to division in scalar algebra. The inverse is given by

$$\mathbf{A}^{-1} = \mathbf{A}^{\dagger}|\mathbf{A}|^{-1}. \tag{A.5}$$

Thus, if

$$\mathbf{A} = \begin{pmatrix} 1 & 2 \\ 3 & 4 \end{pmatrix},$$

then

$$\mathbf{A}^{\dagger} = \begin{pmatrix} 4 & -2 \\ -3 & 1 \end{pmatrix} \quad \text{and} \quad |\mathbf{A}| = (1 \times 4) - (3 \times 2) = -2,$$

so that

$$\mathbf{A}^{-1} = -\frac{1}{2}\begin{pmatrix} 4 & -2 \\ -3 & 1 \end{pmatrix},$$

and to check

$$\mathbf{A}\mathbf{A}^{-1} = -\frac{1}{2}\begin{pmatrix} 1 & 2 \\ 3 & 4 \end{pmatrix}\begin{pmatrix} 4 & -2 \\ -3 & 1 \end{pmatrix} = \begin{pmatrix} 1 & 0 \\ 0 & 1 \end{pmatrix} = \mathbf{1}.$$

Finally, for products of matrices,

$$(\mathbf{ABC\ldots D})^T = \mathbf{D}^T\ldots\mathbf{C}^T\mathbf{B}^T\mathbf{A}^T, \tag{A.6}$$

and, if **A, B, C, ... , D** are all square nonsingular matrices,

$$(\mathbf{ABC\ldots D})^{-1} = \mathbf{D}^{-1}\ldots\mathbf{C}^{-1}\mathbf{B}^{-1}\mathbf{A}^{-1}. \tag{A.7}$$

A point worth remarking is that in practice equations (A.1) and (A.5) are only useful for the practical evaluation of the determinant and inverse of a matrix in simple cases of low dimensionality. For example, in the least-squares method where the matrix of the normal equations (which is positive definite) has to be inverted, the most efficient methods in common use are those based either on the so-called Choleski's decomposition of a positive definite matrix or on Golub's factorization by orthogonal matrices, the details of which may be found in any modern textbook on numerical methods.

A.2. CLASSICAL THEORY OF MINIMA

If $f(x)$ is a function of the single variable x which in a certain interval possesses continuous derivatives

$$f^{(j)}(x) \equiv \frac{d^j f(x)}{dx^j}, \quad (j = 1, 2, \ldots, n+1),$$

then *Taylor's Theorem* states that if x and $(x+h)$ belong to this interval then

$$f(x+h) = \sum_{j=0}^{n} \frac{h^j}{j!} f^{(j)}(x) + R_n,$$

where $f^{(0)}(x) = f(x)$, and the remainder term is given by

$$R_n = \frac{h^{n+1}}{(n+1)!} f^{(n+1)}(x + \theta h), \quad 0 < \theta < 1$$

For a function of p-variables, Taylor's expansion becomes

$$f(\mathbf{x} + t\mathbf{h}) = \sum_{j=0}^{n} \frac{t^j}{j!} (\mathbf{h}\nabla)^j f(\mathbf{x}) + R_n,$$

where $\mathbf{h}$ is the row vector $(h_1, h_2, \ldots, h_p)$, ∇^T is the row vector

$$\left(\frac{\partial}{\partial x_1} \quad \frac{\partial}{\partial x_2} \quad \ldots \quad \frac{\partial}{\partial x_p} \right)$$

and

$$R_n = \frac{t^{n+1}}{(n+1)!} (\mathbf{h}\nabla)^{n+1} f(\mathbf{x} + \theta t\mathbf{h}), \quad 0 < \theta < 1$$

A *necessary* condition for a *turning point* (maximum, minimum or saddle point) of $f(\mathbf{x})$ to exist is that

$$\frac{\partial f(\mathbf{x})}{\partial x_i} = 0$$

for all $i = 1, 2, \ldots, p$. A *sufficient* condition for this point to be a minimum is that the second partial derivatives exist, and that $D_i > 0$ for all $i = 1, 2, \ldots, p$, where

$$D_i = \begin{vmatrix} \dfrac{\partial^2 f}{\partial x_1^2} & \dfrac{\partial^2 f}{\partial x_1 \partial x_2} & \cdots & \dfrac{\partial^2 f}{\partial x_1 \partial x_i} \\ \dfrac{\partial^2 f}{\partial x_2 \partial x_1} & \dfrac{\partial^2 f}{\partial x_2^2} & \cdots & \dfrac{\partial^2 f}{\partial x_2 \partial x_i} \\ \vdots & \vdots & & \vdots \\ \dfrac{\partial^2 f}{\partial x_i \partial x_1} & \dfrac{\partial^2 f}{\partial x_i \partial x_2} & \cdots & \dfrac{\partial^2 f}{\partial x_i^2} \end{vmatrix}.$$

If we seek a minimum of $f(\mathbf{x})$, subject to the s *equality constraints*

$$e_j(\mathbf{x}) = 0, \quad j = 1, 2, \ldots, s,$$

then the quantity to consider is the *Lagrangian form*

$$L(\mathbf{x}, \boldsymbol{\lambda}) = f(\mathbf{x}) + \sum_{j=1}^{s} \lambda_j e_j(\mathbf{x}),$$

where the constants λ_j are the so-called *Lagrange multipliers*. If the first partial derivatives of $e_j(\mathbf{x})$ exist, then the required minimum is the unconstrained solution of the equations

$$e_j(\mathbf{x}) = 0, \quad j = 1, 2, \ldots, s$$

and

$$\frac{\partial f(\mathbf{x})}{\partial x_i} + \sum_{j=1}^{s} \lambda_j \frac{\partial e_j(\mathbf{x})}{\partial x_i} = 0, \quad i = 1, 2, \ldots, p.$$

This technique has been used in several places, and extensively in the discussion of the least-squares method of estimation in chapter 8.

APPENDIX B

Optimization of Nonlinear Functions[1]

OUTLINE

B.1 General Principles	249	B.3.1 Direct Search Methods	253
B.2 Unconstrained Minimization of Functions of One Variable	252	B.3.2 Gradient Methods	254
		B.4 Constrained Optimization	255
B.3 Unconstrained Minimization of Multivariable Functions	253		

In Chapters 7 and 8 we encountered the problem of finding the maxima, or minima, of nonlinear functions, sometimes of several variables. These are examples of a more general class of optimization problems, which although occurring frequently in statistical estimation procedures, are not of a statistical nature. In practice, computer codes exist to tackle these problems, and it is not suggested that the reader write their own optimization code except in the simplest of circumstances, but nevertheless it is useful to know a little of the theory on which the methods are based to better appreciate their limitations. The discussion of this appendix is therefore confined to the main ideas involved, illustrated by one or two examples. Fuller accounts are given in the books cited in the bibliography, from which these brief notes draw extensively.

B.1. GENERAL PRINCIPLES

We consider only *minimization* problems since
$$\min f(\mathbf{x}) = \max [-f(\mathbf{x})].$$

[1] This appendix makes extensive use of matrix notations. These are reviewed briefly in Appendix A.

The general problem to be solved is then to minimize the function $f(x_1, x_2, \ldots, x_p) \equiv f(\mathbf{x})$, subject to the *m inequality constraints*

$$c_i(\mathbf{x}) \geq 0, \quad i = 1, 2, \ldots, m$$

and the *s equality constraints*,

$$e_j(\mathbf{x}) = 0, \quad j = 1, 2, \ldots, s.$$

All other constraints can be reduced to either of the above forms by suitable transformations. We will discuss first the features of methods of optimization in general and then describe in more detail a few of the most successful methods in current use.

Any point that satisfies all the constraints is called *feasible*, and the entire set of such points is called the *feasible region*. Points lying outside the feasible region are said to be *nonfeasible*. Nearly all practical methods of optimization are *iterative* in the sense that an initial feasible vector $\mathbf{x}^{(0)}$ must be specified from which the method will generate a series of vectors $\mathbf{x}^{(1)}, \mathbf{x}^{(2)}, \ldots, \mathbf{x}^{(n)}$ etc., which represent improved approximations to the solution.

The iterative procedure may be expressed by the equation

$$\mathbf{x}^{(n+1)} = \mathbf{x}^{(n)} + h_n \mathbf{d}_n, \tag{B.1}$$

where $\mathbf{d}_n$ is a *p*-dimensional *directional vector*, and h_n is the distance moved along it. The basic problem is to determine the most suitable vector $\mathbf{d}_n$, since once it is chosen the function $f(\mathbf{x})$ can be calculated and a suitable value of h_n found. Iterative techniques fall naturally into two classes, (a) *direct search* methods and (b) *gradient* methods.

Direct search methods are based on a sequential examination of a series of trial solutions produced from an initial feasible point. On the basis of the examinations, the strategy for further searching is determined. These methods are characterized by the fact they only explicitly require values of the function, and knowledge of the derivatives of $f(\mathbf{x})$ is not required. The latter fact is both a strength and a weakness of the methods, for although in problems involving many variables the calculation of derivatives can be difficult, and/or time consuming, it is clear that more efficient methods should be possible if the information contained in the derivatives is used. In practice, direct search methods are most useful for situations involving a few parameters, or where the calculation of derivatives is very difficult, or for finding promising regions in the parameter space where optima might reasonably be located.

Gradient methods make explicit use of the partial derivatives of the function, in addition to values of the function itself. The *gradient direction* at any point is that direction whose components are proportional to the first-order partial derivatives of the function at the point. The importance of this quantity will be seen as follows. If we make small perturbations $\delta \mathbf{x}$ from the current point $\mathbf{x}$, then to first order in $\delta \mathbf{x}$,

$$\delta f = \sum_{j=1}^{p} \frac{\partial f}{\partial x_j} \delta x_j. \tag{B.2}$$

To obtain the perturbation giving the greatest change in the function, we need to consider the Lagrangian form

$$F(\mathbf{x}, \lambda) = \delta f + \lambda \left(\sum_{j=1}^{p} \delta x_j^2 - \Delta^2 \right), \tag{B.3}$$

where λ is a Lagrange multiplier and Δ is the magnitude of the perturbations, i.e.,

$$\Delta = \left[\sum_{j=1}^{p} \delta x_j^2 \right]^{1/2}.$$

Using (B.2) in (B.3), and forming the differential with respect to δx_j, gives

$$\frac{\partial f}{\partial x_j} + 2\lambda \delta x_j = 0, \quad j = 1, 2, \ldots, p$$

and hence

$$\frac{\delta x_1}{\partial f / \partial x_1} = \frac{\delta x_2}{\partial f / \partial x_2} = \ldots = \frac{\delta x_p}{\partial f / \partial x_p}.$$

That is, for any Δ the greatest value of δf is obtained if the perturbations δx_j are chosen to be proportional to $\partial f / \partial x_j$, and that, further, if $\delta f < 0$, i.e., the search is to converge to a minimum, the constant of proportionality must be negative. This direction is called the *direction of steepest descent*. It follows that the function can always be reduced by following the direction of steepest descent, although this may only be true for a short distance.

One remark that is worth making about gradient methods concerns the actual calculation of the derivatives. Although gradient methods are in general more efficient than direct search methods, their efficiency can drop considerably if the derivatives are not obtained analytically, and so if numerical methods are used to calculate these quantities, great care has to be taken to ensure that inaccuracies do not result.

So far we have not specified the form of the function to be minimized, except that it is nonlinear in its variables. However, in many practical problems involving unconstrained functions it is found that the function can be well approximated by a quadratic form in the neighborhood of the minimum. There is therefore considerable interest in methods that *guarantee* to find the minimum of a quadratic in a specified number of steps. Such methods are said to be *quadratically convergent*, and the hope is that problems that are not strictly quadratic may still be tractable by such methods, a hope that is borne out rather well in practice.

The most useful of the methods having the property of quadratic convergence are those making use of the so-called *conjugate directions*, defined as follows. Two direction vectors $\mathbf{d}_1$ and $\mathbf{d}_2$ are said to be conjugate with respect to the positive definite matrix $\mathbf{G}$ if $\mathbf{d}_1^T \mathbf{G}\, \mathbf{d}_2 = 0$. The importance of conjugate directions in optimization problems stems from the following theorem. If $\mathbf{d}_i (i = 1, 2, \ldots, p)$ is a set of vectors mutually conjugate with respect to the positive definite matrix $\mathbf{G}$, the minimum of the quadratic form

$$f(\mathbf{x}) = \frac{1}{2} \mathbf{x}^T \mathbf{G} \mathbf{x} + \mathbf{b}^T \mathbf{x} + a, \tag{B.4}$$

where a is a constant and $\mathbf{b}$ a constant vector, can be found from an arbitrary point $\mathbf{x}^{(0)}$ by a finite descent calculation in which each of the vectors $\mathbf{d}_i$ is used as a descent direction only once, their order of use being arbitrary. The proof of this important result may be found in the books listed in the bibliography.

Although methods having the property of quadratic convergence will guarantee to converge to the exact minimum of a quadratic in p steps, where p is the dimensionality of the problem, when applied to functions that are not strictly quadratic the problem arises of determining when convergence has taken place. A suitable practical criterion is to consider that convergence has been achieved if, for given small values of ε and ε'

$$f[\mathbf{x}^{(n)}] - f[\mathbf{x}^{(n+1)}] < \varepsilon,$$

and/or

$$|\mathbf{x}^{(n)} - \mathbf{x}^{(n+1)}| < \varepsilon',$$

for a sequence of q successive iterations, where q is a number which will vary with the type of function being minimized. A generous overestimate is $q \sim p$, the number of variables and a considerably smaller number of values are usually sufficient.

Finally, it should be mentioned that all present techniques for optimizing nonlinear functions locate only *local optima*, i.e., points $\mathbf{x}_m$ at which $f(\mathbf{x}_m) < f(\mathbf{x})$ for all $\mathbf{x}$ in a region in the neighborhood of $\mathbf{x}_m$. For multivariate problems there may well be better local optima located at some distance from $\mathbf{x}_m$. At present, there are no general methods for locating the *global optimum* (i.e., the absolute optimum) of a function, and so it is essential to restart the search procedure from different initial points $\mathbf{x}^{(0)}$ to ensure that the full p-dimensional space has been explored.

B.2. UNCONSTRAINED MINIMIZATION OF FUNCTIONS OF ONE VARIABLE

The problem of minimizing a function of one variable is very important in practice, because many methods for optimizing multivariate functions proceed by a series of searches along a line in the parameter space, and each of these searches is equivalent to a univariate search. The latter fall into two groups (a) those which specify an interval within which the minimum lies and (b) those which specify the minimum by a point approximating it. The latter methods are the most useful in practice and we shall only consider them here. The basic procedure is as follows. Proceeding from an initial point $\mathbf{x}^{(0)}$, a systematic search technique is applied to find a region containing the minimum. This bracket is then refined by fitting a quadratic interpolation polynomial to the three points making up the bracket, and the minimum of this polynomial found. As a result of this evaluation a new bracket is formed, and the procedure is repeated. The method is both simple and very safe in practice.

A practical implementation of this procedure is as follows. The function is first evaluated at $x^{(0)}$ and $(x^{(0)} + h)$. If $f(x^{(0)} + h) \leq f(x^{(0)})$, then $f(x^{(0)} + 2h)$ is evaluated. This doubling of the step length h is repeated until a value of $f(x)$ is found such that $f(x^{(0)} + 2^n h) > f(x^{(0)} + 2^{n-1} h)$. At this point the step length is halved and a step again taken from the last successive point, i.e., the $(n-1)$th. This procedure produces four points equally spaced along the axis of search, at each of which the function has been evaluated. The end point farthest from the point corresponding to the smallest function value is rejected, and the remaining three points used for quadratic interpolation. Had the first step failed, then the search is continued by reversing the sign of the step length. If the first step in this direction also fails, then the minimum has been bracketed and the interpolation may be made. If the three points used for the interpolation are x_1, x_2, x_3 with $x_1 < x_2 < x_3$ and $x_3 - x_2 = x_2 - x_1 = l$, then the minimum of the fitted quadratic is at

$$x_m = x_2 + \frac{l[f(x_1) - f(x_3)]}{2[f(x_1) - 2f(x_2) + f(x_3)]}.$$

An iteration is completed by evaluating $f(x_m)$. Convergence tests are now applied and, if required, a further iteration is performed, with a reduced step length, using as the initial point whichever of x_2 or x_m corresponds to the smaller function value.

B.3. UNCONSTRAINED MINIMIZATION OF MULTIVARIABLE FUNCTIONS

Many methods for locating optima of multivariate functions are based on a series of linear searches along a line in the parameter space. By a *linear method*, we will therefore mean any technique which uses a set of direction vectors in the search, and which proceeds by explorations along these directions, deciding future strategy by the results obtained in previous searches.

B.3.1. Direct Search Methods

The simplest of all possible *direct search* methods would be to keep $(p-1)$ of the parameters fixed and find a minimum with respect to the pth parameter, doing this in turn for each variable. The progress of such an *alternating variable* search is in general very inefficient because the contours of equal function value will be aligned along the so-called *principal axes*, which are not parallel to the coordinate axes, so only very small steps will be taken at each stage. Moreover, the inefficiency increases as the number of variables increases. It would clearly be very much more efficient to re-orientate the direction vectors along more advantageous directions and this is done in several techniques, the most successful of which is due to Powell.

The method uses conjugate directions and utilizes the fact that, for a positive definite quadratic form, if searches for minima are made along p conjugate directions then the join of these minima is conjugate to all of those directions, a result that follows from the definition of conjugate directions. The procedure is to start from $\mathbf{x}^{(0)}$ and locate the minimum in the direction $\mathbf{d}_1^{(n)}$. Then from the new minimum point $\mathbf{x}^{(1)}$ locate the minimum in the direction $\mathbf{d}_2^{(n)}$ etc. until the minimum in the direction $\mathbf{d}_p^{(n)}$ is found. The direction of total progress made during this cycle is then

$$\mathbf{d} = \mathbf{x}^{(p)} - \mathbf{x}^{(0)}$$

New search directions are now constructed and care must be taken to ensure that the new direction vectors are always linearly independent. Powell showed that for the quadratic form of (B.4), if $\mathbf{d}_i^{(n)}$ is scaled so that

$$\mathbf{d}_i^{(n)T} \mathbf{G} \mathbf{d}_i^{(n)} = 1, \quad i = 1, 2, \ldots, p$$

then the determinant D of the matrix whose columns are $\mathbf{d}_i^{(n)}$ has a maximum if, and only if, the vectors $\mathbf{d}_i^{(n)}$ are mutually conjugate with respect to $\mathbf{G}$. Thus, the direction $\mathbf{d}$ only replaces an existing search direction if by so doing D is increased. In this case, the minimum in the direction $\mathbf{d}$ is found and used as a starting point in the next iteration, the list of direction vectors being updated as follows:

$$(\mathbf{d}_1^{(n+1)}, \mathbf{d}_2^{(n+1)}, \ldots, \mathbf{d}_p^{(n+1)}) = (\mathbf{d}_1^{(n)}, \mathbf{d}_2^{(n)}, \ldots, \mathbf{d}_{j-1}^{(n)}, \mathbf{d}_{j+1}^{(n)}, \ldots, \mathbf{d}_p^{(n)}, \mathbf{d}),$$

where $\mathbf{d}_j^{(n)}$ is that direction vector along which the greatest reduction in the function value occurred during the nth stage.

B.3.2. Gradient Methods

The simplest technique using gradients is that of steepest descent mentioned above. In this method, the normalized gradient vector at the current point is found, and using a step length h_i a new point is generated via the general iterative equation. This procedure is continued until a function value is found which has not decreased. The step length is then reduced and the search restarted from the best previous point. If the actual minimum along each search direction is located, then the performance of this method is similar in appearance to an alternating variable search, and is rather erratic, the search directions oscillating about the principal axes. A method that in principle is far better is based on an examination of the second derivatives of the function.

a. Newton's Method

A second-order Taylor expansion of the function $f(\mathbf{x})$ about the minimum point $\mathbf{x}_{min}$ is

$$f(\mathbf{x}) = f(\mathbf{x}_{min}) + \sum_{j=1}^{p} h_j \left(\frac{\partial f}{\partial x_j} \right)_{\mathbf{x}=\mathbf{x}_{min}} + \frac{1}{2} \sum_{j=1}^{p} \sum_{k=1}^{p} h_j h_k \left(\frac{\partial^2 f}{\partial x_j \partial x_k} \right)_{\mathbf{x}=\mathbf{x}_{min}}.$$

Differentiating this equation gives

$$g_l \equiv \frac{\partial f}{\partial x_l} = \sum_{j=1}^{p} h_j \left(\frac{\partial^2 f}{\partial x_j \partial x_l} \right)_{\mathbf{x}=\mathbf{x}_{min}}, \qquad i = 1, 2, \ldots, p \qquad (B.11)$$

The minimum is therefore obtained in one step by the move $\mathbf{x}_{min} = \mathbf{x} - \mathbf{h}$, where the components of $\mathbf{h}$ are found by solving the p linear equations (B.11). If we define

$$G_{jk} \equiv \left(\frac{\partial^2 f}{\partial x_j \partial x_k} \right).$$

then

$$\mathbf{x}_{min} = \mathbf{x} - \mathbf{G}_{min}^{-1} \mathbf{g},$$

where, again, $\mathbf{G}_{min}$ means that $\mathbf{G}$ is evaluated at $\mathbf{x}_{min}$. Since $\mathbf{G}_{min}^{-1}$ will not of course be known, it is usual to replace it by $\mathbf{G}^{-1}$ evaluated at the current point $\mathbf{x}^{(i)}$ and use the iterative equation

$$\mathbf{x}^{(n+1)} = \mathbf{x}^{(n)} - \mathbf{G}_n^{-1} \mathbf{g}_n. \qquad (B.12)$$

The method is clearly quadratically convergent, but suffers from severe difficulties.

First, there is the numerical problem of calculating the inverse matrix of second derivatives, and second, and more seriously, for a general function, $\mathbf{G}^{-1}$ is not guaranteed to be positive definite, and in this case the method will diverge. Thus, while Newton's method is efficient in the immediate neighborhood of a minimum, away from this point it has little to recommend it, the method of steepest descent being far preferable.

In view of the above remarks, an efficient method would be one that starts by using the method of steepest descent and, at a later stage, uses Newton's method. A method that does this automatically is due to Davidon, and represents the most powerful method currently available for optimizing unconstrained functions.

b. *Davidon's method*

This method is an iterative scheme based on successive approximations to the matrix $\mathbf{G}_{min}^{-1}$. The best approximation to this matrix, say $\mathbf{H}_n$, is used to define a new search direction by a modification of equation (B.12), i.e.,

$$\mathbf{x}^{(n+1)} = \mathbf{x}^{(n)} - h_n \mathbf{H}_n \mathbf{g}_n,$$

where $\mathbf{g}_n$ is the vector of first derivatives of $f(\mathbf{x}^{(n)})$ with respect to $\mathbf{x}^{(n)}$. The step length h_n is that necessary to find the minimum in the search direction $\mathbf{d}_n = -\mathbf{H}_n \mathbf{g}_n$, and may be found by any univariate search procedure. If the sequence $\{\mathbf{H}_n\}$ is positive definite, it can be shown that the convergence of this method is guaranteed. Furthermore, if the search directions $\mathbf{d}_n$ are mutually conjugate, then the method is quadratically convergent.

Davidon has shown that both of these conditions can be met if, at each stage of the iteration, the matrix $\mathbf{H}_n$ is updated according to the relation

$$\mathbf{H}_{n+1} = \mathbf{H}_n + \mathbf{A}_n + \mathbf{B}_n,$$

where the matrices $\mathbf{A}_n$ and $\mathbf{B}_n$ are given by

$$\mathbf{A}_n = \frac{-h_n \left[\mathbf{H}_n \mathbf{g}_n \mathbf{g}_n^T \mathbf{H}_n^T\right]}{(\mathbf{H}_n \mathbf{g}_n)^T \mathbf{V}} \quad \text{and} \quad \mathbf{B}_n = \frac{-\mathbf{H}_n \mathbf{V} \mathbf{V}^T \mathbf{H}_n^T}{\mathbf{V}^T \mathbf{H}_n \mathbf{V}},$$

with $\mathbf{V} = \mathbf{g}_{n+1} - \mathbf{g}_n$, where $\mathbf{g}_{n+1}$ is the gradient at $\mathbf{x}_{n+1}$. It is usual to start the iteration from the unit matrix $\mathbf{H}_0 = 1$. The matrix $\mathbf{A}_n$ ensures that the sequence $\{\mathbf{H}_n\}$ converges to $\mathbf{G}_{min}^{-1}$, and $\mathbf{B}_n$ ensures that each $\mathbf{H}_n$ is positive definite. The derivation of these expressions may be found in the book by Kowalik and Osborne, cited in the bibliography.

B.4. CONSTRAINED OPTIMIZATION

Constrained optimization, not surprisingly, is a more difficult problem than unconstrained optimization, and only a very brief discussion will be given here.

First, an obvious remark: if the constraints can be removed by suitable transformations then this should be done. For example, many problems involve simple constraints on the parameters that can be expressed in the form

$$l \leq x \leq u,$$

which can be removed completely by the transformation

$$x = l + (u - l)\sin^2 y,$$

thereby enabling an unconstrained minimization to be performed with respect to y. Such transformations *cannot* produce additional local optima. If the constraints cannot be removed, then one of the simplest ways of incorporating them is to arrange that the production of nonfeasible points is unattractive. This is the basis of a practical technique involving *penalty functions*, where the function to be minimized is modified by additional terms designed to achieve this.

The general problem was stated in Section B.1. Using the notation for constraints given there, we consider the function

$$F(\mathbf{x}) = f(\mathbf{x}) + \sum_{i=1}^{n} \lambda_i c_i^2(\mathbf{x}) S[c_i(\mathbf{x})] + \sum_{j=1}^{s} \lambda_i' e_j^2(\mathbf{x}), \qquad (B.13)$$

where $S(q)$ is the function

$$S(q) = \begin{cases} 0, & q \geq 0 \\ 1, & q < 0 \end{cases}$$

and λ_i, λ_i' are positive scale factors, chosen so that the contributions of the various terms to (B.13) are approximately equal. The 'penalty', i.e., the sum of the second and third terms on the right-hand side of (B.13), is thus the weighted sum of squares of the amounts by which the constraints are violated.

This method works reasonably well in practice, but has the disadvantage of requiring that values of $f(\mathbf{x})$ be calculated at nonfeasible points, and this may not always be possible, leading to program failure. A method which restricts the search to feasible points is due to Carroll, and is known as *Carroll's created response surface technique*. In this method, if the constraints are inequalities, the surface

$$F(\mathbf{x}, k) = f(\mathbf{x}) + k \sum_{i=1}^{m} \frac{w_i}{c_i(\mathbf{x})},$$

is considered, where $k > 0$, and the w_i are positive constants. A minimum is found as a function of $\mathbf{x}$ and this is then used as the starting value for a new minimization for a reduced value of k, and the procedure repeated until $k = 0$ is reached. In all minimizations, nonfeasible points are excluded. The theoretical development of this method and its extension to incorporate equality constraints may be found in the book of Kowalik and Osborne.

APPENDIX C

Statistical Tables

OUTLINE

C.1 Normal Distribution	257	C.7 Signed-Rank Test 283
C.2 Binomial Distribution	259	C.8 Rank-Sum Test (See footnote to Table C.7.) 284
C.3 Poisson Distribution	266	C.9 Runs Test 285
C.4 Chi-Squared Distribution	273	C.10 Rank Correlation Coefficient 286
C.5 Student's t Distribution	275	
C.6 F Distribution	277	

C.1. NORMAL DISTRIBUTION

The table gives values of the standardized cumulative distribution function

$$F(x) = \frac{1}{(2\pi)^{1/2}} \int_{-\infty}^{x} \exp\left(-\frac{t^2}{2}\right) dt.$$

Note that $F(-x) = 1 - F(x)$.

x	.00	.01	.02	.03	.04	.05	.06	.07	.08	.09
.0	.5000	.5040	.5080	.5120	.5160	.5199	.5239	.5279	.5319	.5359
.1	.5398	.5438	.5478	.5517	.5557	.5596	.5636	.5675	.5714	.5753
.2	.5793	.5832	.5871	.5910	.5948	.5987	.6026	.6064	.6103	.6141
.3	.6179	.6217	.6255	.6293	.6331	.6368	.6406	.6443	.6480	.6517
.4	.6554	.6591	.6628	.6664	.6700	.6736	.6772	.6808	.6844	.6879

(*Continued*)

x	.00	.01	.02	.03	.04	.05	.06	.07	.08	.09
.5	.6915	.6950	.6985	.7019	.7054	.7088	.7123	.7157	.7190	.7224
.6	.7257	.7291	.7324	.7357	.7389	.7422	.7454	.7486	.7517	.7349
.7	.7580	.7611	.7642	.7673	.7704	.7734	.7764	.7794	.7823	.7852
.8	.7881	.7910	.7939	.7967	.7995	.8023	.8051	.8078	.8106	.8133
.9	.8159	.8186	.8212	.8238	.8264	.8289	.8315	.8340	.8365	.8389
1.0	.8413	.8438	.8461	.8485	.8508	.8531	.8554	.8577	.8599	.8621
1.1	.8643	.8665	.8686	.8708	.8729	.8749	.8770	.8790	.8810	.8830
1.2	.8849	.8869	.8888	.8907	.8925	.8944	.8962	.8980	.8997	.9015
1.3	.9032	.9049	.9066	.9082	.9099	.9115	.9131	.9147	.9162	.9177
1.4	.9192	.9207	.9222	.9236	.9251	.9265	.9279	.9292	.9306	.9319
1.5	.9332	.9345	.9357	.9370	.9382	.9394	.9406	.9418	.9429	.9441
1.6	.9452	.9463	.9474	.9484	.9495	.9505	.9515	.9525	.9535	.9545
1.7	.9554	.9564	.9573	.9582	.9591	.9599	.9608	.9616	.9625	.9633
1.8	.9641	.9649	.9656	.9664	.9671	.9678	.9686	.9693	.9699	.9706
1.9	.9713	.9719	.9726	.9732	.9738	.9744	.9750	.9756	.9761	.9767
2.0	.9773	.9778	.9783	.9788	.9793	.9798	.9803	.9808	.9812	.9817
2.1	.9821	.9826	.9830	.9834	.9838	.9842	.9846	.9850	.9854	.9857
2.2	.9861	.9864	.9868	.9871	.9875	.9878	.9881	.9884	.9887	.9890
2.3	.9893	.9896	.9898	.9901	.9904	.9906	.9909	.9911	.9913	.9916
2.4	.9918	.9920	.9922	.9925	.9927	.9929	.9931	.9932	.9934	.9936
2.5	.9938	.9940	.9941	.9943	.9945	.9946	.9948	.9949	.9951	.9952
2.6	.9953	.9955	.9956	.9957	.9959	.9960	.9961	.9962	.9963	.9964
2.7	.9965	.9966	.9967	.9968	.9969	.9970	.9971	.9972	.9973	.9974
2.8	.9974	.9975	.9976	.9977	.9977	.9978	.9979	.9980	.9980	.9981
2.9	.9981	.9982	.9983	.9983	.9984	.9984	.9985	.9985	.9986	.9986
3.0	.9987	.9987	.9987	.9988	.9988	.9989	.9989	.9989	.9990	.9990
3.1	.9990	.9991	.9991	.9991	.9992	.9992	.9992	.9992	.9993	.9993
3.2	.9993	.9993	.9994	.9994	.9994	.9994	.9994	.9995	.9995	.9995
3.3	.9995	.9995	.9996	.9996	.9996	.9996	.9996	.9996	.9996	.9997
3.4	.9997	.9997	.9997	.9997	.9997	.9997	.9997	.9997	.9997	.9998
x		1.282	1.645	1.960	2.326	2.576	3.090	3.291	3.891	4.417
$F(x)$		.90	.95	.975	.99	.995	.999	.9995	.99995	.999995
$2[1 - F(x)]$		.20	.10	.05	.02	.01	.002	.001	.0001	.00001

C.2. BINOMIAL DISTRIBUTION

The table gives values of F, the cumulative binomial distribution, i.e., the probability of obtaining s or more successes in n independent Bernoulli trials,

$$F = \sum_{r=r'}^{n} \binom{n}{r} p^r q^{n-r},$$

for specified values of n, r' and p, where the probability of a success in a single trial is equal to p. If $p > 0.5$ the values for F are obtained from

$$1 - \sum_{r=n-r'+1}^{n} \binom{n}{r} p^r q^{n-r}$$

						p				
n	r'	.05	.10	.15	.20	.25	.30	.35	.40	.45	.50
2	1	.0975	.1900	.2775	.3600	.4375	.5100	.5775	.6400	.6975	.7500
	2	.0025	.0100	.0225	.0400	.0625	.0900	.1225	.1600	.2025	.2500
3	1	.1426	.2710	.3859	.4880	.5781	.6570	.7254	.7840	.8336	.8750
	2	.0072	.0280	.0608	.1040	.1562	.2160	.2818	.3520	.4252	.5000
	3	.0001	.0010	.0034	.0080	.0156	.0270	.0429	.0640	.0911	.1250
4	1	.1855	.3439	.4780	.5904	.6836	.7599	.8215	.8704	.9085	.9375
	2	.0140	.0523	.1095	.1808	.2617	.3483	.4370	.5248	.6090	.6875
	3	.0005	.0037	.0120	.0272	.0508	.0837	.1265	.1792	.2415	.3125
	4	.0000	.0001	.0005	.0016	.0039	.0081	.0150	.0256	.0410	.0625
5	1	.2262	.4095	.5563	.6723	.7627	.8319	.8840	.9222	.9497	.9688
	2	.0226	.0815	.1648	.2627	.3672	.4718	.5716	.6630	.7438	.8125
	3	.0012	.0086	.0266	.0579	.1035	.1631	.2352	.3174	.4069	.5000
	4	.0000	.0005	.0022	.0067	.0156	.0308	.0540	.0870	.1312	.1875
	5	.0000	.0000	.0001	.0003	.0010	.0024	.0053	.0102	.0185	.0312
6	1	.2649	.4686	.6229	.7379	.8220	.8824	.9246	.9533	.9723	.9844
	2	.0328	.1143	.2235	.3447	.4661	.5798	.6809	.7667	.8364	.8906
	3	.0022	.0158	.0473	.0989	.1694	.2557	.3529	.4557	.5585	.6562
	4	.0001	.0013	.0059	.0170	.0376	.0705	.1174	.1792	.2553	.3438
	5	.0000	.0001	.0004	.0016	.0046	.0109	.0223	.0410	.0692	.1094
	6	.0000	.0000	.0000	.0001	.0002	.0007	.0018	.0041	.0083	.0156

(*Continued*)

		\multicolumn{10}{c}{p}
n	r'	.05	.10	.15	.20	.25	.30	.35	.40	.45	.50
7	1	.3017	.5217	.6794	.7903	.8665	.9176	.9510	.9720	.9848	.9922
	2	.0444	.1497	.2834	.4233	.5551	.6706	.7662	.8414	.8976	.9375
	3	.0038	.0257	.0738	.1480	.2436	.3529	.4677	.5801	.6836	.7734
	4	.0002	.0027	.0121	.0333	.0706	.1260	.1998	.2898	.3917	.5000
	5	.0000	.0002	.0012	.0047	.0129	.0288	.0556	.0963	.1529	.2266
	6	.0000	.0000	.0001	.0004	.0013	.0038	.0090	.0188	.0357	.0625
	7	.0000	.0000	.0000	.0000	.0001	.0002	.0006	.0016	.0037	.0078
8	1	.3366	.5695	.7275	.8322	.8999	.9424	.9681	.9832	.9916	.9961
	2	.0572	.1869	.3428	.4967	.6329	.7447	.8309	.8936	.9368	.9648
	3	.0058	.0381	.1052	.2031	.3215	.4482	.5722	.6846	.7799	.8555
	4	.0004	.0050	.0214	.0563	.1138	.1941	.2936	.4059	.5230	.6367
	5	.0000	.0004	.0029	.0104	.0273	.0580	.1061	.1737	.2604	.3633
	6	.0000	.0000	.0002	.0012	.0042	.0113	.0253	.0498	.0885	.1445
	7	.0000	.0000	.0000	.0001	.0004	.0013	.0036	.0085	.0181	.0352
	8	.0000	.0000	.0000	.0000	.0000	.0001	.0002	.0007	.0017	.0039
9	1	.3698	.6126	.7684	.8658	.9249	.9596	.9793	.9899	.9954	.9980
	2	.0712	.2252	.4005	.5638	.6997	.8040	.8789	.9295	.9615	.9805
	3	.0084	.0530	.1409	.2618	.3993	.5372	.6627	.7682	.8505	.9102
	4	.0006	.0083	.0339	.0856	.1657	.2703	.3911	.5174	.6386	.7461
	5	.0000	.0009	.0056	.0196	.0489	.0988	.1717	.2666	.3786	.5000
	6	.0000	.0001	.0006	.0031	.0100	.0253	.0536	.0994	.1658	.2539
	7	.0000	.0000	.0000	.0003	.0013	.0043	.0112	.0250	.0498	.0898
	8	.0000	.0000	.0000	.0000	.0001	.0004	.0014	.0038	.0091	.0195
	9	.0000	.0000	.0000	.0000	.0000	.0000	.0001	.0003	.0008	.0020
10	1	.4013	.6513	.8031	.8926	.9437	.9718	.9865	.9940	.9975	.9990
	2	.0861	.2639	.4557	.6242	.7560	.8507	.9140	.9536	.9767	.9893
	3	.0115	.0702	.1798	.3222	.4744	.6172	.7384	.8327	.9004	.9453
	4	.0010	.0128	.0500	.1209	.2241	.3504	.4862	.6177	.7340	.8281
	5	.0001	.0016	.0099	.0328	.0781	.1503	.2485	.3669	.4956	.6230

C.2. BINOMIAL DISTRIBUTION

							p				
n	r'	.05	.10	.15	.20	.25	.30	.35	.40	.45	.50
	6	.0000	.0001	.0014	.0064	.0197	.0473	.0949	.1662	.2616	.3770
	7	.0000	.0000	.0001	.0009	.0035	.0106	.0260	.0548	.1020	.1719
	8	.0000	.0000	.0000	.0001	.0004	.0016	.0048	.0123	.0274	.0547
	9	.0000	.0000	.0000	.0000	.0000	.0001	.0005	.0017	.0045	.0107
	10	.0000	.0000	.0000	.0000	.0000	.0000	.0000	.0001	.0003	.0010
11	1	.4312	.6862	.8327	.9141	.9578	.9802	.9912	.9964	.9986	.9995
	2	.1019	.3026	.5078	.6779	.8029	.8870	.9394	.9698	.9861	.9941
	3	.0152	.0896	.2212	.3826	.5448	.6873	.7999	.8811	.9348	.9673
	4	.0016	.0185	.0694	.1611	.2867	.4304	.5744	.7037	.8089	.8867
	5	.0001	.0028	.0159	.0504	.1146	.2103	.3317	.4672	.6029	.7256
	6	.0000	.0003	.0027	.0117	.0343	.0782	.1487	.2465	.3669	.5000
	7	.0000	.0000	.0003	.0020	.0076	.0216	.0501	.0994	.1738	.2744
	8	.0000	.0000	.0000	.0002	.0012	.0043	.0122	.0293	.0610	.1133
	9	.0000	.0000	.0000	.0000	.0001	.0006	.0020	.0059	.0148	.0327
	10	.0000	.0000	.0000	.0000	.0000	.0000	.0002	.0007	.0022	.0059
	11	.0000	.0000	.0000	.0000	.0000	.0000	.0000	.0000	.0002	.0005
12	1	.4596	.7176	.8578	.9313	.9683	.9862	.9943	.9978	.9992	.9998
	2	.1184	.3410	.5565	.7251	.8416	.9150	.9576	.9804	.9917	.9968
	3	.0196	.1109	.2642	.4417	.6093	.7472	.8487	.9166	.9579	.9807
	4	.0022	.0256	.0922	.2054	.3512	.5075	.6533	.7747	.8655	.9270
	5	.0002	.0043	.0239	.0726	.1576	.2763	.4167	.5618	.6956	.8062
	6	.0000	.0005	.0046	.0194	.0544	.1178	.2127	.3348	.4731	.6128
	7	.0000	.0001	.0007	.0039	.0143	.0386	.0846	.1582	.2607	.3872
	8	.0000	.0000	.0001	.0006	.0028	.0095	.0255	.0573	.1117	.1938
	9	.0000	.0000	.0000	.0001	.0004	.0017	.0056	.0153	.0356	.0730
	10	.0000	.0000	.0000	.0000	.0000	.0002	.0008	.0028	.0079	.0193
	11	.0000	.0000	.0000	.0000	.0000	.0000	.0001	.0003	.0011	.0032
	12	.0000	.0000	.0000	.0000	.0000	.0000	.0000	.0000	.0001	.0002
13	1	.4867	.7458	.8791	.9450	.9762	.9903	.9963	.9987	.9996	.9999
	2	.1354	.3787	.6017	.7664	.8733	.9363	.9704	.9874	.9951	.9983
	3	.0245	.1339	.2704	.4983	.6674	.7975	.8868	.9421	.9731	.9888
	4	.0031	.0342	.0967	.2527	.4157	.5794	.7217	.8314	.9071	.9539
	5	.0003	.0065	.0260	.0991	.2060	.3457	.4995	.6470	.7721	.8666

(Continued)

		\multicolumn{10}{c}{p}
n	r'	.05	.10	.15	.20	.25	.30	.35	.40	.45	.50
	6	.0000	.0009	.0053	.0300	.0802	.1654	.2841	.4256	.5732	.7095
	7	.0000	.0001	.0013	.0070	.0243	.0624	.1295	.2288	.3563	.5000
	8	.0000	.0000	.0002	.0012	.0056	.0182	.0462	.0977	.1788	.2905
	9	.0000	.0000	.0000	.0002	.0010	.0040	.0126	.0321	.0698	.1334
	10	.0000	.0000	.0000	.0000	.0001	.0007	.0025	.0078	.0203	.0461
	11	.0000	.0000	.0000	.0000	.0000	.0001	.0003	.0013	.0041	.0112
	12	.0000	.0000	.0000	.0000	.0000	.0000	.0000	.0001	.0005	.0017
	13	.0000	.0000	.0000	.0000	.0000	.0000	.0000	.0000	.0000	.0001
14	1	.5123	.7712	.8972	.9560	.9822	.9932	.9976	.9992	.9998	.9999
	2	.1530	.4154	.6433	.8021	.8990	.9525	.9795	.9919	.9971	.9991
	3	.0301	.1584	.3521	.5519	.7189	.8392	.9161	.9602	.9830	.9935
	4	.0042	.0441	.1465	.3018	.4787	.6448	.7795	.8757	.9368	.9713
	5	.0004	.0092	.0467	.1298	.2585	.4158	.5773	.7207	.8328	.9102
	6	.0000	.0015	.0115	.0439	.1117	.2195	.3595	.5141	.6627	.7880
	7	.0000	.0002	.0022	.0116	.0383	.0933	.1836	.3075	.4539	.6047
	8	.0000	.0000	.0003	.0024	.0103	.0315	.0753	.1501	.2586	.3953
	9	.0000	.0000	.0000	.0004	.0022	.0083	.0243	.0583	.1189	.2120
	10	.0000	.0000	.0000	.0000	.0003	.0017	.0060	.0175	.0426	.0898
	11	.0000	.0000	.0000	.0000	.0000	.0002	.0011	.0039	.0114	.0287
	12	.0000	.0000	.0000	.0000	.0000	.0000	.0001	.0006	.0022	.0065
	13	.0000	.0000	.0000	.0000	.0000	.0000	.0000	.0001	.0003	.0009
	14	.0000	.0000	.0000	.0000	.0000	.0000	.0000	.0000	.0000	.0001
15	1	.5367	.7941	.9126	.9648	.9866	.9953	.9984	.9995	.9999	1.0000
	2	.1710	.4510	.6814	.8329	.9198	.9647	.9858	.9948	.9983	.9995
	3	.0362	.1841	.3958	.6020	.7639	.8732	.9383	.9729	.9893	.9963
	4	.0055	.0556	.1773	.3518	.5387	.7031	.8273	.9095	.9576	.9824
	5	.0006	.0127	.0617	.1642	.3135	.4845	.6481	.7827	.8796	.9408
	6	.0001	.0022	.0168	.0611	.1484	.2784	.4357	.5968	.7392	.8491
	7	.0000	.0003	.0036	.0181	.0566	.1311	.2452	.3902	.5478	.6964
	8	.0000	.0000	.0006	.0042	.0173	.0500	.1132	.2131	.3465	.5000
	9	.0000	.0000	.0001	.0008	.0042	.0152	.0422	.0950	.1818	.3036
	10	.0000	.0000	.0000	.0001	.0008	.0037	.0124	.0338	.0769	.1509

C.2. BINOMIAL DISTRIBUTION

							p				
n	r'	.05	.10	.15	.20	.25	.30	.35	.40	.45	.50
	11	.0000	.0000	.0000	.0000	.0001	.0007	.0028	.0093	.0255	.0592
	12	.0000	.0000	.0000	.0000	.0000	.0001	.0005	.0019	.0063	.0176
	13	.0000	.0000	.0000	.0000	.0000	.0000	.0001	.0003	.0011	.0037
	14	.0000	.0000	.0000	.0000	.0000	.0000	.0000	.0000	.0001	.0005
	15	.0000	.0000	.0000	.0000	.0000	.0000	.0000	.0000	.0000	.0000
16	1	.5599	.8147	.9257	.9719	.9900	.9967	.9990	.9997	.9999	1.0000
	2	.1892	.4853	.7161	.8593	.9365	.9739	.9902	.9967	.9990	.9997
	3	.0429	.2108	.4386	.6482	.8029	.9006	.9549	.9817	.9934	.9979
	4	.0070	.0684	.2101	.4019	.5950	.7541	.8661	.9349	.9719	.9894
	5	.0009	.0170	.0791	.2018	.3698	.5501	.7108	.8334	.9147	.9616
	6	.0001	.0033	.0235	.0817	.1897	.3402	.5100	.6712	.8024	.8949
	7	.0000	.0005	.0056	.0267	.0796	.1753	.3119	.4728	.6340	.7228
	8	.0000	.0001	.0011	.0070	.0271	.0744	.1594	.2839	.4371	.5982
	9	.0000	.0000	.0002	.0015	.0075	.0257	.0671	.1423	.2559	.4018
	10	.0000	.0000	.0000	.0002	.0016	.0071	.0229	.0583	.1241	.2272
	11	.0000	.0000	.0000	.0000	.0003	.0016	.0062	.0191	.0486	.1051
	12	.0000	.0000	.0000	.0000	.0000	.0003	.0013	.0049	.0149	.0384
	13	.0000	.0000	.0000	.0000	.0000	.0000	.0002	.0009	.0035	.0106
	14	.0000	.0000	.0000	.0000	.0000	.0000	.0000	.0001	.0006	.0021
	15	.0000	.0000	.0000	.0000	.0000	.0000	.0000	.0000	.0001	.0003
	16	.0000	.0000	.0000	.0000	.0000	.0000	.0000	.0000	.0000	.0000
17	1	.5189	.8332	.9369	.9775	.9925	.9977	.9993	.9998	1.0000	1.0000
	2	.2078	.5182	.7475	.8818	.9499	.9807	.9933	.9979	.9994	.9999
	3	.0503	.2382	.4802	.6904	.8363	.9226	.9673	.9877	.9959	.9988
	4	.0088	.0826	.2444	.4511	.6470	.7981	.8972	.9536	.9816	.9936
	5	.0012	.0221	.0987	.2418	.4261	.6113	.7652	.8740	.9404	.9755
	6	.0001	.0047	.0319	.1057	.2347	.4032	.5803	.7361	.8529	.9283
	7	.0000	.0008	.0083	.0377	.1071	.2248	.3812	.5522	.7098	.8338
	8	.0000	.0001	.0017	.0109	.0402	.1046	.2128	.3595	.5257	.6855
	9	.0000	.0000	.0003	.0026	.0124	.0403	.0994	.1989	.3374	.5000
	10	.0000	.0000	.0000	.0005	.0031	.0127	.0383	.0919	.1834	.3145

(*Continued*)

						p					
n	r'	.05	.10	.15	.20	.25	.30	.35	.40	.45	.50
	11	.0000	.0000	.0000	.0001	.0006	.0032	.0120	.0348	.0826	.1662
	12	.0000	.0000	.0000	.0000	.0001	.0007	.0030	.0106	.0301	.0717
	13	.0000	.0000	.0000	.0000	.0000	.0001	.0006	.0025	.0086	.0245
	14	.0000	.0000	.0000	.0000	.0000	.0000	.0000	.0005	.0019	.0064
	15	.0000	.0000	.0000	.0000	.0000	.0000	.0000	.0001	.0003	.0012
	16	.0000	.0000	.0000	.0000	.0000	.0000	.0000	.0000	.0000	.0001
	17	.0000	.0000	.0000	.0000	.0000	.0000	.0000	.0000	.0000	.0000
18	1	.6028	.8499	.9464	.9820	.9944	.9984	.9996	.9999	1.0000	1.0000
	2	.2265	.5497	.7759	.9009	.9605	.9858	.9954	.9987	.9997	.9999
	3	.0581	.2662	.5203	.7287	.8647	.9400	.9764	.9918	.9975	.9993
	4	.0109	.0982	.2798	.4990	.6943	.8354	.9217	.9672	.9880	.9962
	5	.0015	.0282	.1206	.2836	.4813	.6673	.8114	.9058	.9589	.9846
	6	.0002	.0064	.0419	.1329	.2825	.4656	.6450	.7912	.8923	.9519
	7	.0000	.0012	.0118	.0513	.1390	.2783	.4509	.6257	.7742	.8811
	8	.0000	.0002	.0027	.0163	.0569	.1407	.2717	.4366	.6085	.7597
	9	.0000	.0000	.0005	.0043	.0193	.0596	.1391	.2632	.4222	.5927
	10	.0000	.0000	.0001	.0009	.0054	.0210	.0597	.1347	.2527	.4073
	11	.0000	.0000	.0000	.0002	.0012	.0061	.0212	.0576	.1280	.2403
	12	.0000	.0000	.0000	.0000	.0002	.0014	.0062	.0203	.0537	.1189
	13	.0000	.0000	.0000	.0000	.0000	.0003	.0014	.0058	.0183	.0481
	14	.0000	.0000	.0000	.0000	.0000	.0000	.0003	.0013	.0049	.0154
	15	.0000	.0000	.0000	.0000	.0000	.0000	.0000	.0002	.0010	.0038
	16	.0000	.0000	.0000	.0000	.0000	.0000	.0000	.0000	.0001	.0007
	17	.0000	.0000	.0000	.0000	.0000	.0000	.0000	.0000	.0000	.0001
	18	.0000	.0000	.0000	.0000	.0000	.0000	.0000	.0000	.0000	.0000
19	1	.6226	.8649	.9544	.9856	.9958	.9989	.9997	.9999	1.0000	1.0000
	2	.2453	.5797	.8015	.9171	.9690	.9896	.9969	.9992	.9998	1.0000
	3	.0665	.2946	.5587	.7631	.8887	.9538	.9830	.9945	.9985	.9996
	4	.0132	.1150	.3159	.5449	.7639	.8668	.9409	.9770	.9923	.9978
	5	.0020	.0352	.1444	.3267	.5346	.7178	.8500	.9304	.9720	.9904

C.2. BINOMIAL DISTRIBUTION

						p					
n	r'	.05	.10	.15	.20	.25	.30	.35	.40	.45	.50
	6	.0002	.0086	.0537	.1631	.3322	.5261	.7032	.8371	.9223	.9682
	7	.0000	.0017	.0163	.0676	.1749	.3345	.5188	.6919	.8273	.9165
	8	.0000	.0003	.0041	.0233	.0775	.1820	.3344	.5122	.6831	.8204
	9	.0000	.0000	.0008	.0067	.0287	.0839	.1855	.3325	.5060	.6762
	10	.0000	.0000	.0001	.0016	.0089	.0326	.0875	.1861	.3290	.5000
	11	.0000	.0000	.0000	.0003	.0023	.0105	.0347	.0885	.1841	.3238
	12	.0000	.0000	.0000	.0000	.0005	.0028	.0114	.0352	.0871	.1796
	13	.0000	.0000	.0000	.0000	.0001	.0006	.0031	.0116	.0342	.0835
	14	.0000	.0000	.0000	.0000	.0000	.0001	.0007	.0031	.0109	.0318
	15	.0000	.0000	.0000	.0000	.0000	.0000	.0001	.0006	.0028	.0096
	16	.0000	.0000	.0000	.0000	.0000	.0000	.0000	.0001	.0005	.0022
	17	.0000	.0000	.0000	.0000	.0000	.0000	.0000	.0000	.0001	.0004
	18	.0000	.0000	.0000	.0000	.0000	.0000	.0000	.0000	.0000	.0000
	19	.0000	.0000	.0000	.0000	.0000	.0000	.0000	.0000	.0000	.0000
20	1	.6415	.8784	.9612	.9885	.9968	.9992	.9998	1.0000	1.0000	1.0000
	2	.2642	.6083	.8244	.9308	.9757	.9924	.9979	.9995	.9999	1.0000
	3	.0755	.3231	.5951	.7939	.9087	.9645	.9879	.9964	.9991	.9998
	4	.0159	.1330	.3523	.5886	.7748	.8929	.9556	.9840	.9951	.9987
	5	.0026	.0432	.1702	.3704	.5852	.7625	.8818	.9490	.9811	.9941
	6	.0003	.0113	.0673	.1958	.3828	.5836	.7546	.8744	.9447	.9793
	7	.0000	.0024	.0219	.0867	.2142	.3920	.5834	.7500	.8701	.9423
	8	.0000	.0004	.0059	.0321	.1018	.2277	.3990	.5841	.7480	.8684
	9	.0000	.0001	.0013	.0100	.0409	.1133	.2376	.4044	.5857	.7483
	10	.0000	.0000	.0002	.0026	.0139	.0480	.1218	.2447	.4086	.5881
	11	.0000	.0000	.0000	.0006	.0039	.0171	.0532	.1275	.2493	.4119
	12	.0000	.0000	.0000	.0001	.0009	.0051	.0196	.0565	.1308	.2517
	13	.0000	.0000	.0000	.0000	.0002	.0013	.0060	.0210	.0580	.1316
	14	.0000	.0000	.0000	.0000	.0000	.0003	.0015	.0065	.0214	.0577
	15	.0000	.0000	.0000	.0000	.0000	.0000	.0003	.0016	.0064	.0207
	16	.0000	.0000	.0000	.0000	.0000	.0000	.0000	.0003	.0015	.0059
	17	.0000	.0000	.0000	.0000	.0000	.0000	.0000	.0000	.0003	.0013
	18	.0000	.0000	.0000	.0000	.0000	.0000	.0000	.0000	.0000	.0002
	19	.0000	.0000	.0000	.0000	.0000	.0000	.0000	.0000	.0000	.0000
	20	.0000	.0000	.0000	.0000	.0000	.0000	.0000	.0000	.0000	.0000

C.3. POISSON DISTRIBUTION

The table gives values of the cumulative Poisson distribution

$$F = \sum_{k=k'}^{\infty} f(k; \lambda)$$

for specified values of λ an k, where

$$f(k; \lambda) = \frac{\lambda^k}{k!} \exp(-\lambda), \quad \lambda > 0, \quad k = 0, 1, \ldots$$

					λ					
k'	0.1	0.2	0.3	0.4	0.5	0.6	0.7	0.8	0.9	1.0
0	1.0000	1.0000	1.0000	1.0000	1.0000	1.0000	1.0000	1.0000	1.0000	1.0000
1	.0952	.1813	.2592	.3297	.3935	.4512	.5034	.5507	.5934	.6321
2	.0047	.0175	.0369	.0616	.0902	.1219	.1558	.1912	.2275	.2642
3	.0002	.0011	.0036	.0079	.0144	.0231	.0341	.0474	.0629	.0803
4	.0000	.0001	.0003	.0008	.0018	.0034	.0058	.0091	.0135	.0190
5	.0000	.0000	.0000	.0001	.0002	.0004	.0008	.0014	.0023	.0037
6	.0000	.0000	.0000	.0000	.0000	.0000	.0001	.0002	.0003	.0006
7	.0000	.0000	.0000	.0000	.0000	.0000	.0000	.0000	.0000	.0001
k'	1.1	1.2	1.3	1.4	1.5	1.6	1.7	1.8	1.9	2.0
0	1.0000	1.0000	1.0000	1.0000	1.0000	1.0000	1.0000	1.0000	1.0000	1.0000
1	.6671	.6988	.7275	.7534	.7769	.7981	.8173	.8347	.8504	.8647
2	.3010	.3374	.3732	.4082	.4422	.4751	.5068	.5372	.5663	.5940
3	.0996	.1205	.1429	.1665	.1912	.2166	.2428	.2694	.2963	.3233
4	.0257	.0338	.0431	.0537	.0656	.0788	.0932	.1087	.1253	.1429
5	.0054	.0077	.0107	.0143	.0186	.0237	.0296	.0364	.0441	.0527
6	.0010	.0015	.0022	.0032	.0045	.0060	.0080	.0104	.0132	.0166
7	.0001	.0003	.0004	.0006	.0009	.0013	.0019	.0026	.0034	.0045
8	.0000	.0000	.0001	.0001	.0002	.0003	.0004	.0006	.0008	.0011
9	.0000	.0000	.0000	.0000	.0000	.0000	.0001	.0001	.0002	.0002

C.3. POISSON DISTRIBUTION

					λ					
k'	2.1	2.2	2.3	2.4	2.5	2.6	2.7	2.8	2.9	3.0
0	1.0000	1.0000	1.0000	1.0000	1.0000	1.0000	1.0000	1.0000	1.0000	1.0000
1	.8775	.8892	.8997	.9093	.9179	.9257	.9328	.9392	.9450	.9502
2	.6204	.6454	.6691	.6916	.7127	.7326	.7513	.7689	.7854	.8009
3	.3504	.3773	.4040	.4303	.4562	.4816	.5064	.5305	.5540	.5768
4	.1614	.1806	.2007	.2213	.2424	.2640	.2859	.3081	.3304	.3528
5	.0621	.0725	.0838	.0959	.1088	.1226	.1371	.1523	.1682	.1847
6	.0204	.0249	.0300	.0357	.0420	.0490	.0567	.0651	.0742	.0839
7	.0059	.0075	.0094	.0116	.0142	.0172	.0206	.0244	.0287	.0335
8	.0015	.0020	.0026	.0033	.0042	.0053	.0066	.0081	.0099	.0119
9	.0003	.0005	.0006	.0009	.0011	.0015	.0019	.0024	.0031	.0038
10	.0001	.0001	.0001	.0002	.0003	.0004	.0005	.0007	.0009	.0011
11	.0000	.0000	.0000	.0000	.0001	.0001	.0001	.0002	.0002	.0003
12	.0000	.0000	.0000	.0000	.0000	.0000	.0000	.0000	.0001	.0001
k'	3.1	3.2	3.3	3.4	3.5	3.6	3.7	3.8	3.9	4.0
0	1.0000	1.0000	1.0000	1.0000	1.0000	1.0000	1.0000	1.0000	1.0000	1.0000
1	.9550	.9592	.9631	.9666	.9698	.9727	.9753	.9776	.9798	.9817
2	.8153	.8288	.8414	.8532	.8641	.8743	.8838	.8926	.9008	.9084
3	.5988	.6201	.6406	.6603	.6792	.6973	.7146	.7311	.7469	.7619
4	.3752	.3975	.4197	.4416	.4634	.4848	.5058	.5265	.5468	.5665
5	.2018	.2194	.2374	.2558	.2746	.2936	.3128	.3322	.3516	.3712
6	.0943	.1054	.1171	.1295	.1424	.1559	.1699	.1844	.1994	.2149
7	.0388	.0446	.0510	.0579	.0653	.0732	.0818	.0919	.1005	.1107
8	.0142	.0168	.0198	.0231	.0267	.0308	.0352	.0401	.0454	.0511
9	.0047	.0057	.0069	.0083	.0099	.0117	.0137	.0160	.0185	.0214
10	.0014	.0018	.0022	.0027	.0033	.0040	.0048	.0058	.0069	.0081
11	.0004	.0005	.0006	.0008	.0010	.0013	.0016	.0019	.0023	.0028
12	.0001	.0001	.0002	.0002	.0003	.0004	.0005	.0006	.0007	.0009
13	.0000	.0000	.0000	.0001	.0001	.0001	.0001	.0002	.0002	.0003
14	.0000	.0000	.0000	.0000	.0000	.0000	.0000	.0000	.0001	.0001

(*Continued*)

C. STATISTICAL TABLES

					λ					
k'	4.1	4.2	4.3	4.4	4.5	4.6	4.7	4.8	4.9	5.0
0	1.0000	1.0000	1.0000	1.0000	1.0000	1.0000	1.0000	1.0000	1.0000	1.0000
1	.9834	.9850	.9864	.9877	.9889	.9899	.9909	.9918	.9926	.9933
2	.9155	.9220	.9281	.9337	.9389	.9437	.9482	.9523	.9561	.9596
3	.7762	.7898	.8026	.8149	.8264	.8374	.8477	.8575	.8667	.8753
4	.5858	.6046	.6228	.6406	.6577	.6743	.6903	.7058	.7207	.7350
5	.3907	.4102	.4296	.4488	.4679	.4868	.5054	.5237	.5418	.5595
6	.2307	.2469	.2633	.2801	.2971	.3412	.3316	.3490	.3665	.3840
7	.1214	.1325	.1442	.1564	.1689	.1820	.1954	.2092	.2233	.2378
8	.0573	.0639	.0710	.0786	.0866	.0951	.1040	.1133	.1231	.1334
9	.0245	.0279	.0317	.0358	.0403	.0451	.0503	.0558	.0618	.0681
10	.0095	.0111	.0129	.0149	.0171	.0195	.0222	.0251	.0283	.0318
11	.0034	.0041	.0048	.0057	.0067	.0078	.0090	.0104	.0120	.0137
12	.0011	.0014	.0017	.0020	.0024	.0029	.0034	.0040	.0047	.0055
13	.0003	.0004	.0005	.0007	.0008	.0010	.0012	.0014	.0017	.0020
14	.0001	.0001	.0002	.0002	.0003	.0003	.0004	.0005	.0006	.0007
15	.0000	.0000	.0000	.0001	.0001	.0001	.0001	.0001	.0002	.0002
16	.0000	.0000	.0000	.0000	.0000	.0000	.0000	.0000	.0001	.0001
k'	5.1	5.2	5.3	5.4	5.5	5.6	5.7	5.8	5.9	6.0
0	1.0000	1.0000	1.0000	1.0000	1.0000	1.0000	1.0000	1.0000	1.0000	1.0000
1	.9939	.9945	.9950	.9955	.9959	.9963	.9967	.9970	.9973	.9975
2	.9628	.9658	.9686	.9711	.9734	.9756	.9776	.9794	.9811	.9826
3	.8835	.8912	.8984	.9052	.9116	.9176	.9232	.9285	.9334	.9380
4	.7487	.7619	.7746	.7867	.7983	.8094	.8200	.8300	.8396	.8488
5	.5769	.5939	.6105	.6267	.6425	.6579	.6728	.6873	.7013	.7149
6	.4016	.4191	.4365	.4539	.4711	.4881	.5050	.5217	.5381	.5543
7	.2526	.2676	.2829	.2983	.3140	.3297	.3456	.3616	.3776	.3937
8	.1440	.1551	.1665	.1783	.1905	.2030	.2159	.2290	.2424	.2560
9	.0748	.0819	.0894	.0974	.1056	.1143	.1234	.1328	.1426	.1528

C.3. POISSON DISTRIBUTION

					λ					
k'	5.1	5.2	5.3	5.4	5.5	5.6	5.7	5.8	5.9	6.0
10	.0356	.0397	.0441	.0488	.0538	.0591	.0648	.0708	.0722	.0839
11	.0156	.0177	.0200	.0225	.0253	.0282	.0314	.0349	.0386	.0426
12	.0063	.0073	.0084	.0096	.0110	.0125	.0141	.0160	.0179	.0201
13	.0024	.0028	.0033	.0038	.0045	.0051	.0059	.0068	.0078	.0088
14	.0008	.0010	.0012	.0014	.0017	.0020	.0023	.0027	.0031	.0036
15	.0003	.0003	.0004	.0005	.0006	.0007	.0009	.0010	.0012	.0014
16	.0001	.0001	.0001	.0002	.0002	.0002	.0003	.0004	.0004	.0005
17	.0000	.0000	.0000	.0001	.0001	.0001	.0001	.0001	.0001	.0002
18	.0000	.0000	.0000	.0000	.0000	.0000	.0000	.0000	.0000	.0001
k'	6.1	6.2	6.3	6.4	6.5	6.6	6.7	6.8	6.9	7.0
0	1.0000	1.0000	1.0000	1.0000	1.0000	1.0000	1.0000	1.0000	1.0000	1.0000
1	.9978	.9980	.9982	.9983	.9985	.9986	.9988	.9989	.9990	.9991
2	.9841	.9854	.9866	.9877	.9887	.9897	.9905	.9913	.9920	.9927
3	.9423	.9464	.9502	.9537	.9570	.9600	.9629	.9656	.9680	.9704
4	.8575	.8658	.8736	.8811	.8882	.8948	.9012	.9072	.9129	.9182
5	.7281	.7408	.7531	.7649	.7763	.7873	.7978	.8080	.8177	.8270
6	.5702	.5859	.6012	.6163	.6310	.6453	.6594	.6730	.6863	.6993
7	.4098	.4258	.4418	.4577	.4735	.4892	.5047	.5201	.5353	.5503
8	.2699	.2840	.2983	.3127	.3272	.3419	.3567	.3715	.3864	.4013
9	.1633	.1741	.1852	.1967	.2084	.2204	.2327	.2452	.2580	.2709
10	.0910	.0984	.1061	.1142	.1226	.1314	.1404	.1498	.1505	.1695
11	.0469	.0514	.0563	.0614	.0668	.0726	.0786	.0849	.0916	.0985
12	.0224	.0250	.0277	.0307	.0339	.0373	.0409	.0448	.0495	.0534
13	.0100	.0113	.0127	.0143	.0160	.0179	.0199	.0221	.0245	.0270
14	.0042	.0048	.0055	.0063	.0071	.0080	.0091	.0102	.0115	.0128
15	.0016	.0019	.0022	.0026	.0030	.0034	.0039	.0044	.0050	.0057
16	.0006	.0007	.0008	.0010	.0012	.0014	.0016	.0018	.0021	.0024
17	.0002	.0003	.0003	.0004	.0004	.0005	.0006	.0007	.0008	.0010
18	.0001	.0001	.0001	.0001	.0002	.0002	.0002	.0003	.0003	.0004
19	.0000	.0000	.0000	.0000	.0001	.0001	.0001	.0001	.0001	.0001

(*Continued*)

	λ									
k'	7.1	7.2	7.3	7.4	7.5	7.6	7.7	7.8	7.9	8.0
0	1.0000	1.0000	1.0000	1.0000	1.0000	1.0000	1.0000	1.0000	1.0000	1.0000
1	.9992	.9993	.9993	.9994	.9994	.9995	.9995	.9996	.9996	.9997
2	.9933	.9939	.9944	.9949	.9953	.9957	.9961	.9964	.9967	.9970
3	.9725	.9745	.9764	.9781	.9797	.9812	.9826	.9839	.9851	.9862
4	.9233	.9281	.9326	.9368	.9409	.9446	.9482	.9515	.9547	.9576
5	.8359	.8445	.8527	.8605	.8679	.8751	.8819	.8883	.8945	.9004
6	.7119	.7241	.7360	.7474	.7586	.7693	.7797	.7897	.7994	.8088
7	.5651	.5796	.5940	.6080	.6218	.6354	.6486	.6616	.6743	.6866
8	.4162	.4311	.4459	.4607	.4754	.4900	.5044	.5188	.5330	.5470
9	.2840	.2973	.3108	.3243	.3380	.3518	.3657	.3796	.3935	.4075
10	.1798	.1904	.2012	.2123	.2236	.2351	.2469	.2589	.2710	.2834
11	.1058	.1133	.1212	.1293	.1378	.1465	.1555	.1648	.1743	.1841
12	.0580	.0629	.0681	.0735	.0792	.0852	.0915	.0980	.1048	.1119
13	.0297	.0327	.0358	.0391	.0427	.0464	.0504	.0546	.0591	.0638
14	.0143	.0159	.0176	.0195	.0216	.0238	.0261	.0286	.0313	.0342
15	.0065	.0073	.0082	.0092	.0103	.0114	.0127	.0141	.0156	.0173
16	.0028	.0031	.0036	.0041	.0046	.0052	.0059	.0066	.0074	.0082
17	.0011	.0013	.0015	.0017	.0020	.0022	.0026	.0029	.0033	.0037
18	.0004	.0005	.0006	.0007	.0008	.0009	.0011	.0012	.0014	.0016
19	.0002	.0002	.0002	.0003	.0003	.0004	.0004	.0005	.0006	.0005
k'	8.1	8.2	8.3	8.4	8.5	8.6	8.7	8.8	8.9	9.0
0	1.0000	1.0000	1.0000	1.0000	1.0000	1.0000	1.0000	1.0000	1.0000	1.0000
1	.9997	.9997	.9998	.9998	.9998	.9998	.9998	.9998	.9999	.9999
2	.9972	.9975	.9977	.9979	.9981	.9982	.9984	.9985	.9987	.9988
3	.9873	.9882	.9891	.9900	.9907	.9914	.9921	.9927	.9932	.9938
4	.9604	.9630	.9654	.9677	.9699	.9719	.9738	.9756	.9772	.9788
5	.9060	.9113	.9163	.9211	.9256	.9299	.9340	.9379	.9416	.9450
6	.8178	.8264	.8347	.8427	.8504	.8578	.8648	.8716	.8781	.8843
7	.6987	.7104	.7219	.7330	.7438	.7543	.7645	.7744	.7840	.7932
8	.5609	.5746	.5881	.6013	.6144	.6272	.6398	.6522	.6643	.6761
9	.4214	.4353	.4493	.4631	.4769	.4906	.5042	.5177	.5311	.5443

C.3. POISSON DISTRIBUTION

					λ					
k'	8.1	8.2	8.3	8.4	8.5	8.6	8.7	8.8	8.9	9.0
10	.2959	.3085	.3212	.3341	.3470	.3600	.3731	.3863	.3994	.4126
11	.1942	.2045	.2150	.2257	.2366	.2478	.2591	.2706	.2822	.2940
12	.1193	.1269	.1348	.1429	.1513	.1600	.1689	.1780	.1874	.1970
13	.0687	.0739	.0793	.0850	.0909	.0971	.1035	.1102	.1171	.1242
14	.0372	.0405	.0439	.0476	.0514	.0555	.0597	.0642	.0689	.0739
15	.0190	.0209	.0229	.0251	.0274	.0299	.0325	.0353	.0383	.0415
16	.0092	.0102	.0113	.0125	.0138	.0152	.0168	.0184	.0202	.0220
17	.0042	.0047	.0053	.0059	.0066	.0074	.0082	.0091	.0101	.0111
18	.0018	.0021	.0023	.0027	.0030	.0034	.0038	.0043	.0048	.0053
19	.0008	.0009	.0010	.0011	.0013	.0015	.0017	.0019	.0022	.0024
k'	9.1	9.2	9.3	9.4	9.5	9.6	9.7	9.8	9.9	10.0
0	1.0000	1.0000	1.0000	1.0000	1.0000	1.0000	1.0000	1.0000	1.0000	1.0000
1	.9999	.9999	.9999	.9999	.9999	.9999	.9999	.9999	1.0000	1.0000
2	.9989	.9990	.9991	.9991	.9992	.9993	.9993	.9994	.9995	.9995
3	.9942	.9947	.9951	.9955	.9958	.9962	.9965	.9967	.9970	.9972
4	.9802	.9816	.9828	.9840	.9851	.9862	.9871	.9880	.9889	.9897
5	.9483	.9514	.9544	.9571	.9597	.9622	.9645	.9667	.9688	.9707
6	.8902	.8959	.9014	.9065	.9115	.9162	.9207	.9250	.9290	.9329
7	.8022	.8108	.8192	.8273	.8351	.8426	.8498	.8567	.8634	.8699
8	.6877	.6990	.7101	.7208	.7313	.7416	.7515	.7612	.7706	.7798
9	.5574	.5704	.5832	.5958	.6082	.6204	.6324	.6442	.6558	.6672
10	.4258	.4389	.4521	.4651	.4782	.4911	.5040	.5168	.5295	.5421
11	.3059	.3180	.3301	.3424	.3547	.3671	.3795	.3920	.4045	.4170
12	.2068	.2168	.2270	.2374	.2480	.2588	.2697	.2807	.2919	.3032
13	.1316	.1393	.1471	.1552	.1636	.1721	.1809	.1899	.1991	.2084
14	.0790	.0844	.0900	.0958	.1019	.1081	.1147	.1214	.1284	.1355
15	.0448	.0483	.0520	.0559	.0600	.0643	.0688	.0735	.0784	.0835
16	.0240	.0262	.0285	.0309	.0335	.0362	.0391	.0421	.0454	.0487
17	.0122	.0135	.0148	.0162	.0177	.0194	.0211	.0230	.0249	.0270
18	.0059	.0066	.0073	.0081	.0089	.0098	.0108	.0119	.0130	.0143
19	.0027	.0031	.0034	.0038	.0043	.0048	.0053	.0059	.0065	.0072

(Continued)

					λ					
k'	9.1	9.2	9.3	9.4	9.5	9.6	9.7	9.8	9.9	10.0
20	.0012	.0014	.0015	.0017	.0020	.0022	.0025	.0028	.0031	.0035
21	.0005	.0006	.0007	.0008	.0009	.0010	.0011	.0013	.0014	.0016
22	.0002	.0002	.0003	.0003	.0004	.0004	.0005	.0005	.0006	.0007
23	.0001	.0001	.0001	.0001	.0001	.0002	.0002	.0002	.0003	.0003
24	.0000	.0000	.0000	.0000	.0001	.0001	.0001	.0001	.0001	.0001

C.4. CHI-SQUARED DISTRIBUTION

The table gives values of χ^2 for values of n and F, where $F(\chi^2;n) = \dfrac{1}{2^{n/2}\Gamma(n/2)} \int_0^{\chi^2} t^{n/2-1}\exp(-t/2)\,dt$.

n	.005	.010	.025	.050	.100	.250	.500	.750	.900	.950	.975	.990	.995
1	.0000393	.000157	.000982	.00393	.0158	.102	.455	1.32	2.71	3.84	5.02	6.63	7.88
2	.0100	.0201	.0506	.103	.211	.575	1.39	2.77	4.61	5.99	7.38	9.21	10.6
3	.0717	.115	.216	.352	.584	1.21	2.37	4.11	6.25	7.81	9.35	11.3	12.8
4	.207	.297	.484	.711	1.06	1.92	3.36	5.39	7.78	9.49	11.1	13.3	14.9
5	.412	.554	.831	1.15	1.61	2.67	4.35	6.63	9.24	11.1	12.8	15.1	16.7
6	.676	.872	1.24	1.64	2.20	3.45	5.35	7.84	10.6	12.6	14.4	16.8	18.5
7	.989	1.24	1.69	2.17	2.83	4.25	6.35	9.04	12.0	14.1	16.0	18.5	20.3
8	1.34	1.65	2.18	2.73	3.49	5.07	7.34	10.2	13.4	15.5	17.5	20.1	22.0
9	1.73	2.09	2.70	3.33	4.17	5.90	8.34	11.4	14.7	16.9	19.0	21.7	23.6
10	2.16	2.56	3.25	3.94	4.87	6.74	9.34	12.5	16.0	18.3	20.5	23.2	25.2
11	2.60	3.05	3.82	4.57	5.58	7.58	10.3	13.7	17.3	19.7	21.9	24.7	26.8
12	3.07	3.57	4.40	5.23	6.30	8.44	11.3	14.8	18.5	21.0	23.3	26.2	28.3
13	3.57	4.11	5.01	5.89	7.04	9.30	12.3	16.0	19.8	22.4	24.7	27.7	29.8
14	4.07	4.66	5.63	6.57	7.79	10.2	13.3	17.1	21.1	23.7	26.1	29.1	31.3
15	4.60	5.23	6.26	7.26	8.55	11.0	14.3	18.2	22.3	25.0	27.5	30.6	32.8

(Continued)

C. STATISTICAL TABLES

F

n	.005	.010	.025	.050	.100	.250	.500	.750	.900	.950	.975	.990	.995
16	5.14	5.81	6.91	7.96	9.31	11.9	15.3	19.4	23.5	26.3	28.8	32.0	34.3
17	5.70	6.41	7.56	8.67	10.1	12.8	16.3	20.5	24.8	27.6	30.2	33.4	35.7
18	6.26	7.01	8.23	9.39	10.9	13.7	17.3	21.6	26.0	28.9	31.5	34.8	37.2
19	6.84	7.63	8.91	10.1	11.7	14.6	18.3	22.7	27.2	30.1	32.9	36.2	38.6
20	7.43	8.26	9.59	10.9	12.4	15.5	19.3	23.8	28.4	31.4	34.2	37.6	40.0
21	8.03	8.90	10.3	11.6	13.2	16.3	20.3	24.9	29.6	32.7	35.5	38.9	41.4
22	8.64	9.54	11.0	12.3	14.0	17.2	21.3	26.0	30.8	33.9	36.8	40.3	42.8
23	9.26	10.2	11.7	13.1	14.8	18.1	22.3	27.1	32.0	35.2	38.1	41.6	44.2
24	9.89	10.9	12.4	13.8	15.7	19.0	23.3	28.2	33.2	36.4	39.4	43.0	45.6
25	10.5	11.5	13.1	14.6	16.5	19.9	24.3	29.3	34.4	37.7	40.6	44.3	46.9
26	11.2	12.2	13.8	15.4	17.3	20.8	25.3	30.4	35.6	38.9	41.9	45.6	48.3
27	11.8	12.9	14.6	16.2	18.1	21.7	26.3	31.5	36.7	40.1	43.2	47.0	49.6
28	12.5	13.6	15.3	16.9	18.9	22.7	27.3	32.6	37.9	41.3	44.5	48.3	51.0
29	13.1	14.3	16.0	17.7	19.8	23.6	28.3	33.7	39.1	42.6	45.7	49.6	52.3
30	13.8	15.0	16.8	18.5	20.6	24.5	29.3	34.8	40.3	43.8	47.0	50.9	53.7

C.5. STUDENT'S t DISTRIBUTION

The table gives values of t for specific values of n and F where

$$F(t;n) = \frac{1}{(\pi n)^{1/2}} \frac{\Gamma[(n+1)/2]}{\Gamma(n/2)} \int_{-\infty}^{t} \left(1 + \frac{x^2}{n}\right)^{-(n+1)/2} dx.$$

Note that $F(-t) = 1 - F(t)$.

					F			
n	.60	.75	.90	.95	.975	.99	.995	.9995
1	.325	1.000	3.078	6.314	12.706	31.821	63.657	636.619
2	.289	.816	1.886	2.920	4.303	6.695	9.925	31.598
3	.277	.765	1.638	2.353	3.182	4.541	5.841	12.921
4	.271	.741	1.533	2.132	2.776	3.747	4.604	8.610
5	.267	.727	1.476	2.015	2.571	3.365	4.032	6.869
6	.265	.718	1.440	1.943	2.447	3.143	3.707	5.959
7	.263	.711	1.415	1.895	2.365	2.998	3.499	5.408
8	.262	.706	1.397	1.860	2.306	2.896	3.355	5.041
9	.261	.703	1.383	1.833	2.262	2.821	3.250	4.781
10	.260	.700	1.372	1.812	2.228	2.764	3.169	4.587
11	.260	.697	1.363	1.796	2.201	2.718	3.106	4.437
12	.259	.695	1.356	1.782	2.179	2.681	3.055	4.318
13	.259	.694	1.350	1.771	2.160	2.650	3.012	4.221
14	.258	.692	1.345	1.761	2.145	2.624	2.977	4.140
15	.258	.691	1.341	1.753	2.131	2.602	2.947	4.073
16	.258	.690	1.337	1.746	2.120	2.583	2.921	4.015
17	.257	.689	1.333	1.740	2.110	2.567	2.898	3.965
18	.257	.688	1.330	1.734	2.101	2.552	2.878	3.922
19	.257	.688	1.328	1.729	2.093	2.539	2.861	3.883
20	.257	.687	1.325	1.725	2.086	2.528	2.845	3.850
21	.257	.686	1.323	1.721	2.080	2.518	2.831	3.819
22	.256	.686	1.321	1.717	2.074	2.508	2.819	3.792
23	.256	.685	1.319	1.714	2.069	2.500	2.807	3.768
24	.256	.685	1.318	1.711	2.064	2.492	2.797	3.745
25	.256	.684	1.316	1.708	2.060	2.485	2.787	3.725

(Continued)

	\multicolumn{8}{c}{F}							
n	.60	.75	.90	.95	.975	.99	.995	.9995
26	.256	.684	1.315	1.706	2.056	2.479	2.779	3.707
27	.256	.684	1.314	1.703	2.052	2.473	2.771	3.690
28	.256	.683	1.313	1.701	2.048	2.467	2.763	3.674
29	.256	.683	1.311	1.699	2.045	2.462	2.756	3.659
30	.256	.683	1.310	1.697	2.042	2.457	2.750	3.646
40	.255	.681	1.303	1.684	2.021	2.423	2.704	3.551
60	.254	.679	1.296	1.671	2.000	2.390	2.660	3.460
120	.254	.677	1.289	1.658	1.980	2.358	2.617	3.373
∞	.253	.674	1.282	1.645	1.960	2.326	2.576	3.291

C.6. F DISTRIBUTION

The table gives values of F such that

$$F(F; m, n) = \frac{\Gamma[(m+n)/2]}{\Gamma(m/2)\Gamma(n/2)} \left(\frac{m}{n}\right)^{m/2} \int_0^F \frac{x^{(m-2)/2}}{[1+(m/n)x]^{(m+n)/2}} dx$$

for specified values of m and n. Values corresponding to $F(F) = 0.10$, 0.05 and 0.025 may be found using the relation $F_{1-\alpha}(n,m) = [F_\alpha(m,n)]^{-1}$.

$F(F; m, n) = 0.90$

n	\multicolumn{16}{c	}{m}
	1	2	3	4	5	6	7	8	9	10	12	15	20	24	30	40	60	120	∞
1	39.86	49.50	53.59	55.83	57.24	58.20	58.81	59.44	59.86	60.19	60.71	61.22	61.74	62.00	62.26	62.53	62.79	63.06	63.33
2	8.53	9.00	9.16	9.24	9.29	9.33	9.35	9.37	9.38	9.39	9.41	9.42	9.44	9.45	9.46	9.47	9.47	9.48	9.49
3	5.54	5.46	5.39	5.34	5.31	5.28	5.27	5.25	5.24	5.23	5.22	5.20	5.18	5.18	5.17	5.16	5.15	5.14	5.13
4	4.54	4.32	4.19	4.11	4.05	4.01	3.98	3.95	3.94	3.92	3.90	3.87	3.84	3.83	3.82	3.80	3.79	3.78	3.76
5	4.06	3.78	3.62	3.52	3.45	3.40	3.37	3.34	3.32	3.30	3.27	3.24	3.21	3.19	3.17	3.16	3.14	3.12	3.10
6	3.78	3.46	3.29	3.18	3.11	3.05	3.01	2.98	2.96	2.94	2.90	2.87	2.84	2.82	2.80	2.78	2.76	2.74	2.72
7	3.59	3.26	3.07	2.96	2.88	2.83	2.78	2.75	2.72	2.70	2.67	2.63	2.59	2.58	2.56	2.54	2.51	2.49	2.47
8	3.46	3.11	2.92	2.81	2.73	2.67	2.62	2.59	2.56	2.54	2.50	2.46	2.42	2.40	2.38	2.36	2.34	2.32	2.29
9	3.36	3.01	2.81	2.69	2.61	2.55	2.51	2.47	2.44	2.42	2.38	2.34	2.30	2.28	2.25	2.23	2.21	2.18	2.16
10	3.29	2.92	2.73	2.61	2.52	2.46	2.41	2.38	2.35	2.32	2.28	2.24	2.20	2.18	2.16	2.13	2.11	2.08	2.06
11	3.23	2.86	2.66	2.54	2.45	2.39	2.34	2.30	2.27	2.25	2.21	2.17	2.12	2.10	2.08	2.05	2.03	2.00	1.97
12	3.18	2.81	2.61	2.48	2.39	2.33	2.28	2.24	2.21	2.19	2.15	2.10	2.06	2.04	2.01	1.99	1.96	1.93	1.90
13	3.14	2.76	2.56	2.43	2.35	2.28	2.23	2.20	2.16	2.14	2.10	2.05	2.01	1.98	1.96	1.93	1.90	1.88	1.85
14	3.10	2.73	2.52	2.39	2.31	2.24	2.19	2.15	2.12	2.10	2.05	2.01	1.96	1.94	1.91	1.89	1.86	1.83	1.80

(Continued)

278 C. STATISTICAL TABLES

n \ m	1	2	3	4	5	6	7	8	9	10	12	15	20	24	30	40	60	120	∞
15	3.07	2.70	2.49	2.36	2.27	2.21	2.16	2.12	2.09	2.06	2.02	1.97	1.92	1.90	1.87	1.85	1.82	1.79	1.76
16	3.05	2.67	2.46	2.33	2.24	2.18	2.13	2.09	2.06	2.03	1.99	1.94	1.89	1.87	1.84	1.81	1.78	1.75	1.72
17	3.03	2.64	2.44	2.31	2.22	2.15	2.10	2.06	2.03	2.00	1.96	1.91	1.86	1.84	1.81	1.78	1.75	1.72	1.69
18	3.01	2.62	2.42	2.29	2.20	2.13	2.08	2.04	2.00	1.98	1.93	1.89	1.84	1.81	1.78	1.75	1.72	1.69	1.66
19	2.99	2.61	2.40	2.27	2.18	2.11	2.06	2.02	1.98	1.96	1.91	1.86	1.81	1.79	1.76	1.73	1.70	1.67	1.63
20	2.97	2.59	2.38	2.25	2.16	2.09	2.04	2.00	1.96	1.94	1.89	1.84	1.79	1.77	1.74	1.71	1.68	1.64	1.61
21	2.96	2.57	2.36	2.23	2.14	2.08	2.02	1.98	1.95	1.92	1.87	1.83	1.78	1.75	1.72	1.69	1.66	1.62	1.59
22	2.95	2.56	2.35	2.22	2.13	2.06	2.01	1.97	1.93	1.90	1.86	1.81	1.76	1.73	1.70	1.67	1.64	1.60	1.57
23	2.94	2.55	2.34	2.21	2.11	2.05	1.99	1.95	1.92	1.89	1.84	1.80	1.74	1.72	1.69	1.66	1.62	1.59	1.55
24	2.93	2.54	2.33	2.19	2.10	2.04	1.98	1.94	1.91	1.88	1.83	1.78	1.73	1.70	1.67	1.64	1.61	1.57	1.53
25	2.92	2.53	2.32	2.18	2.09	2.02	1.97	1.93	1.89	1.87	1.82	1.77	1.72	1.69	1.66	1.63	1.59	1.56	1.52
26	2.91	2.52	2.31	2.17	2.08	2.01	1.96	1.92	1.88	1.86	1.81	1.76	1.71	1.68	1.65	1.61	1.58	1.54	1.50
27	2.90	2.51	2.30	2.17	2.07	2.00	1.95	1.91	1.87	1.85	1.80	1.75	1.70	1.67	1.64	1.60	1.57	1.53	1.49
28	2.89	2.50	2.29	2.16	2.06	2.00	1.94	1.90	1.87	1.84	1.79	1.74	1.69	1.66	1.63	1.59	1.56	1.52	1.48
29	2.89	2.50	2.28	2.15	2.06	1.99	1.93	1.89	1.86	1.83	1.78	1.73	1.68	1.65	1.62	1.58	1.55	1.51	1.47
30	2.88	2.49	2.28	2.14	2.05	1.98	1.93	1.88	1.85	1.82	1.77	1.72	1.67	1.64	1.61	1.57	1.54	1.50	1.46
40	2.84	2.44	2.23	2.09	2.00	1.93	1.87	1.83	1.79	1.76	1.71	1.66	1.61	1.57	1.54	1.51	1.47	1.42	1.38
60	2.79	2.39	2.18	2.04	1.95	1.87	1.82	1.77	1.74	1.71	1.66	1.60	1.54	1.51	1.48	1.44	1.40	1.35	1.29
120	2.75	2.35	2.13	1.99	1.90	1.82	1.77	1.72	1.68	1.65	1.60	1.55	1.48	1.45	1.41	1.37	1.32	1.26	1.19
∞	2.71	2.30	2.08	1.94	1.85	1.77	1.72	1.67	1.63	1.60	1.55	1.49	1.42	1.38	1.34	1.30	1.24	1.17	1.00

C.6. F DISTRIBUTION

$F(F; m, n) = 0.95$

n \ m	1	2	3	4	5	6	7	8	9	10	12	15	20	24	30	40	60	120	∞
1	161.4	199.5	215.7	224.6	230.2	234.0	236.8	238.9	240.5	241.9	243.9	245.9	248.0	249.1	250.1	251.1	252.2	253.3	254.3
2	18.51	19.00	19.16	19.25	19.30	19.33	19.35	19.37	19.38	19.40	19.41	19.43	19.45	19.45	19.46	19.47	19.48	19.49	19.50
3	10.13	9.55	9.28	9.12	9.01	8.94	8.89	8.85	8.81	8.79	8.74	8.70	8.66	8.64	8.62	8.59	8.57	8.55	8.53
4	7.71	6.94	6.59	6.39	6.26	6.16	6.09	6.04	6.00	5.96	5.91	5.86	5.80	5.77	5.75	5.72	5.69	5.66	5.63
5	6.61	5.79	5.41	5.19	5.05	4.95	4.88	4.82	4.77	4.74	4.68	4.62	4.56	4.53	4.50	4.46	4.43	4.40	4.36
6	5.99	5.14	4.76	4.53	4.39	4.28	4.21	4.15	4.10	4.06	4.00	3.94	3.87	3.84	3.81	3.77	3.74	3.70	3.67
7	5.59	4.74	4.35	4.12	3.97	3.87	3.79	3.73	3.68	3.64	3.57	3.51	3.44	3.41	3.38	3.34	3.30	3.27	3.23
8	5.32	4.46	4.07	3.84	3.69	3.58	3.50	3.44	3.39	3.35	3.28	3.22	3.15	3.12	3.08	3.04	3.01	2.97	2.93
9	5.12	4.26	3.86	3.63	3.48	3.37	3.29	3.23	3.18	3.14	3.07	3.01	2.94	2.90	2.86	2.83	2.79	2.75	2.71
10	4.96	4.10	3.71	3.48	3.33	3.22	3.14	3.07	3.02	2.98	2.91	2.85	2.77	2.74	2.70	2.66	2.62	2.58	2.54
11	4.84	3.98	3.59	3.36	3.20	3.09	3.01	2.95	2.90	2.85	2.79	2.72	2.65	2.61	2.57	2.53	2.49	2.45	2.40
12	4.75	3.89	3.49	3.26	3.11	3.00	2.91	2.85	2.80	2.75	2.69	2.62	2.54	2.51	2.47	2.43	2.38	2.34	2.30
13	4.67	3.81	3.41	3.18	3.03	2.92	2.83	2.77	2.71	2.67	2.60	2.53	2.46	2.42	2.38	2.34	2.30	2.25	2.21
14	4.60	3.74	3.34	3.11	2.96	2.85	2.76	2.70	2.65	2.60	2.53	2.46	2.39	2.35	2.31	2.27	2.22	2.18	2.13
15	4.54	3.68	3.29	3.06	2.90	2.79	2.71	2.64	2.59	2.54	2.48	2.40	2.33	2.29	2.25	2.20	2.16	2.11	2.07
16	4.49	3.63	3.21	3.01	2.85	2.74	2.66	2.59	2.54	2.49	2.42	2.35	2.28	2.24	2.19	2.15	2.11	2.06	2.01
17	4.45	3.59	3.20	2.96	2.81	2.70	2.61	2.55	2.49	2.45	2.38	2.31	2.23	2.19	2.15	2.10	2.06	2.01	1.96
18	4.41	3.55	3.16	2.93	2.77	2.66	2.58	2.51	2.46	2.41	2.34	2.27	2.19	2.15	2.11	2.06	2.02	1.97	1.92
19	4.38	3.52	3.13	2.90	2.74	2.63	2.54	2.48	2.42	2.38	2.31	2.23	2.16	2.11	2.07	2.03	1.98	1.93	1.88

(Continued)

n \ m	1	2	3	4	5	6	7	8	9	10	12	15	20	24	30	40	60	120	∞
20	4.35	3.49	3.10	2.87	2.71	2.60	2.51	2.45	2.39	2.35	2.28	2.20	2.12	2.08	2.04	1.99	1.95	1.90	1.84
21	4.32	3.47	3.07	2.84	2.68	2.57	2.49	2.42	2.37	2.32	2.25	2.18	2.10	2.05	2.01	1.96	1.92	1.87	1.81
22	4.30	3.44	3.05	2.82	2.66	2.55	2.46	2.40	2.34	2.30	2.23	2.15	2.07	2.03	1.98	1.94	1.89	1.84	1.78
23	4.28	3.42	3.03	2.80	2.64	2.53	2.44	2.37	2.32	2.27	2.20	2.13	2.05	2.01	1.96	1.91	1.86	1.81	1.76
24	4.26	3.40	3.01	2.78	2.62	2.51	2.42	2.36	2.30	2.25	2.18	2.11	2.03	1.98	1.94	1.89	1.84	1.79	1.73
25	4.24	3.39	2.99	2.76	2.60	2.49	2.40	2.34	2.28	2.24	2.16	2.09	2.01	1.96	1.92	1.87	1.82	1.77	1.71
26	4.23	3.37	2.98	2.74	2.59	2.47	2.39	2.32	2.27	2.22	2.15	2.07	1.99	1.95	1.90	1.85	1.80	1.75	1.69
27	4.21	3.35	2.96	2.73	2.57	2.46	2.37	2.31	2.25	2.20	2.13	2.06	1.97	1.93	1.88	1.84	1.79	1.73	1.67
28	4.20	3.34	2.95	2.71	2.56	2.45	2.36	2.29	2.24	2.19	2.12	2.04	1.96	1.91	1.87	1.82	1.77	1.71	1.65
29	4.18	3.33	2.93	2.70	2.55	2.43	2.35	2.28	2.22	2.18	2.10	2.03	1.94	1.90	1.85	1.81	1.75	1.70	1.64
30	4.17	3.32	2.92	2.69	2.53	2.42	2.33	2.27	2.21	2.16	2.09	2.01	1.93	1.89	1.84	1.79	1.74	1.68	1.62
40	4.08	3.23	2.84	2.61	2.45	2.34	2.25	2.18	2.12	2.08	2.00	1.92	1.84	1.79	1.74	1.69	1.64	1.58	1.51
60	4.00	3.15	2.76	2.53	2.37	2.25	2.17	2.10	2.04	1.99	1.92	1.84	1.75	1.70	1.65	1.59	1.53	1.47	1.39
120	3.92	3.07	2.68	2.45	2.29	2.17	2.09	2.02	1.96	1.91	1.83	1.75	1.66	1.61	1.55	1.50	1.43	1.35	1.25
∞	3.84	3.00	2.60	2.37	2.21	2.10	2.01	1.94	1.88	1.83	1.75	1.67	1.57	1.52	1.46	1.39	1.32	1.22	1.00

C.6. F DISTRIBUTION

$F(F; m, n) = 0.975$

n \ m	1	2	3	4	5	6	7	8	9	10	12	15	20	24	30	40	60	120	∞
1	647.8	799.5	864.2	899.6	921.8	937.1	948.2	956.7	963.3	968.6	976.7	984.9	993.1	997.2	1001	1006	1010	1014	1018
2	38.51	39.00	39.17	39.25	39.30	39.33	39.36	39.37	39.39	39.40	39.41	39.43	39.45	39.46	39.46	39.47	39.48	39.49	39.50
3	17.44	16.04	15.44	15.10	14.88	14.73	14.62	14.54	14.47	14.42	14.34	14.25	14.17	14.12	14.08	14.04	13.99	13.95	13.90
4	12.22	10.65	9.98	9.60	9.36	9.20	9.07	8.98	8.90	8.84	8.75	8.66	8.56	8.51	8.46	8.41	8.36	8.31	8.26
5	10.01	8.43	7.76	7.39	7.15	6.98	6.85	6.76	6.68	6.62	6.52	6.43	6.33	6.28	6.23	6.18	6.12	6.07	6.02
6	8.81	7.26	6.60	6.23	5.99	5.82	5.70	5.60	5.52	5.46	5.37	5.27	5.17	5.12	5.07	5.01	4.96	4.90	4.85
7	8.07	6.54	5.89	5.52	5.29	5.12	4.99	4.90	4.82	4.76	4.67	4.57	4.47	4.42	4.36	4.31	4.25	4.20	4.14
8	7.57	6.06	5.42	5.05	4.82	4.65	4.53	4.43	4.36	4.30	4.20	4.10	4.00	3.95	3.89	3.84	3.78	3.73	3.67
9	7.21	5.71	5.08	4.72	4.48	4.32	4.20	4.10	4.03	3.96	3.87	3.77	3.67	3.61	3.56	3.51	3.45	3.39	3.33
10	6.94	5.46	4.83	4.47	4.24	4.07	3.95	3.85	3.78	3.72	3.62	3.52	3.42	3.37	3.31	3.26	3.20	3.14	3.08
11	6.72	5.29	4.63	4.28	4.04	3.88	3.76	3.66	3.59	3.53	3.43	3.33	3.23	3.17	3.12	3.06	3.00	2.94	2.88
12	6.55	5.10	4.47	4.12	3.89	3.73	3.61	3.51	3.44	3.37	3.28	3.18	3.07	3.02	2.96	2.91	2.85	2.79	2.72
13	6.41	4.97	4.35	4.00	3.77	3.60	3.48	3.39	3.31	3.25	3.15	3.05	2.95	2.89	2.84	2.78	2.72	2.66	2.60
14	6.30	4.86	4.24	3.89	3.66	3.50	3.38	3.29	3.21	3.15	3.05	2.95	2.84	2.79	2.73	2.67	2.61	2.55	2.49
15	6.20	4.77	4.15	3.80	3.58	3.41	3.29	3.20	3.12	3.06	2.96	2.86	2.76	2.70	2.64	2.59	2.52	2.46	2.40
16	6.12	4.69	4.08	3.73	3.50	3.34	3.22	3.12	3.05	2.99	2.89	2.79	2.68	2.63	2.57	2.51	2.45	2.38	2.32
17	6.04	4.62	4.01	3.66	3.44	3.28	3.16	3.06	2.98	2.92	2.82	2.72	2.62	2.56	2.50	2.44	2.38	2.32	2.25
18	5.98	4.56	3.95	3.61	3.38	3.22	3.10	3.01	2.93	2.87	2.77	2.67	2.56	2.50	2.44	2.38	2.32	2.26	2.19
19	5.92	4.51	3.90	3.56	3.33	3.17	3.05	2.96	2.88	2.82	2.72	2.62	2.51	2.45	2.39	2.33	2.27	2.20	2.13

(Continued)

C. STATISTICAL TABLES

n \ m	1	2	3	4	5	6	7	8	9	10	12	15	20	24	30	40	60	120	∞
20	5.87	4.46	3.86	3.51	3.29	3.13	3.01	2.91	2.84	2.77	2.68	2.57	2.46	2.41	2.35	2.29	2.22	2.16	2.09
21	5.83	4.42	3.82	3.48	3.25	3.09	2.97	2.87	2.80	2.73	2.64	2.53	2.42	2.37	2.31	2.25	2.18	2.11	2.04
22	5.79	4.38	3.78	3.44	3.22	3.05	2.93	2.84	2.76	2.70	2.60	2.50	2.39	2.33	2.27	2.21	2.14	2.08	2.00
23	5.75	4.35	3.75	3.41	3.18	3.02	2.90	2.81	2.73	2.67	2.57	2.47	2.36	2.30	2.24	2.18	2.11	2.04	1.97
24	5.72	4.32	3.72	3.38	3.15	2.99	2.87	2.78	2.70	2.64	2.54	2.44	2.33	2.27	2.21	2.15	2.08	2.01	1.94
25	5.69	4.29	2.69	3.35	3.13	2.97	2.85	2.75	2.68	2.61	2.51	2.41	2.30	2.24	2.18	2.12	2.05	1.98	1.91
26	5.66	4.27	3.67	3.33	3.10	2.94	2.82	2.73	2.65	2.59	2.49	2.39	2.28	2.22	2.16	2.09	2.03	1.95	1.88
27	5.63	4.24	3.65	3.31	3.08	2.92	2.80	2.71	2.63	2.57	2.47	2.36	2.25	2.19	2.13	2.07	2.00	1.93	1.85
28	5.61	4.22	3.63	3.29	3.06	2.90	2.78	2.69	2.61	2.55	2.45	2.34	2.23	2.17	2.11	2.05	1.98	1.91	1.83
29	5.59	4.20	3.61	3.27	3.04	2.88	2.76	2.67	2.59	2.53	2.43	2.32	2.21	2.15	2.09	2.03	1.96	1.89	1.81
30	5.57	4.18	3.59	3.25	3.03	2.87	2.75	2.65	2.57	2.51	2.41	2.31	2.20	2.14	2.07	2.01	1.94	1.87	1.79
40	5.42	4.05	3.46	3.13	2.90	2.74	2.62	2.53	2.45	2.39	2.29	2.18	2.07	2.01	1.94	1.88	1.80	1.72	1.64
60	5.29	3.93	3.34	3.01	2.79	2.63	2.51	2.41	2.33	2.27	2.17	2.06	1.94	1.88	1.82	1.74	1.67	1.58	1.48
120	5.15	3.80	3.23	2.89	2.67	2.52	2.39	2.30	2.22	2.16	2.05	1.94	1.82	1.76	1.69	1.61	1.53	1.43	1.31
∞	5.02	3.69	3.12	2.79	2.57	2.41	2.29	2.19	2.11	2.05	1.94	1.83	1.71	1.64	1.57	1.48	1.39	1.27	1.00

C.7. SIGNED-RANK TEST

The table gives critical values of w_+ for a one-tailed signed-rank test for samples of size n. For a two-tailed test use the statistic w at a 2α value.

n	One-tailed $\alpha = 0.01$ Two-tailed $\alpha = 0.02$	One-tailed $\alpha = 0.025$ Two-tailed $\alpha = 0.05$	One-tailed $\alpha = 0.05$ Two-tailed $\alpha = 0.10$
5			0
6		0	2
7	0	2	3
8	1	3	5
9	3	5	8
10	5	8	10
11	7	10	13
12	9	13	17
13	12	17	21
14	15	21	25
15	19	25	30
16	23	29	35
17	27	34	41
18	32	40	47
19	37	46	53
20	43	52	60
21	49	58	67
22	55	65	75
23	62	73	83
24	69	81	91
25	76	89	100
26	84	98	110
27	92	107	119
28	101	116	130
29	110	126	140
30	120	137	151

Footnote: Formulas for the calculation of this table are given in F. Wilcoxon, S.K. Katti and R.A. Wilcox, 'Critical values and probability levels for the Wilcoxon rank-sum test and the Wilcoxon signed-rank test', *Selected Tables in Mathematical Statistics*, Vol. 1 (1973), American Mathematical Society, Providence, Rhode Island, U.S.A.

C.8. RANK-SUM TEST (See footnote to Table C.7.)

Values of $u'_{1,2}$ such that $P[u_{1,2} \leq u'_{1,2}] \leq \alpha$, for samples of sizes n_1 and $n_2 > n_1$. For a two-tailed test use the statistic u at a 2α value.

										n_2									
n_1	3	4	5	6	7	8	9	10	11	12	13	14	15	16	17	18	19	20	
2									1	1	1	1	1	2	2	2	2	2	One-tailed test,
3			1	1	2	2	3	3	4	4	5	5	6	6	7	7	8		$\alpha = 0.025$; or
4		1	2	3	4	4	5	6	7	8	9	10	11	11	12	13	14		two-tailed test,
5		2	3	5	6	7	8	9	11	12	13	14	15	17	18	19	20		$\alpha = 0.05$
6			5	6	8	10	11	13	14	16	17	19	21	22	24	25	27		
7				8	10	12	14	16	18	20	22	24	26	28	30	32	34		
8					13	15	17	19	22	24	26	29	31	34	36	38	41		
9						17	20	23	26	28	31	34	37	39	42	45	48		
10							23	26	29	33	36	39	42	45	48	52	55		

										n_2									
n_1	3	4	5	6	7	8	9	10	11	12	13	14	15	16	17	18	19	20	
2					1	1	1	1	2	2	3	3	3	3	4	4	4		One-tailed test,
3		1	2	2	3	4	4	5	5	6	7	7	8	9	9	10	11		$\alpha = 0.05$; or
4	1	2	3	4	5	6	7	8	9	10	11	12	14	15	16	17	18		two-tailed test,
5		4	5	6	8	9	11	12	13	15	16	18	19	20	22	23	25		$\alpha = 0.10$
6			7	8	10	12	14	16	17	19	21	23	25	26	28	30	32		
7				11	13	15	17	19	21	24	26	28	30	33	35	37	39		
8					15	18	20	23	26	28	31	33	36	39	41	44	47		
9						21	24	27	30	33	36	39	42	45	48	51	54		
10							27	31	34	37	41	44	48	51	55	58	62		

C.9. RUNS TEST

The table gives lower and upper critical values for r, the number of runs, in the form (a,b), for given values of n and $m > n$ for use in a one-tailed test at significance level $\alpha = 0.05$, or a two-tailed test at significance level $\alpha = 0.1$. A dash means there is no value that satisfies the required conditions.

n	4	5	6	7	8	9	10	11	12	13	14	15	16	17	18	19	20
2					(2,−)	(2,−)	(2,−)	(2,−)	(2,−)	(2,−)	(2,−)	(2,−)	(2,−)	(2,−)	(2,−)	(2,−)	(2,−)
3	(−,7)	(2,−)	(2,−)	(2,−)	(2,−)	(2,−)	(3,−)	(3,−)	(3,−)	(3,−)	(3,−)	(3,−)	(3,−)	(3,−)	(3,−)	(3,−)	(3,−)
4	(2,8)	(2,9)	(3,9)	(3,9)	(3,−)	(3,−)	(3,−)	(4,−)	(4,−)	(4,−)	(4,−)	(4,−)	(4,−)	(4,−)	(4,−)	(4,−)	(4,−)
5		(3,9)	(3,10)	(3,10)	(3,11)	(4,11)	(4,11)	(4,−)	(4,−)	(4,−)	(5,−)	(5,−)	(5,−)	(5,−)	(5,−)	(5,−)	(5,−)
6			(3,11)	(4,11)	(4,12)	(4,12)	(5,12)	(5,13)	(5,13)	(5,13)	(5,13)	(6,−)	(6,−)	(6,−)	(6,−)	(6,−)	(6,−)
7				(4,12)	(4,13)	(5,13)	(5,13)	(5,14)	(6,14)	(6,14)	(6,14)	(6,15)	(6,15)	(6,15)	(7,15)	(7,15)	(7,−)
8					(5,13)	(5,14)	(6,14)	(6,15)	(6,15)	(6,15)	(7,16)	(7,16)	(7,16)	(7,16)	(8,16)	(8,16)	(8,17)
9						(6,14)	(6,15)	(6,15)	(7,16)	(7,16)	(7,17)	(8,17)	(8,17)	(8,17)	(8,18)	(8,18)	(9,18)
10							(6,16)	(7,16)	(7,17)	(8,17)	(8,17)	(8,18)	(8,18)	(9,18)	(9,19)	(9,19)	(9,19)
11								(7,17)	(8,17)	(8,18)	(8,18)	(9,19)	(9,19)	(9,19)	(10,20)	(10,20)	(10,20)
12									(8,18)	(9,18)	(9,19)	(9,19)	(10,20)	(10,20)	(10,21)	(10,21)	(11,21)
13										(9,19)	(9,20)	(10,20)	(10,21)	(10,21)	(11,21)	(11,22)	(11,22)
14											(10,20)	(10,21)	(11,21)	(11,22)	(11,22)	(12,23)	(12,23)
15												(11,21)	(11,22)	(11,22)	(12,23)	(12,23)	(12,24)
16													(11,23)	(12,23)	(12,24)	(13,24)	(13,25)
17														(12,24)	(13,24)	(13,25)	(13,25)
18															(13,25)	(14,25)	(14,26)
19																(14,26)	(14,27)
20																	(15,27)

C.10. RANK CORRELATION COEFFICIENT

The table gives critical values of Spearman's rank correlation coefficient for specified values of α, the significance level.

n	$\alpha = 0.05$	$\alpha = 0.025$	$\alpha = 0.01$
5	0.900		
6	0.829	0.886	0.943
7	0.714	0.786	0.893
8	0.643	0.738	0.833
9	0.600	0.683	0.783
10	0.564	0.648	0.745
11	0.523	0.623	0.736
12	0.497	0.591	0.703
13	0.475	0.566	0.673
14	0.457	0.545	0.646
15	0.441	0.525	0.623
16	0.425	0.507	0.601
17	0.412	0.490	0.582
18	0.399	0.476	0.564
19	0.388	0.462	0.549
20	0.377	0.450	0.534
21	0.368	0.438	0.521
22	0.359	0.428	0.508
23	0.351	0.418	0.496
24	0.343	0.409	0.485
25	0.336	0.400	0.475
26	0.329	0.392	0.465
27	0.323	0.385	0.456
28	0.317	0.377	0.448
29	0.311	0.370	0.440
30	0.305	0.364	0.432

APPENDIX

D

Answers to Odd-Numbered Problems

OUTLINE

Problems 1	287	Problems 7	289
Problems 2	288	Problems 8	290
Problems 3	288	Problems 9	290
Problems 4	288	Problems 10	290
Problems 5	289	Problems 11	291
Problems 6	289		

PROBLEMS 1

1.1 Possible events are 1H, 1T, 3H, 3T, 5H, 5T, 2HT, 2TH, 2TT, 4HH, 4HT, 4TH, 4TT, 6HH, 6HT, 6TH, and 6TT; 18 elements in the sample space.

1.3 The bins are $0 \leq x < 12.5$, $12.5 \leq x < 25.0$, etc. with widths 12.5. The frequencies are 0, 1, 5, 7, 9, 11, 4, and 3. The cumulative distribution is

Bin	< 12.5	< 25	< 37.5	< 50	< 62.5	< 75	< 87.5	< 100
Cumulative frequency	0	1	6	13	22	33	37	40

1.5 For the unbinned data, $\bar{x} = 58.375$ and $s = 18.62$. For the binned data $\bar{x} = 58.75$ and $s = 18.82$. Shepard's correction reduces this to 18.47, but would not alter the estimate for the mean. The percentage of unbinned data that falls within the range $\bar{x} - 2s < x < \bar{x} + 2s$ is 97.5%. If the data followed a normal distribution, the percentage would be 95.4%.

1.7 Solve the quadratic equation for μ that follows from $f_{max}(x) = 1/\sigma\sqrt{2\pi}$.

1.9 The correlation coefficient $r = -0.64$.

PROBLEMS 2

2.1 Probability for a current to flow is $P = [(1 \cap 2) \cup (3 \cup 4)] = 2p - 2p^3 + p^4$, and for no current to flow is $\overline{P} = P[\overline{1 \cup 2}]P[\overline{3 \cap 4}] = 1 - 2p + 2p^3 - p^4$. $P + \overline{P} = 1$, which checks.

2.3 If F denotes the event where the component is faulty and B the event where the component is part of a faulty batch, then $P[F|B] = 0.94$ and the technician should now be 94% certain that the component is the problem.

2.5 The number of possible arrangements is 103680.

2.7 If + denotes a positive (guilty) test result and $\overline{G}$ innocence, then $P[\overline{G}|+] = 0.254$

2.9 If W_2 means the second draw has resulted in a white ball, etc., then
(a) $P[R_1 \cap W_2 \cap B_3] = 0.022$ and (b) $P[R_1 \cap W_2 \cap B_3] = 0.018$.

PROBLEMS 3

3.1 $\mu'_1 = \gamma/a$ and $\mu'_2 = \gamma(\gamma+1)/a^2$.

3.3 Expected number is 4.095.

3.5 (a) $P[x < y] = 0.25$. (b) $P[x > 1, y < 2] = 0.367$.

3.7 $f^C(x|y) = 12x^2(1 + x - y)/(7 - 4y)$

3.9 $f(u) = u$, $(0 \leq u < 1)$ and $f(u) = 2 - u$, $(1 \leq u < 2)$, i.e., a triangle with a peak at $u = 1$.

PROBLEMS 4

4.1 Without medical intervention, $P[n = 14] = 0.0369$, i.e., 3.7%. This is to be compared with 70% who recover if they use the drug. So, on the basis of this very small sample, the drug is effective.

4.3 $P = 0.0087$.

4.5 Using the binomial, $P[r \geq 5] = 0.564$; using the Poisson, $P[r \geq 5] = 0.559$.

4.7 $P[k \geq 3] = 0.047$.

4.9 $P[W > 6 \times 10^5] \approx 0.09$

PROBLEMS 5

5.1 An unbiased estimator for I is $\bar{x}_n - 1/2$ and $E\left[(\bar{x}_n - \mu)^2\right] = 1/12n + 1/4$.

5.3 $P[x < 7.5] = 0.1056$.

5.5 Any more than 22 exposures.

5.7 Approximately 96.

5.9 The percentage error on F is 12%.

PROBLEMS 6

6.1 $\chi^2 = 82.3$.

6.3 Use the properties of the characteristic function of a variate χ_j^2 having a chi-squared distribution with n_j degrees of freedom.

6.5 $n = 7$

6.7 $k = -2.093$

6.9 The probability of a difference in the variances of the measured size is between 0.02 and 0.05 and hence the machines do not appear to be consistent.

PROBLEMS 7

7.1 The ML estimator $\hat{\tau}$ of the lifetime τ is the mean of the measured times $\bar{t}$. To show that it is an unbiased estimator for τ, show that $E[\hat{\tau}(t_1, t_2, \ldots, t_n)] = \tau$. The variance is $\sigma^2(\hat{\tau}) = \tau^2/n$.

7.3 The mean $\bar{k}$ is an unbiased estimator $\hat{\lambda}$ for the parameter λ and $\sigma^2(\bar{k}) = \lambda/n$.

7.5 $\hat{k} = -2 - 3/\sum \ln x_i$

7.7 Find the ML estimators for E_0 in the case where $|E_i - E_0| \ll \Gamma$ and $\Gamma > 0$.

PROBLEMS 8

8.1 From the χ^2 probabilities only the cubic fit is acceptable.

8.3 $\hat{\lambda}_1 = 1.04$ and $\hat{\lambda}_2 = -1.06$, with an error matrix

$$\mathbf{E} = \begin{pmatrix} 1.6 & -0.4 \\ -0.4 & 0.6 \end{pmatrix} \times 10^{-2}.$$

8.5 $x_1 = h_1\left(1 - \dfrac{\theta_1}{\theta_2}\right)$, $x_2 = h_2\left(1 - \dfrac{\theta_1}{\theta_2}\right)$, $x_3 = h_3\left(1 + \dfrac{\theta_1}{\theta_3}\right)$,

where

$$\theta_1 = h_1^2 + h_2^2 - h_3^2;\ \theta_2 = 2\left(h_1^2 + h_2^2 + h_3\sqrt{h_1^2 + h_2^2}\right);$$

and

$$\theta_3 = 2\left(h_3^2 + h_3\sqrt{h_1^2 + h_2^2}\right).$$

8.7 The Bayes' estimator is

$$\hat{p} = \frac{(\sum r) + 1}{n + 2}.$$

This differs slightly from the ML estimator found in Problem 7.7, which was $\hat{p} = (\sum r)/n$.

PROBLEMS 9

9.1 $0.329 < p < 0.371$

9.3 $7.5 \le S \le 12.5$

9.5 $137.0 < \tau < 324.7$

9.7 credible interval: $780.5 < \mu < 782.7$: confidence interval: $774.1 < \mu < 785.9$.

PROBLEMS 10

10.1 The test statistic is $z = 2.24$ and $z_{0.025} = 1.96$. Therefore, the hypothesis must be rejected.

10.3 The test statistic is $\chi^2 = 43.9$. As this is greater than $\chi^2_{0.1}(19)$, the null hypothesis must be rejected.

10.5 The test statistic is $F = 0.738$ for 11 and 8 degrees of freedom. Since $F < F_{0.95}(11,8) \approx 3.3$, we cannot reject the alternative hypothesis.

10.7 Using ANOVA, $F = s_B^2/s_W^2 = 1.39$. Since $F < F_{0.05}(5,18) = 2.77$, the hypothesis cannot be rejected at this significance level.

10.9 $P[k \geq 63] \approx 0.033 < 0.05$, so the supplier's claim is rejected.

PROBLEMS 11

11.1 Using chi-squared to compare the values of the expected number of intervals e_i with a given number of counts with the observations, gives $\chi^2 = 10.51$ for 12 degrees of freedom, and since $\chi_{0.5}^2 = 11.3$ and $\chi_{0.75}^2 = 8.4$, the hypothesis is acceptable at any reasonable significance level.

11.3 From (11.10) $D = 0.356$, and from (11.11) $D^* = 1.18$. Thus, from (11.12) the hypothesis would accepted at a 10% significance level.

11.5 The calculated $\chi^2 = 6.03$ for 4 degrees of freedom, and since $\chi^2 < \chi_{0.01}^2(4) = 7.78$, we accept the null hypothesis of independence.

11.7 From the rank numbers, $w_+ = 37.5$, $w_- = 15.0$ and hence $w = \min\{w_+, w_-\} = 15.0$. The critical region for $n = 10$ in a two-tailed test (because we are testing an equality) with $\alpha = 0.05$ is $w \leq 8$, so we accept the null hypothesis at this significance level.

11.9 Assigning + and − signs to whether the value is greater than or less than the mean, we have $n = 8$, $m = 10$ and $r = 9$ runs. The critical regions at a 10% significance level are $r \leq 6$ and $r \geq 14$, so the hypothesis is accepted.

Bibliography

The short list below is not a comprehensive bibliography. Rather, it contains a number of relevant works at the introductory and intermediate level, together with a few classic books on the subject.

PROBABILITY THEORY

H. Cramèr *The Elements of Probability Theory and some of its Applications*, John Wiley and Sons (1955) pp281. A very readable introduction to the frequency theory of probability and probability distributions.

H. Cramèr *Random Variables and Probability Distributions*, 3rd edn, Cambridge University press (1970) pp118. A short, more advanced, discussion of probability distributions.

GENERAL PROBABILITY AND STATISTICS

M. Kendall and A. Stuart *The Advanced theory of Statistics*, 4th edn., Griffen (1976). An immense, definitive three-volume work (earlier editions were in one or two volumes) on most aspects of mathematical statistics. Written in an authoritative style, with many examples, both worked and as exercises.

A.M. Mood and F.A. Graybill *Introduction to the Theory of Statistics*, McGraw-Hill (1963) pp443. A good intermediate level book. Numerous problems, but no solutions.

The following three substantial volumes provide good general introductions to probability and statistics, with solved examples and numerous additional problems.

R. Deep *Probability and Statistics,* Elsevier (2006) pp686.
S.M. Ross *Probability and Statistics for Engineers and Scientists* 4th edn., Elsevier (2009) pp664.
R.E. Walpole *Introduction to Statistics*, 3rd edn., Maxwell Macmillan (1990) pp520.

The following two books give clear introductions to statistics from the Bayesian viewpoint, with worked examples.

G. Antelman *Elementary Bayesian Statistics*, Edward Elgar (1997) pp459.
D.A. Berry *Statistics. A Bayesian Perpsective,* Wadsworth Publishing Company (1996) pp518.

STATISTICS FOR PHYSICAL SCIENCES

R.J. Barlow *Statistics*, John Wiley and Sons (1989) pp204. A short informal guide to the use of statistical methods in the physical sciences. Contains some worked examples and problems with solutions.

P.R. Bevington and D.K. Robinson *Data Reduction and Error Analysis for the Physical Sciences*, 3rd edn. McGraw Hill pp320. Mainly about the least-squares method of estimation. Contains some worked examples and problems with answers.

B.E. Cooper *Statistics for Experimentalists*, Pergamon Press (1969) pp336. General overview with some emphasis on tests of significance and analysis of variance. Some problems with hints for their solution.

G. Cowan *Statistical Data Analysis*, Oxford University Press (1998) p197. Similar to the book by Barlow above, but more advanced. The examples are taken largely from particle physics. No solved problems.

S. Brandt *Data Analysis, 3rd edn*, Springer (1998) pp652. An extensive book on statistical and computational methods for scientists and engineers, with many Fortran computer codes. It contains a useful chapter on optimization methods. Some worked examples and solved problems.

F.E. James *Statistical Methods in Experimental Physics, 2nd edn.*, World Scientific (2006) pp345. An advanced work with many practical applications, mainly taken from particle physics.

L. Lyons *Statistics for Nuclear and Particle Physicists*, Cambridge University Press (1986) pp227. Mainly about errors and estimation methods, but with a short introduction to general statistics. Contains a long section on the Monte Carlo technique.

L. Lyons *Data Analysis for Physical Science Students*, Cambridge University Press (1991) pp95. A very short practical guide to experimental errors, and estimation by the least-squares method.

S. Meyer *Data Analysis for Scientists and Engineers*, John Wiley and Sons (1975 pp510. A substantial volume covering most aspects of relevant statistics, with many worked examples.

B.P. Roe *Probability and Statistics in Experimental Physics*, Springer-Verlag (1992) pp208. Similar in level to the book by James above; some problems and worked examples.

OPTIMIZATION THEORY

M.J. Box, D. Davies and W.H. Swann *Non-Linear Optimization Techniques,* Oliver and Boyd (1969) pp60. A short, very clear introduction, without detailed proofs.

B.D. Bunday *Basic Optimization Methods* Edward Arnold (1984) pp128. A short book covering constrained and unconstrained optimization, and containing proofs, exercises and sample computer codes in BASIC.

J. Kowalik and M.R. Osborne *Methods for Unconstrained Optimization Problems,* American Elsevier (1968) pp148. A more detailed book than that of Box *et. al.* Despite the title, it also discusses constrained problems.

STATISTICAL TABLES

W.H. Beyer (ed) *Handbook of Tables for Probability and Statistics,* Chemical Rubber Company (1966). A large reference work.

D.V. Lindley and W.F. Scott *New Cambridge Statistical Tables, 2nd Edition,* Cambridge University Press (1984). A collection of the most useful tables.

Index

Note: Page numbers followed by *f* indicate figures, *t* indicate tables.

A

Acceptance region, 194
Accuracy, 17
Additive property of χ^2, 110
Additive rule of probability, 23
Algebra, matrix, 243–247
Algebraic moment, 41
Alternative hypothesis, 194
Analysis of variance (ANOVA), 215–218
    multi-way analysis, 216
    one-way analysis, 215–216
    sum of squares and, 216, 217t
ANOVA. *See* Analysis of variance
Arithmetic mean, 8

B

Bar chart, 4–5, 5f
Bayes' estimators, 167–171
    loss function and, 168
Bayes' postulate, 29–31
Bayes' theorem, 29–30, 47
Bayesian confidence intervals, 189–190
Bernoulli number, 11
Bernoulli trial, 69–72
Best fit curve, 143–144
Bias, 87
Binomial probability distributions, 69–74
    Bernoulli trial and, 69–72
    characteristic function, 72–73
    examples, 71b–72b
    hypothesis testing and, 214
    limiting form of, 70f, 72–73
    moment distribution function, 72
    normal approximations, 73
    Poisson distribution and, 75–76, 78t
    probability function, 69–71
    sampling distributions related to, 121f
    tables, 259–265
Bins, 4–5
Bivariate normal, 65–66
Breit-Wigner formula, 69

C

Cauchy probability distribution, 68–69
    central limit theorem and, 95
    Lorentz distribution and, 68–69

Central interval, 174, 174f
Central limit theorem
    Cauchy distribution and, 95
    sampling and, 93–97
Central moment, 11, 41
Characteristic function (cf)
    binomial distribution, 73
    chi squared distribution, 109
    normal distribution, 61
    Poisson distribution, 76–79
    random variables and, 42–44
    sampling distributions and, 86
Chebyshev's inequality, 12
    weak law of large numbers and, 93
Chi-square test, Pearson's, 224
Chi-squared distribution
    additive property of, 110
    characteristic function, 109
    critical values, 108, 108f
    degrees of freedom and, 111
    density and distribution functions of, 107f, 108
    independence tests and, 231
    mgf, 109
    normal approximations, 109–110, 110t
    normal distribution and, 105–111
    normality convergence and, 109–110, 110t
    percentage points of, 108, 108f
    population distributions related to, 121f
    quality of fit and, 152
    Student's t distribution and F distribution and, 119–121, 121f, 121t
    tables, 272–274
Classical theory of minima, 247–248
Composite hypotheses, 194, 201–204
    likelihood ratio test, 198, 202, 203b–204b
    UMP test and, 201
Conditional density function, 45–49
Conditional probability, 23, 25f
Confidence belt, 177–178
Confidence coefficient, 174
Confidence intervals, 174–190
    Bayesian, 189–190
    central interval, 174, 174f
    confidence belt, 177–178
    confidence coefficient, 174
    confidence region, 176

Confidence intervals (*Continued*)
   credible intervals, 189
   general method, 177–179, 177f, 178f
   for mean and variance, 183–184
   large samples, 186–187
   near boundaries, 187–189
   normal distribution, 179–184
     for mean, 180–182, 181t
     for mean and variance, 183–184
     for variance, 182–183
   one-tailed, 176, 176b
   Poisson distribution, 184–186, 185t
   two-tailed, 175
Confidence level, 174
Confidence regions, 176, 183–184
Consistent estimator, 86–87
Constrained optimization, 255–256
Contingency table, 231, 231t
Continuous single variate probability distribution, 37
Converge in probability, 86–87
Convolutions, 53
Correlation, 12–14. *See also* Spearman rank correlation coefficient
   binned data and, 14
   coefficient, 50
   interpretation, 13–14
   Pearson's correlation coefficient, 13
   rank correlation coefficient, 239–241
   scatter plot, 13, 13f
Covariance, 50
   estimators for, 90–93
   sample, 12–13
Cramér-Rao inequality, 137–140
Credible intervals, 189
Critical region, 194
Critical values, 108, 108f
Cumulants, 43
Cumulative distribution function, 37, 39f
Cumulative frequency, 4–5, 5f

D

Data
   numerically summarized, 7–15
     location measures, 8–9
     more than one variable, 12–15
     spread measures, 9–12
   representations
     bar chart, 4–5, 5f
     frequency table, 4–5
     histograms, 4–5, 6f, 7
     lego and scatter plots, 7, 7f
Davidon's method of minimization, 225
Degrees of freedom, chi-squared distribution and, 111

Density function. *See* Probability distributions
Descriptive statistics
   defined, 1
   displaying data, 4–7
     bar chart, 4–5, 5f
     frequency table, 4–5
     histograms, 4–5, 6f, 7
     lego and scatter plots, 7, 7f
   experimental errors, 17–19
   experiments and observations, 2–4
   large samples and, 15–17
Discrete single variate, 36
Dispersion, 9–10
Displaying data
   bar chart, 4–5, 5f
   frequency table, 4–5
   histograms, 4–5, 6f, 7
   lego and scatter plots, 7, 7f
Distribution. *See* Probability distributions
Distribution-free tests. *See* Nonparametric tests

E

Equation of regression curve of best fit, 143–144
Error, 17–19
   bars, 17–18
   contributions to, 17
   experimental
     descriptive statistics and, 17–19
     estimation and, 103–104
     outliers and, 98–99
     probable error, 97
     propagation of errors law, 99–103
     random, 17
     sampling and, 103–104
     statistical, 17
     systematic errors, 102
   matrix, 149–150
   mean squared, 88–89
   precision and accuracy, 17–18, 18f
   probabilities, 194–195
   standard, 62
     of mean, 91–92
   statistical hypotheses and, 194
   systematic, 18, 18f
   type I or II, 194
Estimation
   central limit theorem and, 93–97
   interval estimation, 173–191
   large number laws and, 93–97
   of mean, variance and covariance, 90–93
   point estimation
     Bayesian, 167–171
     least-squares, 143–162

maximum likelihood, 123–135
method of moments, 165–166
minimum chi-squared, 163–164
minimum variance, 136–140
Event, 2
Expectation values, 40–41, 49–51
Experimental errors, 17–19
outliers and, 98–99
probable error and, 97
propagation of errors law, 99–103
sampling and estimation and, 103–104
systematic errors, 102
variance matrix, 102–103
Experiments and observations, 2–4
Exponential probability distributions, 66–68
memory and, 67
Extended likelihood function, 128

F

F distribution
chi-squared and Student's t distributions and, 119–121, 121f, 121t
constructing form of, 116
linear hypotheses and, 230–231
mgf, 117
noncentral, 230–231
normal distribution and, 116–119
pdf, 117
percentage points of, 117, 118f
population distributions related to, 121f
quality of fit and, 152
tables, 276–282
Fréchet inequality, 137–140
Frequency interpretation of probability, 27–29
table, 4–5
Full width at half maximum height (FWHM), 17
Functions of random variables. *See* Random variables

G

Gamma distributions, 66
Gaussian distribution, 16, 16f
General hypotheses: likelihood ratios, 198–204
composite hypotheses, 201–204
likelihood ratio test, 198
Neymane-Pearson lemma and, 199
simple hypothesis, 198–201
Generalized likelihood ratio test, 198, 202
Goodness-of-fit tests, 221–231
continuous distributions, 225–228
discrete distributions, 222–225
independence tests as, 231–232

Kolmogorov-Smirnov test, 226–227, 227f
linear hypotheses, 228–231
Pearson's chi-squared test, 224–225, 225b
rank correlation coefficient, 239–241
runs test, 237–239
signed-rank test, 234–236
Gradient method of optimization, 254–255
Davidon's method, 255
Newton's method, 254–255

H

Half-width, 17
Histograms, 4, 6f, 7
Hypothesis testing
acceptance region, 194
alternative hypothesis, 205
analysis of variance, 215–218
ANOVA, 215–218
composite hypothesis, 194
critical region, 194
error probabilities, 194
generalized likelihood ratio, 202–204
goodness-of-fit tests, 222–231
likelihood ratios, 198–201
Neyman-Pearson lemma, 199
nonparametric tests, 233–241
normal distribution, 204–214
table of tests, 213
tests of means, 206–210
tests on variances, 210–214
null hypothesis, 194
OC curve, 196–197, 197f
one-tailed test, 195
p-value, 195
Poisson distribution, 214–215
power of test, 195
rejection region, 194
significance level, 194
simple hypothesis, 84–85
tests for independence, 231–233
two-tailed test, 195
type I errors, 194
type II errors, 194
uniformly most powerful test, 201–202

I

Independence tests
chi-squared procedure and, 231
as goodness-of-fit tests, 231–232
contingency table, 231, 231t
Independent measurements, 2–3
Inference, statistical, 29
Interval estimation. *See* Confidence intervals

Inverse lifetime, 67
Inversion theorem, 43

J

Joint marginal distribution, 64
Joint moments, 49–51
Joint probability density, 45

K

Kolmogorov-Smirnov test, 226–227, 227f
Kurtosis, 60–61

L

Large numbers laws, 93–97
Large samples, 15–17, 186–187
Law of total probability, 25–26
Laws of large numbers, 93–97
Least squares estimation
    constrainted, 159–162
        variance matrix, 161
    nonlinear, 162–163
        linearization procedure and, 163
    unconstrained linear, 143–159
        best fit curve, 143–144
        binned data, 147–148
        combining experiments, 158–159
        error matrix, 149–150
        minimum variance properties, 148–149
        normal equations, 145–147
        orthogonal polynomials, 152–154
        parameters estimates errors, 149–151, 151b
        parameters general solution, 145–149
        quality of fit, 151–152
        residuals, 145
        straight line fit, 154–158
        weight matrix, 149–150
Lego plots, 7, 7f
Likelihood, 32
    ratios, 198–204
        composite hypotheses, 201–204
        Neymane-Pearson lemma and, 199
        simple hypothesis, 198–201
        test, 198, 202
Linear hypotheses
    goodness-of-fit tests and, 228–231
    noncentral F distribution and, 230–231
    power of test and, 230–231
Location measures, 8–9
    mean, 8
    mode and median, 8
    quantile and percentile, 8–9
Lorentz distribution, 68–69
Loss function, 168

M

Mann-Whitney rank-sum test, 236–237
    summarized, 237t
Marginal density function, 45–49, 47f
Marginal probability, 25–26
Mass function, 37, 39f
Matrix algebra, 243–247
Maximum likelihood (ML)
    Bayes' theorem, 170–171
    estimation
        approximate methods, 130–133
        binned data and, 124–125
        combining experiments, 135
        defined, 123–124
        disadvantages, 135
        extended likelihood function, 128
        graphical method, 131, 131f
        interpretation of, 135
        minimum variance bound and, 137–140
        several parameters, 133–135
        single parameter, 123–128
        unbiased estimator and, 126
        variance of estimator, 128–133, 137–140
    normality of large samples, 125–126
    principle, 32
Mean
    arithmetic, 8
    confidence intervals for, 180–182
    error of, 91–92
    estimators for, 90–93
    linear combinations of, 96
    location measures, 8
    point estimators for, 90–91
    population, 8
    squared error, 88–89
Median, 8
Memory, 67
Method of moment, 165–167
mgf. *See* Moment generating function
Minima, classical theory of, 247–248
Minimum chi-square, 163–165
Minimum variance, 136–140. *See also* Least squares
    bound, 137–140
    parameter estimation and, 136–137
    properties, 148–149
    Schwarz inequality and, 138
ML. *See* Maximum likelihood
Mode, 8
Moment generating function (mgf), 42–43
    chi-squared distribution and, 109
    F distribution, 117
    multivariate normal distribution, 66

Poisson distribution, 76–79
    sampling distributions, 86
    Student's t distribution, 114
Moments, 11
    algebraic, 41
    central, 41
    expectation values relation to, 41, 49–51
    generating function, 72
    joint, 49–51
Monte Carlo method, 85–86
Multinomial probability distributions, 74–75
Multiplicative rule, probability, 23–24
Multivariate distributions
    conditional density function, 45–49
    joint density function, 45, 46f
    joint distribution function, 45
    normal probability distributions, 63–66
        bivariate normal, 65–66
        independent variable, 64
        joint marginal distribution, 64
        joint mgf, 66
        quadratic form, 63–64
Multi-way analysis, 216

N

Newton's method of minimization, 254–255
Neyman-Pearson lemma, 199
Noncentral F distribution, 230–231
Nonlinear functions, optimization of
    constrained optimization, 255–256
    general principles, 249–252
    unconstrained minimization
        direct search methods, 253–254
        of multivariable functions, 253–255
        of one variable functions, 252–253
Nonlinear least squares, 162–163
    linearization procedure and, 163
Nonparametric tests, 233–241
    rank correlation coefficient, 239–241
    rank-sum test, 236–237
        summarized, 237t
    runs test, 237–239
    sign test, 233–234
    signed-rank test, 234–236
        summarized, 235t
Normal approximations
    binomial, 73–74
    chi-squared, 109–110
    student's t, 114–115
Normal density function, 59–63, 60f
Normal distribution, 16, 16f
    bivariate, 65–66
    characteristic function, 61

chi-squared distribution and, 105–111
confidence intervals, 179–184
    for mean, 180–182, 181t
    for mean and variance, 183–184
    for variance, 182–183
Gaussian distribution and, 59–63, 60f
hypothesis testing: parameters, 204–214
    table of tests, 213t
    tests on means, 206–210
    tests on variances, 210–214
inverse of standardized
    N^{-1} table, 180–181, 181t
multivariate, 66
sampling distributions related to, 121f
standard form, 61
Student's t distribution and, 111–116
tables, 257–258
univariate, 59–63, 60f
Null hypothesis, 194

O

OC. See Operating characteristic curve
One-tailed confidence intervals, 176
One-tailed test, 204–205
One-way analysis, 215–216
Operating characteristic (OC) curve, 196–197, 197f
Optimal estimator, 88–89
Optimization of nonlinear functions. See Nonlinear functions, optimization of
Orthogonal polynomials, 153
    ill-conditioning and, 152–153
    least squares and, 152–154
    recurrence relations and, 153–154
Outliers, 98–99

P

Parameter, 7
Parameter estimation. See Estimation
Parametric statistics, 86
pdf. See Probability density function
Pearson's chi-squared test, 224–225
Pearson's correlation coefficient, 13
Percentage points, 108–109, 108f
    chi-squared distribution, 108–109, 108f
    Student's t distribution, 113f
Percentile, 8–9
Permutations theorem, 26
Point estimators, 86
    Bayesian, 167–171
    for mean, 90–91
    least squares method, 143–158
    minimum variance and, 136–140
        bound, 137–140

Point estimators (*Continued*)
    parameter estimation, 136—137
        Schwarz inequality, 138
    ML method, 123—136
    moments method, 165—167
Poisson probability distributions, 75—80
    binomial distribution and, 75—76, 78t
    cf and mgf for, 76—79
    hypothesis testing and, 215
    interval estimation and, 184—186, 185t
    normal approximation, 76—79
    sampling distributions related to, 121f
    tables, 266—272
Population
    defined, 2—3
    distributions. *See* Probability distributions
    mean, 8
    variance, 9—10
Posterior probability density, 167—168
Power, 194—195
Power of test
    linear hypotheses and, 230—231
    normal distribution and, 204—205, 205f
    UMP and, 201
Prior probability density, 168
Probability
    additive rule, 23
    axioms of, 21—23
    calculus of, 23—27
    conditional probability, 23—24, 25f
    frequency interpretation, 27—29
    intervals, 189
    law of total probability, 25—26
    marginal, 25—26
    multiplicative rule, 23—24
    permutations theorem, 26
    posterior probability density, 167—168
    prior probability density, 168
    subjective interpretation, 27—32
    Venn diagrams, 22, 22f
Probability density function (pdf). *See* Probability distributions
    F distribution and, 117
    sampling distributions and, 86
Probability distributions
    binomial, 69—74
    bivariate normal, 65—66
    Cauchy, 68—69
    chi-squared, 105—111
    continuous single variate, 37
    cumulative distribution function, 37, 39f
    discrete single variate, 36
    exponential, 66—68
    F distribution, 116—119
    gamma, 66
    marginal density function, 45—49, 47f
    mass function, 37, 39f
    multinomial, 74—75
    multivariate conditional density function, 45—49
    multivariate joint density function, 45, 46f
    multivariate joint distribution function, 45
    multivariate normal, 63—66
    Poisson, 75—80
    Student's t distribution, 111—116
    uniform, 57—58
    univariate normal, 59—63, 60f
    Weibull, 68
Probability mass function, 37, 39f
Probable error, 97
Propagation of errors law, 99—103
p-value, 195—196

Q

Quadratic form, 63—64
Quality of fit
    least squares and, 151—152
        chi-squared test, 152
        F distribution test, 152
Quantile, 8—9

R

Random error, 17
Random number, 58
    generators, 85—86
Random samples
    central limit theorem and, 93—97
    definition regarding, 3—4, 84
    estimators and, 83—90
    large numbers laws and, 93—97
    point estimators, 86—90
    with and without replacement, 90
    sampling distributions, 84—86
        cf, pdf, and mgf and, 86
        Monte Carlo method and, 85—86
    simple sampling, 3—4
Random variables
    cf, 42—44
    convolutions, 53
    cumulative distribution function, 37, 39f
    density function, 37, 40f
    expectation values, 40—41
    functions of, 51—55
    marginal density function, 45—49, 47f
    mass function, 37, 39f
    mgf, 42—44

moments and expectation values, 41, 49–51
multivariate
  conditional density function, 45–49
  joint density function, 45, 46f
  joint distribution function, 45
single variates, 36–44
Rank correlation coefficient, 239–240
  tables, 286
Rank-sum test, 236–237
  summarized, 237t
  tables, 284
Region of acceptance, 194
Region of rejection, 194
Regression curve, 144
Rejection region, 194
Risk function, 168
Runs test, 237–239
  tables, 284–285

S

Sample
  covariance, 12–13
  space, 2–3
  variance, 9–10
Sampling
  central limit theorem and, 93–97
  experimental errors and, 103–104
  large numbers laws and, 93–97
  linear combinations of means and, 96
  random samples and estimators, 83–90
Sampling distributions, 84–86
  cf, pdf, and mgf and, 86
  chi-squared distribution, 105–111
  F distribution, 116–119
  Monte Carlo method and, 85–86
  parametric statistics and, 86
  population distributions related to, 121f
  statistic and, 84–85
  Student's t distribution, 111–116
Scatter plots, 7, 7f
Schwarz inequality, 138
Sheppard's corrections, 11
Sign test, 233–234
Signed-rank test, 234–236
  summarized, 235t
  tables, 283
Significance level, 194
Simple hypothesis: one alternative, 198–201
  likelihood ratios and, 198–204
    likelihood ratio test, 198
    Neymane-Pearson lemma and, 199
Simple random sampling, 3–4
Single variates, 36–44

cf, 42–44
continuous, 37
cumulative distribution function, 37, 39f
discrete, 36
expectation values, 40–41
mgf, 42–44
probability density function, 37, 40f
probability distributions, 36–40
probability mass function, 37, 39f
Skewness, 12
Spearman rank correlation coefficient, 239–241
Spread measures, 9–12
SSB. *See* Sum of squares between groups
SST. *See* Total sum of squares
SSW. *See* Sum of squares within groups
Standard bivariate normal density function, 65–66
Standard deviation, 9–10
Standard error, 62
  of mean, 91–92
Standard normal density function, 61
Standard normal distribution function, 61
Statistic, 7
Statistical error, 17
Statistical hypotheses, 194–198
  error probabilities and, 194
  OC curve and, 196–197, 197f
  one-tailed test, 204–205
  power and, 194–195
  p-value and, 195–196
  two-tailed test and, 195, 196f
Statistical independence, 48
Statistical tables
  binomial distribution, 259–265
  chi-squared distribution, 272–274
  F distribution, 276–282
  normal distribution, 257–258
  Poisson distribution, 266–272
  rank correlation coefficient, 286
  rank-sum test, 284
  runs test, 284
  signed-rank test, 283
  Student's t distribution, 275–276
Statistics
  applications, 1–2
  definitions, 1
Stratified sampling, 4
Strong law of large numbers, 93–97
Student's t distribution
  asymptotic behavior of, 114–115
  background, 111–112
  derivation, 112
  mgf and, 114

Student's t distribution (*Continued*)
  normal distribution and, 111–116
  percentage points of, 113f
  population distributions related to, 121f
  tables, 275–276
Subjective interpretation of probability, 29–30
Sufficient estimator, 89
Sum of squares between groups (SSB), 216, 217, 217t
Sum of squares within groups (SSW), 216, 217t
Systematic error, 18, 18f
  propagation of errors law and, 102
Systematic sampling, 4

T

Tables. *See* Statistical tables
Taylor's series, 247–248
Tests for independence, 231–233
Total sum of squares (SST), 216, 217, 217t, 218b
Two-tailed confidence intervals, 175b, 175–176
Two-tailed test, 195, 196f
Type I or II errors, 194

U

UMP. *See* Uniformly most powerful test
Unbiased estimator, 126
Unbiased point estimators, 87–88
Unconstrained linear least squares, 143–159
  best fit curve, 143–144
  binned data, 147–148
  combining experiments, 158–159, 159b
  error matrix, 149–150
  minimum variance properties, 148–149
  normal equations, 145–147
  orthogonal polynomials, 152–154
  parameters estimates errors, 149–151, 151b
  parameters general solution, 145–149
  quality of fit, 151–152
  residuals, 145
  straight line fit, 154–158
  weight matrix, 149–150
Unconstrained minimization
  of multivariable functions, 253–255
  direct search methods, 253–254
  gradient methods, 254–255
  of one variable functions, 252–253
Uniform probability distribution, 57–58
Uniformly most powerful (UMP) test, 201
Univariate normal distribution function, 59–63, 60f
Univariate normal probability distributions, 59–63, 60f
  linear sum distribution, 62–63

V

Variance
  confidence intervals, 182–184
  of estimator, 128–133
    approximate methods, 130–133
    graphical method and, 131, 131f
  estimators for, 90–93
  matrix, 102–103
  minimum, 136–140
    bound, 137–140
    parameter estimation and, 136–137
    Schwarz inequality and, 138
  ML estimator and, 128–133, 137–140
  point estimators for, 91–92
  population, 9–10
  sample, 9–10
  unconstrained linear least squares and, 148–149
Venn diagrams, 22, 22f

W

Weak law of large numbers, 93–97
  Chebyshev's inequality and, 93
Weibull distribution, 68
Weighted mean, 130, 158
Weight matrix, 149–150
Wilcoxon rank-sum test, 236–237
  summarized, 237t
Wilcoxon signed-rank test, 234–236
  summarized, 235t